Early Life on Earth

Early Life on Earth
Evolution, Diversification, and Interactions

Kenichiro Sugitani

CRC Press
Taylor & Francis Group
Boca Raton London New York

CRC Press is an imprint of the
Taylor & Francis Group, an **Informa** business

First edition published 2022
by CRC Press
6000 Broken Sound Parkway NW, Suite 300, Boca Raton, FL 33487-2742

and by CRC Press
2 Park Square, Milton Park, Abingdon, Oxon, OX14 4RN

CRC Press is an imprint of Taylor & Francis Group, LLC

© 2022 Taylor & Francis Group, LLC

Library of Congress Cataloging-in-Publication Data
Names: Sugitani, Kenichiro, author.
Title: Early life on earth : evolution, diversification, and interactions /
Kenichiro Sugitani.
Description: First edition. | Boca Raton, FL : CRC Press, 2021. |
Includes bibliographical references and index. |
Identifiers: LCCN 2021040198 (print) | LCCN 2021040199 (ebook) |
ISBN 9780367425647 (hardback) | ISBN 9781032198484 (paperback) |
ISBN 9780367855208 (ebook)
Subjects: MESH: Biological Evolution | Biodiversity | Archaea | Origin of
Life | Fossils
Classification: LCC QH366.2 (print) | LCC QH366.2 (ebook) | NLM QH 366.2 |
DDC 576.8—dc23
LC record available at https://lccn.loc.gov/2021040198
LC ebook record available at https://lccn.loc.gov/2021040199

ISBN: 978-0-367-42564-7 (hbk)
ISBN: 978-1-032-19848-4 (pbk)
ISBN: 978-0-367-85520-8 (ebk)

DOI: 10.1201/9780367855208

Typeset in Times
by codeMantra

Contents

Preface

My first field trip to the Pilbara Craton, an outback in Western Australia, now widely known as a hotspot of early life studies, was in summer of 1989. This trip was led by Dr. Ryuichi Sugisaki (deceased) and Dr. Mamoru Adachi; they are now emeritus professors at Nagoya University, Japan. Dr. Sugisaki was a supervisor of my PhD project, focusing on the geochemistry of Archean chemical sediments. Although we had arrived at Perth, Western Australia with excitement, we could not take an airplane from to Port Hedland, a gateway to the Pilbara Craton, due to a booking trouble and instead had to drive on the Great Northern Highway to Port Hedland, through Geraldton, Canarrvon, and Karratha. This long trip was really fun for me. During the 2 week-long field trip, we visited a variety of places in the Pilbara Craton, including Marble Bar, Tom Price, and Point Samson and collected various types of rocks including schist, basalt, komatiite, tuff, dolerite, banded iron formation, shale, chert, conglomerate, and jasper. I really enjoyed this field trip and have always been impressed by everything I saw. Particularly, I was impressed with the Japanese cemeteries found in some coastal towns, in which young deceased individuals were sleeping. They came to this distant foreign land more than 100 years ago for pearl hunting and died young, probably due to diving illness. At the time, I had never thought that I would have nearly 10 times of field trips to the Pilbara Craton until now and fortunately found 3.0 Ga- and 3.4 Ga-old microfossils I would like to talk here about the story of fossil-discovery in place of a typical preface for this book.

Based on geochemical analyses of cherts, jaspers, and shales collected during this first field trip, I published my first paper on Archean rocks in a scientific journal, Precambrian Research, entitled "Geochemical characteristics of Archean cherts and other sedimentary rocks in the Pilbara Block, Western Australia: evidence for Archean seawater enriched in hydrothermally derived iron and silica." Based on my paper published in Precambrian Research, I was awarded my Doctor of Science (DSc) in 1992. My DSc can be said a present from the Pilbara Craton. After my DSc achievement, I resumed fieldwork in the Pilbara Craton in 1993, focusing on the Goldsworthy greenstone belt, 100 km east of Port Hedland. Several times fieldwork had given me a certain amount of geological and analytical data and the whole picture of this area, which brought various scientific questions and research plans to me. I began to be interested in black chert in the sedimentary succession at the Goldsworthy greenstone belt. Chert generally refers to sedimentary rock composed of microcrystalline quartz. Its black color is attributed to the fine particles of organic matter contained therein. Organic matter in the sedimentary rock is generally derived from organisms. Thus, the organic matter in the black chert of the Goldsworthy greenstone belt is expected to have information about ancient life, assumed to be 3.0 Ga-old. I have planned systematic analyses of trace elements and carbon isotopic compositions of the black cherts from various stratigraphic levels from the Goldsworthy greenstone belt.

In the 2001 survey, I had asked my old friend working for local government to support my field trip because my collaborator, Dr. Koichi Mimura at Nagoya University was so busy and had no time for fieldtrip at that time. Days available for the field trip were limited, because my friends took a paid leave. My plan was that we performed a detailed collection of samples at Mt. Goldsworthy and a preliminary investigation at Mt. Grant, focusing on the collection of black carbonaceous cherts.

To save funds and time, we stayed overnight at Perth International Airport, and took a flight to Port Hedland in the early morning. At Port Hedland Airport, we immediately headed to the Goldsworthy greenstone belt, 100 km away from Port Hedland. We arrived around 3:00 p.m. and thus did not have enough time for field-work. The outcrop at Mount Goldsworthy was not directly accessible from the unpaved but well-maintained, Marble Bar-Goldsworthy Road. We had to drive a nar-row track (approximately 2 km) and then walk 5 minutes to the outcrop. I had visited the targeted outcrop several times and had never experienced trouble. This field trip was in July, which was in the midst of the dry season in Pilbara. The track appeared to be dry and hard, and I had no worries about traffic. However, in short order, our car got stuck at the mouth of the track. Tires of our car were deeply buried in dark mud. If our car had a four-wheel drive, this accident was not be serious at all, but our car had a two-wheel drive. We did not see any other cars on the Marble Bar-Goldsworthy. There was no water and no food, just two stupid guys and some cows. Fortunately, we escaped from the mud trap, but it was too late to restart the fieldwork. We dejectedly returned to the hotel without any results. It was a really poor first day in Pilbara. Anyway, after this experience, I decided to avoid driving on tracks to access the outcrops, which meant a long walk with a heavy backpack. On the final day, we conducted a fieldwork session at Mount Grant, which extends to the west of Mount Goldsworthy. Mount Grant had the appearance of a huge surf swell, although it is only 100 m above sea level even at the highest point. We found a track that con-tinued 7 km from the Marble Bar – Goldsworthy Road to a radio tower on the sum-mit of Mount Grant, but placed our car near its entrance and started walking toward the west, mainly along the ridge, while collecting black chert samples. Black cherts were easily found, and I collected randomly them. Consequently, the weight of my backpack was increasingly growing. We started sampling in the morning and arrived at a point, about 6 km from the car park, around the noon. There, beds of sandstone and chert rose vertically like a folding screen on the southern flank of Mount Grant, probably due to their relatively higher resistance against weathering and erosion (the cover photo with three images of extracted microfossils from the 3.0 Ga-old rocks). We had lunch there. The sun was intense, but the wind was pleasant. A small hawk was hovering on the updraft generated along the southern flank. The hawk was stay-ing in the air, at the same level as my eyes, as close as 10 m in distance.

After a short lunch, we climbed up dozens of meters from this "folding screen" rock to reach a flat summit of white coarse-grained sandstone. There we found two layers composed of columnar crystals, up to 30 cm long, forming fan-shaped aggregates at the uppermost portion of this sandstone layer. These crystals were supposed to be silicified evaporite minerals, and the same deposits had already been identified at Mount Goldsworthy. I was confident that sedimentary succes-sion at Mount Goldsworthy and that at Mount Grant were indeed identical. The evaporite beds were overlain by a thin unit composed of tuff, silt, and sandstone. At the uppermost portion of this unit, I found a beautiful jet-black chert layer about 20 cm thick. Although this was to be collected, I was hesitant to collect samples. I don't know why I had such hesitation. One reason might have been the fatigue of the previous days, which was not completely gone from my body. Also, at that time, I had already collected significant amounts of black cherts, and my backpack was heavy.

I asked my friend, "This seems good, but...do you think we should collect this?" My friend is neither a geobiologist nor an astrobiologist, so it didn't matter to him if I collected the beautiful jet-black chert or not. My friend replied coolly, "You've come all this way from Japan, so you might as well take them." The words pushed me to collect two small samples possibly around 200 g in weight. One was a slightly grayish chert, and the other was a jet-black chert, with the names ORW4A and ORW4B, respectively. After collecting these, we got off Mount Grant and walked back on a terribly hot track to our car.

The planned sampling had been completed successfully, and the rest of the work comprised only sending the samples from the post office to Japan. It has been routine work since my first visit to Pilbara. However, I was caught in an unexpected situation when packing the collected rocks in cardboard boxes in front of the post office at Port Hedland. Customs officers came and asked, "What are you doing?" I then replied, "We are packing rock samples to export to Japan." The customs officers then asked, "Do they contain fossils?" I replied, "Maybe" as a joke. However, this response had made the officers nervous. They had suspected that we were going to export stromatolites. I explained in a hurry, "I mean microfossils: fossils of microorganisms. However, the possibility is very low." Actually my samples did not contain stromatolites, and I did not expect at all that they contained microfossils at that time. Organic matter was enough for me. I am not sure if they were convinced with my explanation, but the officers released us.

Despite further twists and turns that cannot be described here, the collected samples arrived safely in Japan and were sent to my laboratory. These samples were intended to be used mainly for analyses of chemical and isotopic compositions. As a part of routine, however, petrographic thin sections were made by a technical staff member. In order to properly understand chemical data, it is very important to examine the petrological characteristics of the analyzed rock and select samples suitable for the research purposes.

ORW4B, a specimen collected from Mount Grant, was carbonaceous as expected from its jet-black color. It had abundant minute carbonaceous particles, but their heterogenous distribution with irregularly shaped chert masses was different from those of carbonaceous black cherts that I had previously examined. When I continued my observation, I found a film-like structure. This film, 100 μm across, appeared to be branched. My suspicion turned into conviction by the discovery of a spindle (lenticular)-shaped structure. It was also large, being up to 60 μm across. It was obvious to me that the structure could not have been produced by the random aggregation of carbon particles. There could not be seen any mineral relicts inside. Rather, inner carbonaceous contents appeared to have been blown out. I muttered to myself, "No way....".

Following this discovery, I decided to start searching for microfossil-like structures in earnest. First, I asked a technical staff member to make more petrographic thin sections from the same hand specimen, in order to confirm that the structures were abundant in the sample and to find more specimens. At the same time, I asked for thin sections of the black cherts collected from Mount Goldsworthy. In fact, the same, and even various morphologies of microfossil-like structures, were found in the newly prepared thin sections made from ORW4B and from the Goldsworthy samples. A few weeks later, I showed Dr. K. Mimura, the petrographic thin sections. He, though an organic geochemist, seemed to intuitively understand what the micro-structures were and said "Amazing!"

I was overjoyed with this discovery and had an expectation that my findings would soon give a powerful impact on scientific societies. However, reality was not sweet. Indeed, it took full six years to publish the first, plainly armed paper with scientific conscience and protection. There were many reasons for this, but the well-known Apex Chert controversy between Dr. J.W. Schopf and Dr. M. Brasier, arising in 2002, definitely affected the publication of my work.

Even if you find what you can call "microfossils," particularly in the Archean, it is a completely different story as to whether it would be accepted by scientific societies. When I found the microfossils, I was completely unknown in the field of Precambrian paleontology. It was obvious to me that my manuscript submitted to the journal would be in the storm of criticisms and finally rejected. Furthermore, I had realized that the amount of identified microfossils and other information were not enough for a convincing argument. In the end, I had conclusion to work with an Australian paleontologist. Fortunately, I found on the internet the name of Dr. Kath Grey, a senior paleontologist at the Geological Survey of Western Australia (GSWA). She had a long record of working on Proterozoic microfossils and stromatolites, so she seemed an appropriate person to consult because some of the microstructures which I had found resembled Proterozoic microfossils.

I wrote a letter to her, with some photographs of the microstructures. I would not have been surprised if she had not replied, because she knew nothing about me. Fortunately, she responded quickly, saying that the structures were interesting and requested if it would be possible for her to examine them. I immediately scheduled my business trip to Perth. The GSWA building was near the Swan River, a few minutes-walk from my hotel and Dr. Kath Grey met me at the reception desk with a welcoming smile. She took me to her office and immediately started to examine the rock specimen and the petrographic thin sections. After several minutes, she stopped her examination. I asked her "What do you think?" She said, "Well, it's a 50/50 chance". I asked her, "Please clarify which one you are referring to." Her answer was, "They are probably not biogenic." At that point, I had probably looked disappointed. Mercifully, she added, "But I need more time to examine them at higher magnification, and there are a lot of criteria for biogenicity to be met. Can you come again tomorrow, so that we may take a closer look?" It was hanging by a hair. After returning to the hotel, I considered the criteria for biogenicity pointed out by Kath and tried to answer them. I prepared a short report and showed that to Dr. Kath Grey in the next morning. She said, "Ok, now that we have both had time to think about them, let's have another look."

After reading my report clarifying the provenance of the chert containing micro-fossils, her reaction had become to be more positive. The more she examined the specimens, the more excited she became about the objects under the microscope, saying "What is this? It looks very much like some of the younger organic-walled microfossils, which are probably planktonic algae...." After having left me watching her anxiously while she made a thorough examination, she showed me some of the specimens she had found so interesting and said, "I think these may be very important." Because of the language difficulty, Kath started drawing diagrams to show me the features that she thought indicated that my specimens were biogenic. After this session, Kath introduced me to her colleague, Dr. Martin Van Kranendonk, who was a senior geologist at GSWA at that time. He later became a professor at the University of New South Wales (UNSW) and a director of the Australian Centre for Astrobiology (ACA). She also introduced me to another key person, Dr. Malcolm Walter, a founding director at ACA, who was also a professor at Macquarie University and a director of ACA at that time.

In collaboration with these researchers and encouragement from my laboratory colleagues in School of Informatics and Sciences, Nagoya University (unfortunately abolished), my quest for ancient life had progressed and I, together with collaborators including Dr. Kath Grey and Dr. Koichi Mimura, published the first paper in Precambrian Research in 2007, overcoming various obstacles. Now that I think about it, I was a bit reckless in my approach. However, Dr. Kath Grey and her colleagues helped me to understand how important it was to provide evidence that supported claims about biogenicity, even though it was a time-consuming exercise. I was so lucky to meet Dr. Kath Grey. Without meeting her, the microfossils may have been forgotten or summarily dismissed, and would have remained unknown to anyone but me, leaving me deeply disappointed.

Also, without the full support of my supervisor, Dr. Ryuichi Sugisaki and of my colleague, Dr. Koichi Mimura for his extensive field support, my Archean studies would have not begun, and I would not have continued this wonderful work. Furthermore, I could not get this difficult task done (writing a single book in nonnative language) without the help of my family. So, I would sincerely like to dedicate this book to Dr. Kath Grey, Dr. Ryuichi Sugisaki, Dr. Koichi Mimura and my family (Midori Sugitani, Takako Kataya, Emiko Sugitani, Hinako Sugitani, Hana). Obviously, many researchers (over 20 people from six countries), technical staff members, and my students have been involved in this project. I do not list all their names here, but I do sincerely acknowledge all of them. Acknowledgment should also be given for the support of JSPS KAKENHI 19H02013, 19340150, 22340149 and 24654162.

I wrote this book as an extended version of my chapter "8.5 Fossils of Ancient Microorganisms" in *Handbook of Astrobiology*, published in 2019. It was intended to introduce the joys of Precambrian geobiology, particularly the Archean ones, and astrobiology to undergraduate students and graduate students of wide disciplines interested in this field. That is the reason why I introduce basic topics in the beginning chapters. This book has various aspects, including a textbook and a review paper, in addition to a monograph. I have also given somewhat detailed reviews for some selected works of particular fascination to me. I hope many readers enjoy this book.

Matsusaka & Nagoya,
Japan, Kenichiro Sugitani

Author

Kenichiro Sugitani is a professor at the Graduate School of Environmental Studies at Nagoya University, Chikusa, Nagoya. He received the 1995 Geochemical Society of Japan encouragement award for contributions to the understanding of geochemistry and origin of ancient siliceous sediments including Archean cherts. He has undertaken fieldwork in the Pilbara Craton and mapped the Goldsworthy greenstone belt and discovered microfossils from the 3.0 Ga Farrel Quartzite. He is an associate member of Australian Centre for Astrobiology at the University of New South Wales and serves as a regional editor of Astrobiology and a member of editorial advisory board of Geobiology.

1 Space, Solar System, and the Earth

1.1 INTRODUCTION

In this first chapter, the origins of elements in the universe, evolution of our solar system and the Earth, and the origins of the oceans and the atmosphere are reviewed. Based on "the Big Bang theory", our universe began 13.8 Ga ago, whereas our solar system had evolved around 4.6 Ga ago, which is assumed to be nearly identical to the age of the Earth (Figure 1.1). Our (not only human beings but also the other organisms) existence is inseparable from the universe, where elements have been generated through dynamic processes. Formation of our habitable planet is closely related to the evolution of the solar system. Habitability for "the Earth-type life" prerequisites the prolonged presence of water (H_2O) and the preceding supply of building blocks of life. These topics are mentioned here.

1.2 ELEMENTS IN THE UNIVERSE AND THEIR ORIGINS

To date, 118 elements including synthetic ones have been identified, from $_1H$ to synthetic $_{118}Og$ (oganesson). Despite such diversity of elements, their abundance in the universe deduced from that of our solar system is extremely biased to elements with small atomic numbers (Figure 1.2). The elemental abundance of our solar system has been deduced based on spectroscopic analyses of the solar sphere for volatile elements and chemical compositions of chondrite for nonvolatile elements: chondrite is a sort of meteorite and is thought to have preserved primitive solid phase compositions of the solar system (see *Column*).

Hydrogen is the most abundant element in our solar system and the universe, and He is the next one. These two dominate more than 98% of all the elements. Elemental

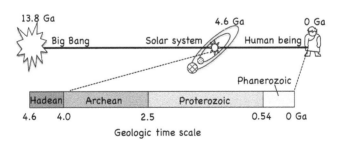

FIGURE 1.1 Time scale from the Big Bang to the present. This book mainly deals with the deep time (Hadean to Archean) of the Earth.

DOI: 10.1201/9780367855208-1

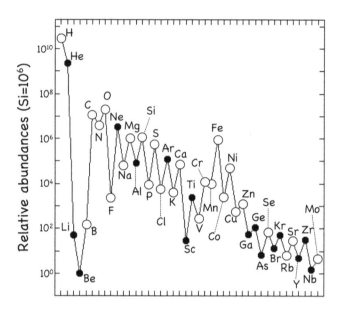

FIGURE 1.2 Relative abundances of elements in the solar system (Si = 10⁷), from H to Mo. White-circled elements are essential elements for plants and animals (Cuss et al., 2020, and references therein).

abundances exponentially decrease to elements with larger atomic numbers, with some irregularities represented by relative depletion of Li, Be, and B and by relative enrichment of Fe (Figure 1.2). The most important essential elements for life such as H, C, N, O, S, and P, comprising carbonhydrates, lipids, nucleic acids, and polyphosphates are major constituents.

Hydrogen, He, Li, and Be were produced soon after the beginning of the universe. In other words, the other elements were not present at that time. Some of the other elements have been produced by stepwise nuclear fusions inside fixed stars composed dominantly of H. The heaviest element produced by this process is Ti, and the size of a star (fixed star) gives constrains on what element is finally generated in its core. In the core of the Sun, four atoms of ^{1}H are fused to produce one atom of ^{4}He. Inside fixed stars larger than the Sun, heavier elements could be produced by further nuclear fusion. This process is followed by a process called neutron capture that could increase the mass number of an element, with an increase of atomic number to a lesser extent, eventually producing ^{56}Fe. Elements heavier than ^{56}Fe are produced also by neutron capture. However, this process occurs only during supernova explosion, which would occur at the end of fixed stars more than ca. 10 times heavier the Sun. This neutron capture process is called "rapid process", whereas the process of neutron capture inside fixed stars is called "slow process".

Associated with nuclear fissions, energy is produced and emitted to the space as the electromagnetic radiation from fixed stars. The Sun has solar radiation estimated to be 3.6×10^{26}W. The Earth receives only 4.55×10^{-8} % of this radiation energy. The solar radiation is dominated by visible light (47%) (380–700 nm), ultraviolet (7%)

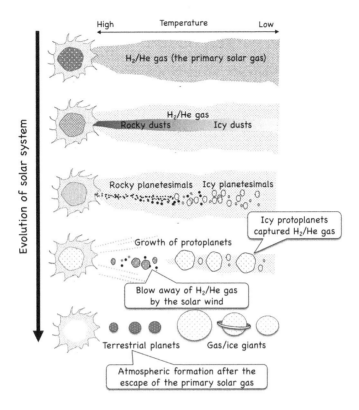

FIGURE 1.3 The evolution of the solar system. Modified from Dig. 14.2 of Genda (2019).

(10–380 nm), and infrared (46%) (700 nm–1 mm) {note that these ranges are somewhat ambiguous (https://photosyn.jp/pwiki/)}. Most photoautotrophic organisms (see Chapter 3) almost exclusively utilize visible light to produce organic matter (OM), and some algae can utilize near infrared (700–750 nm) (e.g., Nürnberg et al., 2018). The terrestrial and the marine surface ecosystems are founded on this solar radiation energy.

1.3 EVOLUTION OF OUR SOLAR SYSTEM

Our solar system had evolved from molecular clouds composed of interstellar dusts and gases (interstellar matter). The interstellar matter was originated largely from previously existed stars and their planetary systems, scattered by for example supernova explosion. These wrecks were raw materials for our solar system and life on the Earth and if present, others. Elements produced directly by the Big Bang (Big Bang nucleosynthesis) were limited to light ones up to Be. Namely, our bodies and those of other organisms are composed largely of elements produced by nucleus syntheses within fixed stars and during their supernova explosions. Elements produced by the rapid process include Cu, Zn, and Mo, which are essential to many of the organisms on the Earth. These elements comprise reaction centers of several important enzymes such as superoxide dismutase, laccase, and nitrogenase.

Interstellar medium refers to interstellar matter and cosmic rays present between solar systems. The region with higher density of interstellar medium, compared with the surroundings, is called "interstellar cloud". Molecular cloud refers to the region with a density of 10^4–10^6 H_2 per cm^{-3}; the major component of molecular cloud is H_2, although many of other molecules, including e.g., carbon monoxide (CO), H_2O, ammonia (NH_3), hydrogen cyanide (HCN), and formaldehyde (HCHO), have been detected. Recently, methylamine (CH_3NH_2), which is a precursor of glycine ($C_2H_5NO_2$), the simplest amino acid, was recently detected (Bøgelund et al., 2019). Within such a molecular cloud, fixed star forms. The fixed star formation is thought to be triggered by fragmentation of high-density area of the cloud (molecular cloud core) or by shock wave generated by, for example, supernova explosion. In the following, the assumed formation process of our solar system is briefly described.

Destruction of gravitational stability resulted in contraction and rotation of the molecular cloud, eventually forming the primordial sun and the surrounding disk (the primitive solar sytem nebula) composed of gases (mainly H_2) and dusts (mainly H_2O ice), with trace amounts of minerals and metals. The formation of this disk was associated with precipitation of dusts to the rotating surface, which released their potential energies. Released potential energy was transformed to thermal energy, which was higher in the region closer to the primitive sun. Obviously, the solar radiation was stronger in the region closer to the primitive sun. Temperature gradient in the primitive solar system expected from these two factors (potential energy of dusts and solar radiation) indicates that within the inner region closer to the primitive sun (~3 astronomical unit: AU), dusts were once entirely evaporated, except for materials with very high boiling points. Once vaporized, materials within the inner zone would subsequently condense to form minerals and metals.

The growth of planets is thought to have been a stepwise dynamic process. Namely, particles (dusts) once formed planetesimals, which had grown through collision coalescence with each other to protoplanets (100–1,000 km in diameter). Compositions of planetesimals were different depending on distance from the primodal sun (Figure 1.3). Planetesimals formed within the inner zone had rocky composition, whereas those in the outer zone had icy H_2O composition, and protoplanets as well. Through further collision coalescence, protoplanets had grown to the primitive planets, direct precursors of the present planets. The planets had grown in the primary solar atmosphere mostly composed of H_2 at least in their early stage of growth. As H_2O was much more abundant compared with rocky materials in the solar system, icy primitive planets could have grown much larger than the rocky planets and captured the primary H_2 atmosphere, resulting in the formation of gas giants such as Jupiter and Saturn. Uranus and Neptune are also gaseous and icy planets (ice giants), although their masses are much smaller than Jupiter and Saturn. One explanation for this is that their primitive planets had formed after the blow-off of the primary atmosphere by strong solar wind. This wind had completely removed the primary atmosphere surrounding the rocky planets, which was closely related to the origins of atmosphere and oceans of the Earth and Mars (e.g., Wurm, 2019 and reference therein).

1.4 EVOLUTION OF THE EARTH'S INNER STRUCTURE

The Earth is characterized by a three-layered concentric structure composed of core, mantle, and crust. The core is composed dominantly of metallic Fe, with some light elements required. It is also divided into the inner solid core and the outer liquid core, which is predicted by analyses of seismic waves and the presence of dipole magnetic field of the Earth. Candidates for light elements contained in the core have been thought to be C, S, O, H, and Si. Recent studies emphasize H. One approach was made by estimation of temperature just above the core (the bottom of the solid mantle) (Nomura et al., 2014). The estimated temperature was much lower than that required for melting of pure metallic Fe. In order to explain this inconsistency, a large amount of H (0.6 weight %) should be contained in the outer core. The authors suggested the possibility that dissolution of H into metallic Fe had occurred within magma ocean (see below). Surprisingly, this amount of H is 80 times of the present seawater as H_2O.

Formation of the core and the Earth's layered structure was triggered by magma ocean, which was formed by elevated surface temperature due to heat generated by radioactive decay, bombardment of planetesimals and greenhouse effect of thick H_2O vapor atmosphere (Figure 1.4). The depth of the magma ocean might have

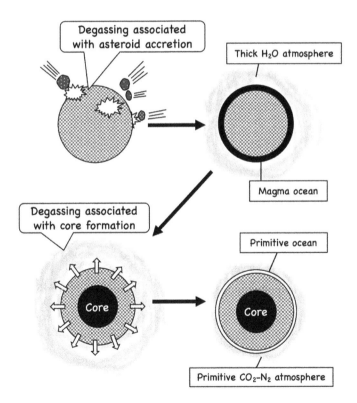

FIGURE 1.4 Evolution of the inner structure of the Earth and related formation of the primitive atmosphere and ocean. Note that compositions of the primitive atmosphere are still in controversial (see the main text).

been up to 2,000 km (Elkins-Tanton, 2012 and reference therein). Within the magma ocean, silicate components and metal components had been separated with each other, just like in a blast furnace. As being heavier than silicates, metallic Fe had accumulated at the bottom of the magma ocean. This process had eventually resulted in gravitational destruction and formation of the metallic Fe core in the center of the Earth and the outer silicate mantle: degassing associated with this event is thought to have contributed to the formation of the Earth's atmosphere. Also, the formation of the earliest crust might have been related to this event. A recent study suggests that the formation of felsic crust characterized by high concentrations of Si, Al, and K, like modern continental crust, could be back to 4.5 Ga (Iizuka et al., 2015), though controversial (see Chapter 5).

1.5   ORIGINS OF THE OCEANS AND THE ATMOSPHERE

Origin of the oceans and the atmosphere is crucial for the origin of life, because it has been widely believed that life had emerged in a close association with the presence of liquid H_2O and that life had emerged in the oceans, regardless of the validity of this hypothesis as discussed in Chapter 3.

The Earth's atmosphere and oceans were once considered to have been gradually oozed out from its interior mainly as volcanic gases and hot springs whose major compositions are H_2O and CO_2 (Rubey, 1951). However, this model has already been discarded. Most of the volcanic gases and hot springs are recycled precipitation and seawater. Precipitated H_2O penetrates underground and reacts with ascending magma (heated rocks) to be volcanic gases or hot springs. Water (H_2O) and CO_2 in seawater are incorporated into oceanic crust by several processes including formation of hydrate minerals, carbonates (e.g., $CaCO_3$: exoskeletons of organisms), and OM. These materials would be decomposed by subduction of the oceanic crust, releasing $H_2O(OH)$, CO_2, and other volatiles (see Chapter 2).

Large-scale degassing in the very early history of the Earth was theoretically suggested by Matsui and Abe (1986). Degassing during collisional accretion of planetesimals with 0.1% H_2O produced a thick H_2O atmosphere, volumetrically consistent with the mass of the present oceans, at the stage when the radius of the proto-Earth reached 0.4 R_0 (40% of the present value). The H_2O atmosphere increased the surface temperature, together with released heat energy from collisional accretion of planetesimals to have produced a magma ocean. After the main accretional event, the surface temperature dropped, and H_2O vapor in the atmosphere condensed to have formed oceans and left mildly oxidizing CO_2 ($+CO$)-N_2 atmosphere, although Hashimoto et al. (2007) suggested that accretions of CI-chondrite-like materials produced the reducing atmosphere enriched with H_2, methane (CH_4), and CO. The primitive reducing atmosphere is supported also from unexpected high concentrations of highly siderophile elements (HSEs, including the platinum-group elements, Re and Au) in the mantle. Theoretically, the core-mantle differentiation process had resulted in extensive separation of HSEs into the core at high fractionation factor. However, the present mantle contains higher amounts of HSEs than theoretically expected (e.g., Day et al., 2016). This issue has been challenged by many researchers; proposed hypotheses include e.g., incomplete separation hypothesis, magma-ocean hypothesis, and late veneer hypothesis. Late veneer hypothesis suggests that after the core-mantle differentiation producing CO_2- and N_2-rich atmosphere, asteroids enriched in HSEs collided and coalesced to the primitive Earth's mantle (e.g., Willbold et al. 2011 and references therein). The collided asteroids are assumed to be compositionally equivalent to enstatite chondrite. If this is the case, the Earth's earliest atmosphere was probably highly reducing, being enriched in H_2, CH_4, and NH_3, although different models have been proposed as represented by Kasting (1993), and the late veneer hypothesis itself is controversial. In any case, the reducing Earth's earliest atmosphere seems to be convenient for chemical evolution and birth of life.

The formation of oceans might have been much earlier than 4.0 Ga ago. The early formation of oceans is evidenced by the presence of 3.7–3.8 Ga-old pillow lavas in the Isua Supracrustal Belt (ISB), West Greenland (Maruyama and Komiya, 2011). Pillow lava, one of the major volcanic structures, is composed of stacked lobate-shaped volcanic masses whose cross sections resemble pillows (Figure 1.5). The structure is formed by eruption of magma (mainly basaltic) in an aqueous setting. Hot magma erupted into cool water is chilled, and its surface quickly solidifies to form a thin skin. Magma is still inside and with additional supply of new magma from

FIGURE 1.5 (a) Pillow lava, approximately 3.5 Ga-old, in the Pilbara Craton, Western Australia. (b) Closer view of the different block, showing alignment of vesicles along the pillow margin. Vesicles are traces of trapped gas, vaporized due to decompression of magma. Scale is the coin above the pillow.

the magma chamber, giving rise to an increase in the internal pressure. Eventually, magma breaks the skin to overflow to form lobate mass. This process repeats to produce characteristic stacked pillow-shaped lavas. As the ISB pillow lava occurs as a significantly large block several km across, its formation should have required the presence of significant mass of water such as ocean. Sedimentological and geochemical evidence for the presence of ocean at least 3.7 Ga ago comes from the ≥3.75 Ga Nuvvuagittuq greenstone belt (NGB) in northern Québec, Canada, where iron formation (IF) occurs. Iron formation refers to chemical sedimentary rocks composed of quartz (SiO_2) and iron minerals represented by hematite (Fe_2O_3) and magnetite (Fe_3O_4) and is generally considered to have deposited in oceans (Chapter 5). Mloszewska et al. (2012) argued seawater origin of the NGB IF, based on trace element characteristics including rare-earth elements (see *Column* in Chapter 8 for details). The presence of liquid H_2O, not always ocean, in the Hadean (>4.0 Ga) has been implied from isotopic features of 4.4 Ga-old detrital zircon ($ZrSiO_4$) from the Jack Hills conglomerate, Western Australia (Wilde et al., 2001; Valley et al., 2002).

As described above, planetesimals formed within the inner zone of the primitive solar system are thought to had been of rocky compositions. Though controversial (Campbell and O'Neill, 2012), one hypothesis is that the Earth's bulk composition is similar to enstatite chondrite (Javoy et al., 2010). As enstatite chondrite is depleted in H_2O (~0.01%), one tenth of model materials used by Matsui and Abe (1986), other materials such as volatile-rich comets may have been required as the sources of Earth's volatiles. Piani et al. (2020), on the other hand, recently suggested the possibility that Earth's water was inherited from enstatite chondrite-like materials. Origins of the Earth's volatiles are still not fully understood and highly controversial.

COLUMN: METEORITE

Meteorite refers to extraterrestrial solid material fallen onto planets. This term has been often confused with the term "asteroid", which refers to a small solar system body other than dwarf planets. Meteorites are classified into three types including stony meteorite, stony-iron meteorite, and iron meteorite. Stony meteorites are further classified into chondrite and achondrite. Chondrites, except for one subtype, are characterized by containing ~1 mm spherules called chondrules. Origins of chondrules have long been controversial, but it is generally accepted that they were formed by rapid condensation of vaporized materials before formation of planetesimals. Chondrites are thought to be derived from primitive mother celestial bodies. Here "primitive" means that the celestial body for chondrite had not experienced evolution to develop inner structure like the Earth. Chondrites have been classified based on petrological and chemical variations, which reflect thermal history and/or redox condition during their formation. Among various chondrite types, carbonaceous chondrites are of special interest in the context of origin of life, because they are volatile-rich and contain amino acids and other organic compounds, building blocks of biopolymers. CI chondrite referred to in the main text is one type of carbonaceous chondrites and is the most oxidized type, contrastive to highly reduced and volatile-poor enstatite chondrite. Achondrite, stony-iron meteorite, and iron meteorite are thought to be derived from the evolved celestial body composed of core and mantle. Such structure was formed by segregation of metallic Fe and silicates by large-scale melting, and subsequent gravitational collapse. Thus, achondrite corresponds to mantle material, whereas iron meteorite to core. One type of stony-iron meteorite (pallasite) may be derived from the boundary between core and mantle.

REFERENCES

Bøgelund, E.G., McGuire, B.A., Hogerheijde, M.R., van Dishoek, E.F., Ligterink, N.F.W. 2019. Methylamine and other simple N-bearing species in the hot cores NGC 6334I MM1-3. *Astrinomy and Astrophysics* 624, A82.

Campbell, I.H., O'Neill, H.S.C. 2012. Evidence against a chondritic Earth. *Nature* 483, 553–558.

Cuss, C.W., Glover, C., Javed, M.B., Nagel, A.H., Shotyk, W. 2020. Geochemical and biological controls on the ecological relevance of total, dissolved, and colloidal forms of trace elements in large boreal rivers: review and case studies. Environmental Reviews 28, 138–163.

Day, J.M.D., Brandon, A.D., Walker, R.J. 2016. Highly siderophile elements in Earth, Mars, the Moon, and asteroids. *Reviews in Mineralogy and Geochemistry* 81, 161–238.

Elkins-Tanton, L.T. 2012. Magma oceans in the inner solar system. *The Annual Review of Earth and Planetary Sciences* 40, 113–139.

Genda, H. 2019. Evolution of early atmosphere. In A. Yamagishi et al., eds. *Astrobiology: From the Origins of Life to the Search for Extraterrestrial Intelligence.* Springer, Singapore. pp. 197–207.

Hashimoto, G.L., Abe, Y., Sugita, S. 2007. The chemical composition of the early terrestrial atmosphere: Formation of a reducing atmosphere from CI-like material. *Journal of Geophysical Research* 112, E05010.

Iizuka, T., Yamaguchi, T., Hibiya, Y., Amelin, Y. 2015. Meteorite zircon constraints on the bulk Lu-Hf isotope composition and early differentiation of the Earth. *Proceedings of the National Academy of Sciences of the United States of America* 112, 5331–5336.

Javoy, M., Kaminski, E., Guyot, F., Andrault, D., Sanloup, C., Moreira, M., Labrosse, S., Jambon, A., Agrinier, P., Davaille, A., Jaupart, C. 2010. The chemical composition of the Earth: Enstatite chondrite models. *Earth and Planetary Science Letters* 293, 259–268.

Kasting, J.F. 1993. Earth's early atmosphere. *Science* 259, 920–926.

Maruyama, S., Komiya, T. 2011. The oldest pillow lavas, 3.8–3.7 Ga from the Isua Supracrustal Belt, SW Greenland: Plate tectonics had already begun by 3.8 Ga. *Journal of Geography* 120, 869–876.

Matsui, T., Abe, Y. 1986. Evolution of an impact-induced atmosphere and magma ocean on the accreting Earth. *Nature* 319, 303–305.

Mloszewska, A.M., Pecoits, E., Cates, N.L., Mojzsis, S.J., O'Neil, J., Robbins, L.J., Konhauser, K.O. 2012. The composition of Earth's oldest iron formations: The Nuvvuagittuq Supracrustal Belt (Québec, Canada). *Earth and Planetary Science Letters* 317–318, 331–342.

Nomura, R., Hirose, K., Uesugi, K., Ohishi, Y., Tsuchiyama, A., Miyake, A., Ueno, Y. 2014. Low core-mantle boundary temperature inferred from the solidus of pyrolite. *Science* 343, 522–525.

Nürnberg, D.J., Morton, J., Santabarbara, S., Telfer, A., Joliot, P., Antonaru, L.A., Ruban, A.V., Cardona, T., Krausz, E., Boussac, A., Fantuzzi, A., Rutherford, A.W. 2018. Photochemistry beyond the red limit in chlorophyll f-containing photosystems. *Science* 360, 1210–1213.

Piani, L., Marrocchi, Y., Rigaudier, T., Vacher, L.G., Thomassin, D., Marty, B. 2020. Earth's water may have been inherited from material similar to enstatite chondrite meteorites. *Science* 369, 1110–1113.

Rubey, W.W. 1951. Geologic history of sea water. *An attempt to state the problem. Bulletin of the Geological Society of America* 62, 1111–1148.

Valley, J.W., Peck, W.H., King, E.M., Wilde, S.A. 2002. A cool early Earth. *Geology* 30, 351–354.

Wilde, S.A., Valley, J.W., Peck, W.H., Graham, C.M. 2001. Evidence from detrital zircons for the existence of continental crust and oceans on the Earth 4.4 Gyr ago. *Nature* 409, 175–178.

Willbold, M., Elliott, T., Moorbath, S. 2011. The tungsten isotopic composition of the Earth's mantle before the terminal bombardment. *Nature* 477, 195–198.

Wurm, G. 2019. Travelling to the origins of the Solar System. *Science* 363, 230–231.

2 Solid Earth

2.1 INTRODUCTION

In this chapter, the solid Earth is overviewed, including basic classification of igneous and sedimentary rocks and their origins and implications, and plate tectonics. As it is now widely accepted that life has been closely interacted with the solid Earth, it is worth introducing terrestrial rocks before entering into the main topic of this book. Information preserved in rocks is often blurred and even misleading, but we have no choice other than rocks to have direct insights into the ancient Earth and life. Rather, rocks occasionally have long preserved traces of early life over billion years.

Terrestrial rocks are classified into three main groups, igneous, sedimentary, and metamorphic rocks. Igneous rocks are direct products of cooling and lithification of magma and lava, whereas sedimentary rocks represent rocks formed by the surface processes including transportation and deposition of detrital materials produced by weathering and erosion of previously existing rocks and deposition of biominerals such as shells and corals. Metamorphic rocks are igneous

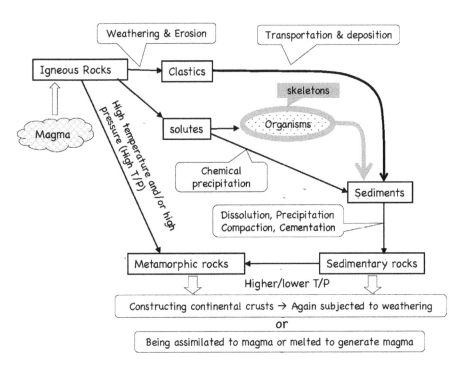

FIGURE 2.1 Schematic model of rock- and sediment-related material flows on the Earth.

DOI: 10.1201/9780367855208-2

and sedimentary rocks subjected to high (occasionally low) pressure and/or temperature. These three types of rocks are closely related with each other as shown in Figure 2.1. This schematic figure is illustrated for imaging a virtual closed system of material cycles, although the formation of magma and thus igneous rock is designated as the starting point. This is in fact true for the early Earth, considering the formation process of the Earth described in Chapter 1. It is also well known that magmatism in the modern Earth is closely related with plate tectonics and it is a great issue when the modern-style plate tectonics started and contributed to the continental growth.

2.2   PLATE TECTONICS, DRIVING FORCE OF DYNAMISM OF THE EARTH

Plate tectonics is a theory proposed by J. Tuzo-Wilson and others in 1968, constructed based on the theory of continental drift proposed by A. Wegener in 1912 and the theory of seafloor spreading proposed by H.H. Hess in the early 1960s. Plate tectonics interprets that the Earth's surface is composed of multiple rigid plates and that their relative motions have driven geotectonic activities. The geotectonic activities include e.g., continental drift, oceanic spreading, volcanism, orogeny, and earthquake, which have been closely related to evolutions and extinctions of life.

   Plate is also called "lithosphere". Lithosphere is composed of crust and mantle (mantle lithosphere) (Figure 2.2). Lithosphere is up to $> 200\,km$ in thickness at the continental region (continental lithosphere), whereas oceanic lithosphere is thinner, generally less than $100\,km$ thick (Conrad and Lithgow-Bertelloni, 2006). Oceanic lithosphere is produced at oceanic spreading center. It increases its thickness and density as it moves away from spreading center, eventually subducting into viscous asthenosphere over $500\,km$ thick. At the subduction zone (trench), huge earthquakes often occur, and parts of oceanic crusts are accreted to continents. Oceanic spreading center and subduction zone refer to divergent and convergent plate boundaries, respectively. These two types of plate boundaries are sites of active magmatism, although their mechanisms are distinct with each other. At the spreading center, two lithospheres move apart, resulting in the deficiency of lithosphere at the axis and thus decompression of rising hot asthenosphere. Decompression lowers the melting point of mantle material (peridotite) to trigger its partial melting to produce primary basaltic magma, whereas subduction-related magmatism is triggered by "water (H_2O, OH)", which could reduce the melting point of rocks. Subducting oceanic crusts have hydrate minerals containing OH and H_2O (e.g., Seyfried and Mottl, 1982), which are formed mainly by hydrothermal alteration of crusts at spreading centers. Hydrate minerals would be decomposed at the subduction zone due to elevation of temperature and pressure, releasing "water". Mantle wedge, being added with "water" released from oceanic crust, could be melted to produce magma. It may be noted that magmatism also occurs at intraplate settings, including, e.g., hot spot magmatism related to plume of mantle material from the boundary with the outer core and continental collision-related magmatism caused by heating due to crustal thickening.

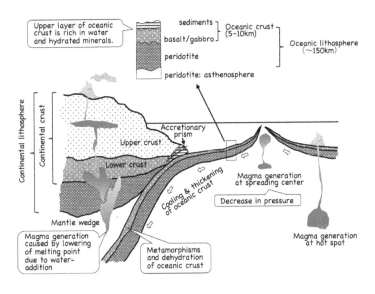

Fig. 2.3 Schematic model of plate tectonics and related magmatism.

FIGURE 2.2   Schematic model of plate tectonics and related magmatism.

2.3   IGNEOUS ROCKS

2.3.1   CLASSIFICATION SCHEME

Igneous rocks can be classified into two categories including plutonic rocks and volcanic rocks, depending on their textural variations, which is function of cooling speed of magma (and lava), and into four categories, depending on their concentrations of SiO_2 (Figure 2.3).

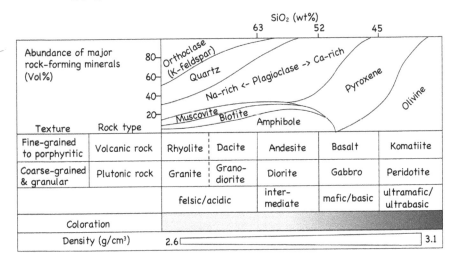

FIGURE 2.3   Igneous rock types and related texture and mineralogy. Based on Figure 4.4 of Grotzinger and Press (2007).

Lava is magma erupted onto the Earth's surfaces (on land and ocean floor). The temperature of lava is generally higher than 800°C. It is thus cooled quickly just after eruption, which results in its rapid consolidation. Rapid cooling prevents the growth of crystals, resulting in fine-grained and even glassy textures. Some minerals with high melting points crystallize to form euhedral crystals, which appears to be floating in the fine-grained matrix. Such crystals are called phenocrysts. Igneous rocks characterized by fine-grained texture occasionally with phenocrysts are called volcanic rocks. If magma is cooled slowly in the subsurface magma chamber, crystallized minerals could grow to be large crystals, forming equigranular texture. Igneous rock of this type is called plutonic rock characterized by interlocking anhedral crystals large enough to be identified by naked eyes. Another classification scheme is chemical composition. Based on concentrations of SiO_2, igneous rocks are classified into ultramafic/ultrabasic ($SiO_2 < 45$ wt%), mafic/basic (45 wt% $< SiO_2 < 52$ wt%), intermediate (52 wt% $< SiO_2 < 63$ wt%), and acidic/feldspathic ($SiO_2 > 63$ wt%). One of the major mechanisms responsible for such the variations is chemical evolution of magma in the chamber, in addition to degree of partial melting. Basically, partial melting of ultramafic mantle material (peridotite) produces relatively SiO_2-enriched primary basaltic magma, which migrates into magma chamber within the crust. With decreasing temperature of magma in the chamber, minerals with high melting points such as olivine and subsequently pyroxene crystallize and deposit to the bottom. As olivine with the formula of $(Fe, Mg)_2SiO_4$ and pyroxene with the formula of $(Fe, Mg) SiO_3$ have high $(Fe, Mg)/SiO_2$ weight ratios than the other silicate minerals, residual magma tends to be enriched in SiO_2. Such the process (only partially described here) is called "fractional crystallization". In addition to mantle peridotite, lower continental crusts and subducting lithosphere (slab) could be partially melted, which is caused by hot ascending magma and to lowering of melting temperature due to hydration, respectively. Chemical compositions of magma could also be modified by contamination of crustal materials. As it is beyond the scope of this book to make descriptions of all types of igneous rocks here, granitic and basaltic rocks in some details, representing igneous rocks comprising continental and oceanic crusts, respectively, are described in this chapter.

2.3.2 Granitic Rocks (Granitoids)

Granite in a strict sense and granitic rocks (e.g., granodiorite and tonalite) with similar texture and composition to granite are important components in continental crusts, particularly the upper ones. Granitic rocks are enriched in quartz, plagioclase and orthoclase (K-feldspar), with lesser amounts of mica and mafic minerals such as hornblende and biotite (Figure 2.4a). Chemically, they are characterized by relative enrichment of Si, Na and K and depletion in Mg, Ca, and Fe, showing highly differentiated composition from parental mantle materials. Their average density is around 2.5~2.7 g/cm^3, smaller than basalt (2.8~3.0 g/cm^3) and peridotite (~3.5 g/cm^3). Thus, once formed, continental crusts have buoyancy against mantle to form stable subaerial exposures. Granitic magma could be produced by fractional crystallization of primary basaltic magma, as final products. However, it is widely accepted that the formation of granitic rocks cannot be explained solely by this process. Several

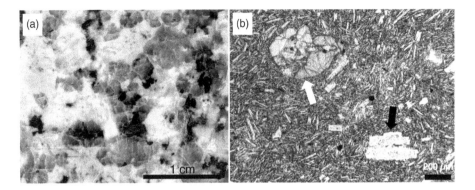

FIGURE 2.4 (a) Granite, unknown locality (scanned image). Light grey somewhat translucent mineral is quartz, whereas white and pink minerals is feldspars. Black minerals are hornblend or biotite. (b) Basalt with phenocrysts of olivine (white arrow) and plagioclase (black arrow: probably Ca-rich), unknown locality. The matrix is composed of fine crystals of relatively Na-rich plagioclase. Open nicol.

classification schemes have been proposed for granitic rocks. Alphabetic classification system was first proposed by Chappell and White (1974), which has been upgraded later into five types, including I-type, S-type, A-type, M-type, and C-type, based on their geochemistry related to sources of granitic magma. For example, the first presented I-type and S-type represent granitoids formed by partial melting of igneous protoliths and that of sedimentary (supracrustal) rocks, respectively. For further descriptions and interpretations of granitic rocks, see, e.g., Bonin et al. (2020).

Growth of continents due to granitic magmatism has several important implications for the early Earth evolution. Although granitic rocks could form at various geodynamic settings, including intraplate setting and plate boundary, juvenile granitic rocks are typically produced in a close association with subduction of oceanic lithosphere. As described above, dehydration of subducted slabs triggers partial melting of mantle to produce basaltic magma, which ascends and reaches the lower continental crust of basaltic composition. Partial melting of the lower basaltic crust, if occurs, would produce more silicic (granitic), magma (Figure 2.2). Namely, granitic magmatism is closely related to plate tectonics. The onset of plate tectonics sometime in the Archean is believed to have facilitated granitic magmatism and growth of continents. Subaerial exposure of granitic and other type of rocks due to continental growth would have accelerated weathering. Weathering (chemical weathering) of silicate minerals releases essential elements such as K, Ca, Mg, and P to the oceans and at the same time reduces CO_2 in the atmosphere (see Chapter 4). Also, exposure of light-colored granitic rocks had likely increased albedo (reflectance of the solar radiation) of the Earth. Therefore, the growth of continents had potentially lowered the surface temperature of the Earth.

2.3.3 BASALTIC ROCKS

Basalt is a representative mafic volcanic rock, composed mostly of fine crystals of plagioclase and pyroxene. Large, ideally euhedral crystals (phenocrysts) of pyroxene, olivine, and/or calcium-rich plagioclase are often observed (Figure 2.4b): These have higher melting temperatures than the other rock-forming minerals and thus crystallize earlier in the magma chamber. Basalt and crystalline counterparts (gabbro) comprise the oceanic crusts, together with sediments.

Basaltic rocks have been classified based on chemical compositions and tectonic settings. Chemically, basaltic rocks are classified into alkaline, calc-alkaline, and tholeiitic, with the other four minor types: Note that these terms also refer to magma suites, from which other rock types would be formed by fractional crystallization. Alkaline basalt is relatively enriched in Na and K, whereas depleted in Si. This magma suite is typically formed at intraplate settings represented by continental rift valleys and ocean islands. Calc-alkaline basalt is relatively rich in Ca and depleted in alkaline elements (Na and K). Tholeiitic basalt is depleted in alkaline elements and is relatively enriched in SiO_2. This magma suite is typically formed at mid-ocean ridges, large igneous provinces (LIPs), and occasionally ocean islands. In tectonic framework, the terms of MORB (mid-ocean ridge basalt), OIB (oceanic island basalt), and IAB (island arc basalt) have been widely used. OIB is produced typically at intraplate setting like Hawaii Islands, whereas IAB at subduction zones. These types can be geochemically characterized and discriminated with each other. In other words, tectonic setting where ancient basaltic rocks were formed could be deduced from their chemical compositions, particularly trace elements (e.g., Xia and Li, 2019) (*Column*).

Basaltic rocks represent volcanic-sedimentary successions called "greenstone belts", comprising Archean cratons together with acidic/felsic plutonic rocks and gneisses (metamorphic rocks) (see Chapter 5). Comparison of basaltic rocks in the Archean greenstone belts with modern ones could provide clues to tectonics during the Archean. Trace element geochemistry related to assimilation of felsic materials has also been used to infer the evolution of continental crusts.

2.4   SEDIMENTARY ROCKS

Sedimentary rocks have generally been classified into four types including volcaniclastic (pyroclastic) rock, terrigenous clastic rock, chemical sedimentary rock, and biogenic sedimentary rock; additionally, a new type of biochemical sedimentary rock is proposed here.

2.4.1   VOLCANICLASTIC (PYROCLASTIC) ROCKS

Volcaniclastic rock, in the strict sense, refers to sedimentary rock formed directly related to volcanic activities, whereas in the wider sense, sediments composed of reworked volcanic materials are included. According to Fisher and Smith (1991), the term "volcaniclastic" refers to clastic materials composed partially or entirely of volcanic fragments. Volcaniclastic particles are formed by various processes and mechanisms, based on which pyroclast, hydroclast, epiclast, and autoclast and alloclast are identified (Fisher and Smith, 1991). Volcaniclastics are, in general, products of explosive volcanisms. Pumice, highly vesicular volcanic glass, would

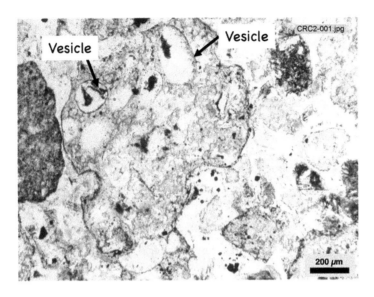

FIGURE 2.5 Photomicrograph of volcaniclastic sandstone from the 3.3 Ga-old Euro Basalt in the Pilbara Craton, Western Australia. Open nicol.

be broken to be glass shards. Hydroclasts are also glassy but poorly vesiculated compared with pyroclasts.

Volcaniclastic rocks are common in the Archean sedimentary records (Figure 2.5). This is not surprising, considering more extensive volcanic activities during this era. In the context of the early evolution of life, volcaniclastic rocks provide two perspectives. One is related to identification of traces of Archean life. Wacey et al. (2018) described some volcaniclastic rocks (tephra) to have vesicles morphologically similar to described microfossils and suggested the possibility that organic-walled (microbially colonized) volcanic vesicles mimic cellularly preserved microfossils, which will be discussed in detail in Chapter 11. Brasier et al. (2011) also proposed a hypothesis that volcaniclastic material, specifically glassy, porous, and gas-rich pumice, was a potential site for emergence of life and the earliest habitat. Glassy volcaniclastic materials are susceptible to chemical weathering and thus could have effectively provided nutrients to early life.

2.4.2 TERRIGENOUS CLASTIC ROCKS

Terrigenous clastic rocks refer to sedimentary rocks that are composed mainly of mineral grains and rock fragments derived from preexisting rocks, which are cemented and lithified by precipitated minerals in pore spaces. They are classified into the main three types based on grain size variation, including mudstone (claystone and siltstone), sandstone, and conglomerate (Figure 2.6). Rocks subjected to weathering would be fragmented to be detrital materials of various sizes. This process is accompanied by dissolution of some rock-forming minerals and formation of secondary minerals, the so-called clay minerals such as kaolinite [$Al_4Si_4O_{10}(OH)_8$], known as

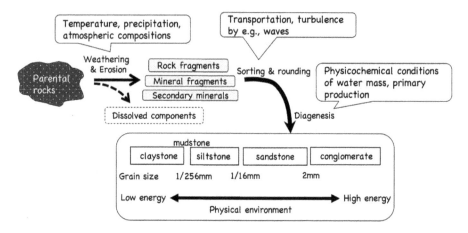

FIGURE 2.6 Schematic model for processes involved in the formation of terrigenous clastic rocks.

raw materials for pottery. Dissolved components without being incorporated into the secondary minerals would be transported by groundwater and river water and finally reach oceans and lakes. River and wind flows transport the eroded detrital materials to other places both in aqueous and subaerial environments. Unconsolidated sediments would be lithified through the process called "diagenesis". Diagenesis refers to the process that occurs in the sediments after deposition until lithification. During diagenesis, unstable detrital materials are dissolved. New materials such as amorphous opaline silica ($SiO_2 \cdot nH_2O$) and carbonate such as calcite ($CaCO_3$) would be precipitated from porewaters, contributing to lithification of sediments and thus formation of "terrigenous clastic rocks". Microbial activities closely related to decomposition of organic matter (OM) is also involved in the early stage of diagenesis, which will be discussed in Chapter 4.

2.4.2.1 Factors Controlling Chemistry and Mineralogy of Terrigenous Clastic Rocks

Chemical and mineralogical compositions of terrigenous clastic rocks are functions of various factors, including provenance, mode of weathering, length of transport, degree of abrasion, and chemistry of water mass at the depositional site (Figure 2.6). Provenance refers to rocks subjected to weathering and erosion and preliminarily gives constrain on chemical and mineralogical compositions of terrigenous clastic rocks. Nature of provenance, namely rock type, is often closely related to tectonic settings. Weathering and erosion are largely controlled by climate (temperature and precipitation) and atmospheric compositions. Chemical weathering (dissolution of rock-forming minerals and formation secondary minerals) is accelerated under high temperature, high precipitation, and high CO_2 concentration. In the extreme case, most of the minerals would be eventually decomposed, leaving residual soils enriched in oxides and/or hydroxides of Al,

Ti, and Fe, which are called laterite. On the other hand, mechanical weathering (physical breaking minerals and rocks) is accelerated by, e.g., daily large temperature fluctuation. Rocks would be broken to be mineral particles or clasts (detrital materials) of various sizes through repeated expansion and contraction by temperature fluctuations, with limited chemical decomposition. Volume expansion from water to ice (~10%) in cracks is one of the major driving forces of mechanical weathering. Obviously, chemical weathering and mechanical one are complementary to each other. Additionally, biological activities are also known to be involved in chemical weathering (e.g., Samuels et al., 2020).

If detrital materials produced by weathering are transported for long distance, by river, they would be sorted due to their size and density differences. In general, silt (1/16–1/256 mm) and clay (<1/256 mm) dominate sediments of downstream areas, whereas various-sized detrital materials including sand (2–1/16 mm) and gravel (>2 mm) are found at upstream areas. Reworking at the depositional site such as beach subjected to high energy wave actions also accelerate mechanical breakdown and sorting of detrital materials. Physicochemical compositions of water mass and primary production at the depositional site would significantly contribute to the mode of diagenesis. For example, if the depositional site with high production of planktonic algae is in a closed basin and thus the water circulation is restricted, the sediments could be rich in OM and occasionally diagenetically produced sulfide minerals such as pyrite (FeS_2) (see Chapter 4).

Terrigenous clastic rocks also have information on crustal growth and ancient terrestrial environment. Taylor and McLennan (1985), for example, compiled the chemical compositions of Archean (> 2.5 Ga ago) and post-Archean fine-grained clastic rocks (shale and siltstone) and demonstrated that the values of Th/Sc and La/Sc are significantly higher in post-Archean clastic rocks. Here, the chemical compositions of fine-grained clastic rocks are regarded to represent those of the upper (continental) crust. Thorium and La tend to be enriched in granitic rocks, major components of the continental crust, whereas Sc in mafic rocks. Thus, the observed secular change is interpreted in the context of evolution of the continental crust. Also, chemical compositions of fine-grained clastic rocks have often been used in order to assume the degree of chemical weathering of source rocks. Indices for chemical weathering such as CIW (Chemical Index of Weathering) (Equation 2.1) and WI (Weathering Index) (Harnois, 1988) have been proposed. CIW is an index using Al_2O_3, CaO, and Na_2O concentrations of sedimentary rocks (and soils), based on the fact that during the weathering, Si, Ma, Ca, and Na of parental rocks are leached, whereas Al and Ti basically remain.

$$CIW = \left[Al_2O_3 / \left(Al_2O_3 + CaO + Na_2O \right) \right] \times 100 \ (\text{molecular ratio}) \qquad (2.1)$$

As the degree of chemical weathering is function of temperature, amount of rainfall, and pH of rain waters (depending on $_pCO_2$ and other acids), those of Archean sedimentary rocks and paleosols are expected to place some constraints on the ancient surface environment. Hessler and Lowe (2006), for example, analyzed alluvial and fluvial deposits (terrigenous sedimentary rocks) of the 3.2 Ga-old Moodies Group in the Barberton greenstone belt, South Africa and suggested that these sediments were derived from aggressively weathered source rocks, being attributed to increased rainfall, higher temperature, and/or high atmospheres $_pCO_2$ in the Archean. Under the presumed absence of O_2 in the Archean atmosphere, it is expected that some minerals easily decomposed under the modern oxygenated surface environment could have been stable as detrital materials. Such minerals include uraninite (UO_2), siderite ($FeCO_3$), and various sulfide minerals represented by pyrite. In other words, detection of these minerals in the terrigenous clastic rock record could be taken as evidence for low $_pO_2$ in the Archean atmosphere (Rasmussen and Buick, 1999), which will be discussed in some detail in Chapter 6.

2.4.2.2   Sedimentary Structures of Terrigenous Clastic Rocks and Their Implications

Terrigenous clastic rocks display various structures, including graded bedding, cross-bedding, ripple mark, mudcracks, sole marking, and raindrop impression. Here, graded bedding, cross-beddings, and raindrop impressions are interpreted briefly, and then examples applying to Archean environmental studies are introduced.

Graded bedding refers to fining-upward trend of particle size observed in a single bed. In general, graded bedding occurs repeatedly. This structure is formed commonly in association with turbidity currents and has long been used as a way-up indicator. Cross-bedding (stratification and lamination) refers to beds inclined against the thicker stratum unit. This structure is formed by moving media such as stream water, wind, wave, and tidal current, in various scales. Various types of cross-bedding are known, including, e.g., planar, hummocky, trough, and herringbone cross-beddings, corresponding to different depositional environments and flow regimes. Cross-bedding is also used to determine way-up direction. Volcanic sedimentary successions in Archean greenstone belts are often significantly deformed and nearly vertically inclined. Thus, the determination of way-up direction using these structures is essential for reconstructing lithostratigraphy. An interesting example is an attempt to decipher ancient tidal cycles from the 3.2 Ga-old terrigenous clastic sediments (the Moodies Group, South Africa), based on measurement of thickness of bundles of cross-bedding sandstone forests (Eriksson and Simpson, 2000) (Figure 2.7).

Though more or less controversial, small circular, crater- or bump-like structures observed at the bedding surface of fine-grained sedimentary rocks are interpreted as raindrop impressions. Fossilized raindrop impressions have been used as an indicator of ancient atmospheric pressure. Som et al. (2012), for example, combined experimental and field studies on raindrop impressions to have inferred the ancient atmospheric density. Measurements of raindrop impressions recorded in tuff of the 2.7 Ga-old Ventersdorp Supergroup, South Africa gave results that the contemporaneous air pressure was below twice modern level and probably below 1.1 bar, although the other estimate suggested higher partial pressure of N_2 around 1.6–2.4 bar in the Archean (Goldblatt et al., 2009).

FIGURE 2.7 Cross-bedding of the 3.2 Ga-old tidal sandstone, the Barberton greenstone belt, South Africa.

2.4.3 BIOGENIC, CHEMICAL, AND BIOCHEMICAL SEDIMENTARY ROCKS

Nonclastic sedimentary rocks have traditionally been classified into two types: chemical sedimentary rock and biogenic sedimentary rock: Obviously clastic materials are more or less contained in many cases. The term "biochemical sedimentary rock" may be confused with "biogenic sedimentary rock". These three terms may be better in strict sense, because the term "chemical sediments" has often been applied to sediments produced directly or indirectly by biological activities. The representative is iron formation (IF), iron-rich sedimentary rocks characterizing Archean to Paleoproterozoic sedimentary successions. Iron formation has traditionally been categorized into chemical sediments. However, it is now widely accepted that some of the IFs were formed in a close association with bacterial activity as discussed in detail in Chapter 5. Therefore, such nonskeletal but biologically induced chemical sediments are better called "biochemical sediments (sedimentary rocks)". This terminology automatically constrains definitions of chemical and biogenic sedimentary rocks. Namely, chemical sedimentary rocks refer to lithified sediments composed of pure chemical precipitates represented by evaporites, whereas biogenic sedimentary rocks refer to sediments composed largely of biogenic skeletons, like coral limestone composed largely of calcite ($CaCO_3$), or organic remains, like coal.

2.4.3.1 Biogenic Sedimentary Rocks

In addition to coral limestone and coal, biogenic sedimentary rocks include diatomite, radiolarite, and chalk. Diatomite is composed largely of siliceous frustules of diatom, unicellular algae widespread in terrestrial, freshwater, and marine environments. Emergence of diatom is back to the Mesozoic. Radiolarite (radiolarian

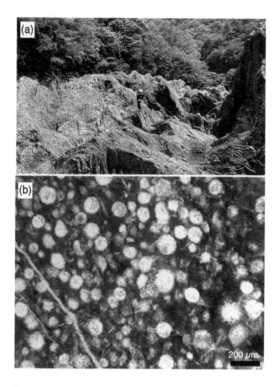

FIGURE 2.8 (a) Spectacular outcrop of the Triassic radiolarian bedded chert at Gifu Prefecture, central Japan. K.S. in the center for scale. (b) Photomicrograph of Triassic radiolarian bedded chert, from Gifu Prefecture, central Japan. Open nicol.

chert) is also composed largely of microcrystalline quartz originated from opaline ($SiO_2 \cdot nH_2O$) skeletons of radiolaria, marine zooplankton (Figure 2.8), which had evolved in the Cambrian. Chalk is a sort of limestone, composed mainly of tiny calcareous ($CaCO_3$) skeletons of unicellular organisms (coccolithophore and foraminifera). Spectacular cliff of chalk facing the Strait of Dover is composed of numerous such skeletons. If marine sediments contain at least 30% such siliceous or calcareous skeletons, they are called oozes. Over 60% of the ocean floor is covered with such oozes. Even siliceous oozes alone cover approximately 25% of the ocean floor (Open University Oceanography Course Team, 1989). This indicates that the productivities of these microorganisms significantly influence ocean chemistry. Indeed, concentrations of dissolved silica [$Si(OH)_4$](DSi) in modern seawater, particularly in shallow seawater, are much lower than those of river waters. Thus, before the evolution of organisms requiring silica as a nutrient, the Earth's oceans were enriched in silica (Siever, 1992). This cannot be overlooked when considering the sedimentation system during the Precambrian as discussed in Chapter 5.

2.4.3.2 Chemical Sedimentary Rocks

Chemical sedimentary rocks refer to lithified sediments that are composed dominantly of chemically precipitated materials, ideally without involvement of biological activities. Evaporites and hydrothermal deposits are well known as chemical sediments (sedimentary rocks).

Evaporites are sediments (sedimentary rocks) formed due to evaporation of water mass (originated from lake or ocean water) and resultant supersaturation to specific minerals. Minerals comprising evaporite are various, depending on water chemistry and degree of evaporation. Carbonates ($CaCO_3$, Na_2CO_3), sulfate (Na_2SO_4, $CaSO_4 \cdot 2H_2O$), borax ($Na_2B_4O_7 \cdot 10H_2O$), and halite (NaCl) represent evaporitic minerals precipitated from evaporated seawater. Chert is also formed by evaporation of sodium carbonate brines, as shown by Magadi-type chert that occurs at Lake Magadi in the Kenyan Rift Valley. Magadi-type chert is not direct precipitation of silica but diagenetic product of magadiite, a hydrous sodium silicate mineral [$NaSi_7O_{13}(OH)_3 \cdot 4H_2O$] (Figure 2.9a) (Eugster, 1967). Evaporites are common in Archean sedimentary successions, although to my knowledge, most of them had been replaced by cherts or dissolved away to leave casts. Unfortunately, therefore, primary mineral phases that could provide important constraints on seawater and atmospheric chemistry cannot be identified. It may also be mentioned here that the deposition of Archean evaporites might have been attributed not only to prolonged dry weather but also an abrupt increase in temperature due to asteroid impact (Lowe and Byerly, 2015).

FIGURE 2.9   (a) Magadi-type chert (Pleistocene), Lake Magadi, Kenya. (b) Marble Bar Chert (3.5-billion-year-old), Pilbara Craton, Western Australia.

Many of hydrothermal deposits are generally produced by precipitation of minerals due to changes in physicochemical conditions, including, e.g., temperature drop and changes of pH and redox potential. Some Phanerozoic metalliferous bedded cherts have been revealed to be hydrothermal in origin (e.g., Yamamoto, 1987). Archean cherts and IFs have often been considered to be of hydrothermal origin (e.g., Sugitani, 1992) (Figure 2.9b). However, their deposition was not always a direct consequence of specific hydrothermal activities. Archean seawater was likely characterized by background enrichment in Si and Fe. Indeed, Archean chert formation, especially in shallow basin, may reflect pH change or salinity increase (Stefurak et al., 2014). Carbonates such as limestone and dolostone composed of dolomite [$CaMg(CO_3)_2$] in the Precambrian, when organisms secreting carbonate minerals were absent, were mostly chemical sediments, although the formation of stromatolitic carbonates is likely triggered by microbial respiration.

2.4.3.3 Biochemical Sedimentary Rocks

Biochemical sedimentary rocks here refer to non-skeletal chemical sediments precipitated in a close association with biological activities. This category includes some IFs, phosphorites, cherts, and carbonates. When IFs are products of oxidation of ferrous iron (Fe^{2+}) directly by O_2 produced by cyanobacteria (oxygenic photosynthesizers), they could be called biogeochemical sedimentary rocks. Phosphorite refers to deposit enriched in phosphate (PO_4^{3-}) minerals represented by fluorapatite [$Ca_5(PO_4)_3F$], although the concentration of P as P_2O_5 varies widely from several % to more than 30%. Its formation generally requires high surface productivity and high flux of organic-P to the seafloor. This may remind readers to better call them biogenic sediments, although as described later, formation of some modern phosphorites is closely related to sulfur bacteria that release enough phosphate for precipitation of hydroxyapatite [$Ca_5(PO_4)_3(OH)$] (Schulz and Schulz, 2005). The oldest phosphorite is known to have been formed around 2.0 Ga ago. This event likely reflected elevated primary productivity at global scale, which was closely related to increase in chemical weathering of continental crusts and consequent increase in phosphorous fluxes to oceans (e.g., Papineau, 2010).

2.5 METAMORPHIC ROCKS

Metamorphism refers to the phenomenon that textures and mineral assemblage of rocks change without melting. This generally occurs under higher pressure and/or higher temperature than those under which rocks were lithified, although it also could occur under lower temperature and pressure conditions (retrograde metamorphism). In case of sedimentary rocks, metamorphism refers to textural and mineralogical rearrangement that occurs at temperature and pressure beyond diagenetic temperature and pressure in the broad sense (around 200°C and 200 MPa). Additionally, fluids occasionally play an important role for metamorphism. Recrystallization (change of grain shape and size), phase change (change of crystal type with the same chemical formula), and neomorphism (formation of new minerals) are the major processes of metamorphism. Metamorphisms could drastically change the appearance of rocks both in outcrops and under the microscope (Figure 2.10a, b).

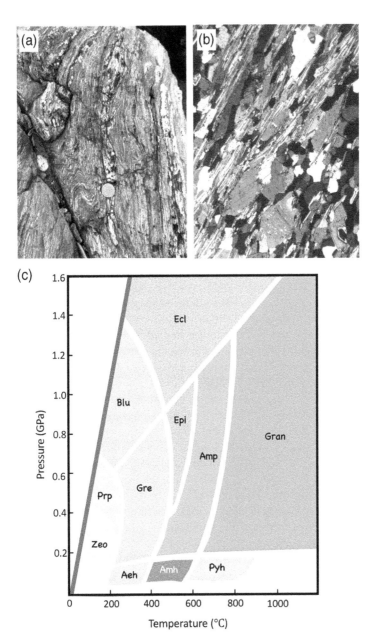

FIGURE 2.10 (a) Outcrop of the metamorphic rock of greenschist to amphibolite facies of the Mesozoic Sanbagawa Metamorphic Belt, in Mie Prefecture, central Japan. (b) Photomicrograph of mica schist {not from the outcrop shown in (a)}. Crossed nicols. (c) Pressure-temperature diagram for metamorphism of various facies represented by metamorphic minerals as follows. {Modified from Sen (2014)}. Zeo: Zeolite, Prp: Prehnite-Pumpellyite, Blu: Blueschist, Ecl: Eclogite, Gre: Greenschist, Epi: Epidote-Amphibolite, Amp: Amphibolite, Gran: Granulite, Aeh: Albite-Eclogite Hornfels, Amh: Amphibole Hornfels, Pyh: Pyroxene Hornfels.

Various types and facies of metamorphisms have been defined, including e.g., contact, regional, cataclastic, hydrothermal, burial, and shock metamorphisms. Facies of metamorphism include zeolite, prehnite-pumpellyite, blueschist, green-schist, epidote-amphibolite, amphibolite, granulite, and eclogite facies. Additionally, hornfels represents facies mainly restricted to contact metamorphism (metamor-phism caused by contact with magma). The temperature-pressure diagram for the metamorphic facies is shown in Figure 2.10c, which describes that the metamor-phic grade changes from the lowest zeolite facies to the highest granulite facies as a series of consistently increasing pressure and temperature, corresponding to increasing depth of burial, to intrusion of magma or hot igneous bodies, and to tectonics such as continental collisions producing extremely high pressure (e.g., Mason, 1990). Metasomatism is sometimes classified as one of the metamorphic types. This is a phenomenon that minerals of the protolith are dissolved and simul-taneously replaced at a solid state by other minerals, related to circulation of hydro-thermal and other fluids.

Metamorphism and metasomatism are critical in searching for traces of ancient life, because these phenomena could destroy or significantly modify biosignatures, which has long been a problem particularly in search for the Archean life. This is obvious, because most organisms live in an environment with very narrow "modest" ranges of temperature and pressure, compared with the ranges for metamorphisms. Deep sea is a high-pressure environment, but the pressure at the seafloor 100,000 m in depth is 100 MPa, half of the maximum pressure of diagenesis.

Complex OM in sediments is a diagenetic product of biopolymers: here the term "diagenesis" is used in the narrow sense, corresponding temperature up to ~60°C. Organic matter resistant to solvents such as dichloromethane (CH_2Cl_2) and ace-tone (C_3H_6O) is called kerogen, whereas solvent-dissolved OM is called bitumen. Kerogen would further degrade and volatilize during katagenesis (~60°C–150°C) and metagenesis (150°C–250°C), which would be followed by metamorphism. Through these processes, H, O, S, and P in kerogen would be preferentially released, resulting in kerogen composed mainly of C. Such thermally matured kerogen rarely preserves biomarkers, organic compounds unambiguously originated from biopoly-mers. ^{12}C-enriched carbon isotopic composition, one of the most widely utilized biosignatures in ancient rocks, could be modified by metamorphism to be heavier values (e.g., Wada et al., 1994). Recrystallization also hinders preservation of biosig-natures, particularly of morphological ones (cellular fossils). Recrystallization basi-cally includes enlargement of crystal sizes, which could modify and even destroy cellular structures. Recrystallization and resultant enlargement of crystals may dis-place and sweep away amorphous organic particles to mimic cellular microfossils (e.g., Brasier et al., 2006).

COLUMN: GEOCHEMISTRY OF IGNEOUS ROCKS AND OF MAGMATIC PROCESSES

As described in the main text, the chemical compositions of igneous rocks are con-trolled by various processes such as the composition of parental rock, degree of partial melting, fractional crystallization, and assimilation. During partial melting, elemental fractionation occurs between liquid and solid phases, leading to generation of magma compositionally different from parental rock, represented by an increase in Si in liquid phase due to retention of olivine and pyroxene group minerals in solid phase. In addition, radius and charge number of ions are critical factors for elemental fractionation. Elements with significantly large ionic radii tend to be fractionated into liquid phase. Such elements are called large ion lithophile elements (LILE), including K, Li, Rb, Cs, Sr, Ba, rare earth elements (REEs), Th, and U. Elements with large charge number also tend to be concentrated into liquid phase. Such elements are called high field strength elements (HFSE), including Hf, Zr, Ti, Nb, and Ta. The term "incompatible elements" integrates LILE and HFSE. Due to repeated partial melting and magmatism, the Earth's mantle has gradually depleted in incompatible elements, which have been accumulated eventually into the continental crusts. Thus, concentrations of incompatible elements should differ between basalts erupted at oceanic setting and those at continental setting. The latter is expected to be enriched in incompatible elements, reflecting contamination of crustal materials (assimilation). In other words, concentrations of incompatible elements could be clue to the tectonic setting of basalt eruption, providing implications for how Archean greenstone belts were formed.

REFERENCES

Bonin, B., Janoušek, V., Moyen, J-F. 2020. Chemical variation, modal composition and classification of granitoids. *Geological Society, London, Special publications* 491, 9–51.

Brasier, M., McLoughlin, N., Green, O., Wacey, D. 2006. A fresh look at the fossil evidence for early Archean cellular life. *Philosophical Transactions of the Royal Society B* 361, 887–902.

Brasier, M.D., Matthewman, R., McMahon, S., Wacey, D. 2011. Pumice as a remarkable substrate for the origin of life. *Astrobiology* 11, 725–735.

Chappell, B.W., White, A.J.R. 1974. Two contrasting granite types. *Pacific Geology* 8, 173–174.

Conrad, C.P., Lithgow-Bertelloni, C. 2006. Influence of continental roots and asthenosphere on plate-mantle coupling, *Geophysical Research Letters* 33, L05312.

Eriksson, K.A., Simpson, E.L. 2000. Quantifying the oldest tidal record: The 3.2 Ga Moodies Group, Barberton Greenstone Belt, South Africa. *Geology* 28, 831–834.

Eugster, H.P. 1967. Hydrous sodium silicates from Lake Magadi, Kenya: Precursors of bedded chert. *Science* 157, 1177–1180.

Fisher, R.V., Smith, G.A. 1991. Volcanism, tectonics and sedimentation. In *Sedimentation in volcanic settings. SEPM (Society for Sedimentary Geology) Special Publication* 45, 1–5.

Goldblatt, C., Claire, M.W., Lenton, T.M., Matthews, A.J., Watson, A.J., Zahnle, K.J. 2009. Nitrogen-enhanced greenhouse warming on early Earth. *Nature Geoscience* 2, 891–896.

Grotzinger, J., Press, S. 2007. *Understanding Earth*. W. H. Freeman and Company, New York. 579pp.

Harnois, L. 1988. The CIW index: A new chemical index of weathering. *Sedimentary Geology* 55, 319–322.

Hessler, A.M. Lowe, D.R. 2006. Weathering and sediment generation in the Archean: An integrated study of the evolution of silcicilastic sedimentary rocks of the 3.2 Ga Moodies Group, Barberton greenstone belt, South Africa. *Precambrian Research* 151, 185–210.

Lowe, D.R., Byerly, G.R. 2015. Geologic record of partial ocean evaporation triggered by giant asteroid impacts, 3.29-3.23 billion years ago. *Geology* 43, 535–538.

Mason, R. 1990. Metamorphism in collision zones. In: *Petrology of the Metamorphic Rocks*. Springer, Dordrecht. doi:10.1007/978-94-010-9603-4_7.

Open University Oceanography Course Team, 1989. *Ocean Chemistry and Deep-Sea Sediments*. Pergamon Press, New York, p. 134.

Papineau, D. 2010. Global biogeochemical change at both ends of the Proterozoic: Insights from phosphorites. *Astrobiology* 10, 165–181.

Rasmussen, B., Buick, R. 1999. Redox state of the Archean atmosphere: Evidence from detrital heavy minerals in ca. 3250–2750 Ma sandstones from the Pilbara Craton, Australia. *Geology* 27, 115–118.

Samuels, T., Bryce, C., Landenmark, H., Marie-Loudon, C., Nicholson, N., Stevens, A.H., Cockell, C. 2020. Microbial weathering of minerals and rocks in natural environments. In K. Dontsova et al., eds. *Biogeochemical Cycles: Ecological Drivers and Environmental Impact, Geophysical Monograph 251*. American Geophysical Union. Wiley & Sons, Inc. pp. 59–79. doi:10.1002/9781119413332.ch3.

Schulz, H.N., Schulz, H.D. 2005. Large sulfur bacteria and the formation of phosphorite. *Science* 307, 416–418.

Sen, G. 2014. Petrology—Principles and practice. p. 260. Springer, Verlag. p. 368.

Seyfried Jr., W.E., Mottl, M.J. 1982. Hydrothermal alteration of basalt by seawater under seawater-dominated conditions. *Geochimica et Cosmochimica Acta* 46, 985–1002.

Siever, R. 1992. The silica cycle in the Precambrian. *Geochimica et Cosmochimica Acta* 56, 3265–3272.

Som, S.M., Catling, D.C., Harnmeijer, J.P., Polivka, P.M., Buick, R. 2012. Air density 2.7 billion years ago limited to less than twice modern levels by fossil raindrop imprints. *Nature* 484, 359–362.

Stefurak, E.J.T., Lowe, D.R., Zentner, D., Fischer, W.W. 2014. Primary silica granules—A new mode of Paleoarchean sedimentation. *Geology* 42, 283–286.

Sugitani, K. 1992. Geochemical characteristics of Archean cherts and other sedimentary rocks in the Pilbara Block, Western Australia: Evidence for Archean seawater enriched in hydrothermally-derived iron and silica. *Precambrian Research* 57, 21–47.

Taylor, S.R., McLennan, S.M. 1985. *The Continental Crust: Its Composition and Evolution*. Blackwell Scientific Publications, Oxford.

Wacey, D., Noffke, N., Saunders, M., Guagliardo, P., Pyle, D.M. 2018. Volcanogenic pseudo-fossils from the ~3.48 Ga Dresser Formation, Pilbara, Western Australia. *Astrobiology* 18, 539–555.

Wada, H., Tomita, T., Matsuura, K., Tuchi, K., Ito, M., Morikiyo, T. 1994. Graphitization of carbonaceous matter during metamorphism with references to carbonate and pelitic rocks of contact and regional metamorphisms, Japan. *Contribution to Mineralogy and Petrology* 118, 217–228.

Xia, L., Li, X. 2019. Basalt geochemistry as a diagnostic indicator of tectonic setting. *Gondwana Research* 65, 43–67.

Yamamoto, K. 1987. Geochemical characteristics and depositional environments of cherts and associated rocks in the Franciscan and Shimanto Terranes. *Sedimentary Geology* 52, 65–108.

3 Life on the Earth 1

3.1 INTRODUCTION

Life is quite difficult to be uniquely defined. However, the following constraints can be placed on cellular life on the Earth. First, life should be composed of at least one compartment called "cell", being isolated from the environment by membrane. Second, the membrane has an ability to take materials such as nutrients from the outer environment into cell lumen and to excrete, e.g., metabolic products. Third, such the cellular life should replicate itself to produce offspring. Many geobiologists and astrobiologists now believe that life on the Earth had emerged in the Hadean (>4.0 Ga ago).

Larsen et al. (2017) estimated that at least 1–6 billion species were present on the Earth and that unlike most previous studies, they were dominated by bacteria (approximately 70%–90% of the total species). The number of already described species was approximately 1,500,000. So if Larsen et al.'s estimate is the case, only less than 15% of species living on our planet have been identified. All the known cellular life from unicellular bacteria less than 5 μm in diameter to multicellular giant marine mammal "blue whale" up to 30 m long have the same system of life. The system has been called "central dogma"; genetic information stored in deoxyribonucleic acid (DNA) is transferred to ribonucleic acid (RNA), based on which proteins carrying metabolisms are synthesized, although this does not always mean that DNA world of life evolved first. Despite more than 80% of species on the Earth have not yet been identified, little scientists appear to consider that distinct life systems could be employed by unidentified species. In either case, surprising is that so diverse organisms have employed the same system of life. Though indirect, this could be taken as evidence for all the extant organisms having a common ancestor, which is called, e.g., Last Universal Common Ancestor (LUCA). This also secures our premise that modern life is a clue to early life and even origin of life. From these points of view, life on the Earth is overviewed in this chapter, including origin of life, classification system of life, and diversity of metabolisms and ecosystems. It may be noted that some researchers have argued the possibility that life had initially evolved on Mars and had been transported to the Earth (e.g., Kirschvink and Weiss, 2002). This is a fascinating hypothesis but not mentioned to here.

3.2 CHEMICAL EVOLUTION AND EMERGENCE OF LIFE ON THE EARTH

Chemical evolution is generally defined as a process involving evolution of biopolymers such as proteins, carbohydrates, nucleic acids, and lipids, preceding evolution of protocells. Organic compounds (OCs) as building blocks of biomolecules were initially produced from simple inorganic molecules such as H_2, CO, CH_4 and NH_3 through abiogenetic processes. Inorganic molecules were reacted with each other,

DOI: 10.1201/9780367855208-3

with appropriate energies and catalysts, to form monomers such as amino acids and nucleobases. Monomers were subsequently polymerized through dehydration condensation to eventually form biopolymers. This theory is called Oparin-Haldane theory, with respect to the two independent advocators, A. Oparin in Russia and J.B.S. Haldane in United Kingdom. A number of experiments on chemical evolution have been conducted since Miller's paper published in 1953 (Miller, 1953). Miller's experiment, extraterrestrial delivery of building blocks of life and the two major conflicting hypotheses of origin of life (deep-sea hydrothermal origin vs. terrestrial hydrothermal origin) are described here.

3.2.1 REVISIT TO MILLER'S EXPERIMENT

In Miller's first short technical paper published in 1953 (Miller, 1953), he said that the experiment was conducted in order to test Urey and others' hypothesis that biopolymers were abiotically formed when the Earth had a strongly reducing atmosphere enriched in CH_4, NH_3, and H_2 (Urey, 1952). Urey postulated that the primitive reducing atmosphere was formed in the consequence of chemical reactions that occurred during collisions of planetesimals containing metallic Fe and H_2O. Here, H_2 produced by reactions between metallic Fe and H_2O is a key component. For example, H_2 reacts with CO to produce C, which further reacts with H_2 to produce CH_4 (Urey, 1952). Based on this, Miller conducted a very simple experiment mimicking reactions presumably occurring in the Earth's primitive atmosphere, where reducing compounds (CH_4, NH_3, H_2) and H_2O have reacted with each other in association with free radicals produced by electrical discharge. Strikingly the products contained three amino acids including glycine, and alanine. Miller had continued the experiments with improved reaction apparatus and procedures, and published comprehensive paper in 1955 (Miller, 1955), which reported production of five well-identified amino acids (glycine, alanine, sarcosine, β-alanine, and α-aminobutyric acid). In addition, volatile acids such as formic acid (HCOOH) and acetic acid (CH_3COOH) were also identified. Abiogenic origin of these compounds was comprehensively discussed in this paper: one of the key points for this is that the produced amino acid (alanine) was racemic. It was also suggested that aldehydes (R-CHO) and hydrogen cyanide (HCN) were direct products of electrical discharges. These two compounds and NH_3 are used as starting materials in the well-known "Strecker Reaction (artificial synthesis of amino acids)". Formaldehyde (HCHO) and HCN detected in interstellar molecular clouds are known to be potential building blocks for sugars and nucleobases.

Although these innovative studies had led many researchers to have conducted experiments of chemical evolution, their premise "the primitive, highly reducing, H_2- and NH_3-containing atmosphere" was later questioned. The primitive atmosphere might have been composed of $CO_2+N_2+H_2O$, (Kasting, 1993 and reference therein); under such neutral ~ mildly oxidizing atmospheric condition, the efficiency of OC synthesis was much lower (e.g., Miyakawa et al., 2002). This view appears to be still persistent and indeed many researchers favor other sites for chemical evolutions, represented by deep-sea hydrothermal vents and extraterrestrial environments as described later. However, we should not overlook the discussion by Cleaves et al.

(2008), who demonstrated that the poor "final" products obtained under the neutral atmospheric compositions have resulted from decomposition of primarily synthesized OCs by oxidizers such as nitrite (NO_2^-) and nitrate (NO_3^-) that were produced by electric discharge. It was also emphasized that the breakdown of primary OCs could be inhibited by reductants such as ferrous iron (Fe^{2+}), likely abundant on the early Earth (but also see Kuwahara et al., 2012). Additionally, we need to keep it in mind that the primitive atmospheric condition is still controversial, and the reducing model involving the presence of significant amounts of H_2, CO, and CH_4 has recently be revived (Catling and Zahnle, 2020; Hashimoto et al., 2007) as discussed in Chapter 1.

3.2.2 DELIVERY OF BUILDING BLOCKS OF LIFE FROM SPACE

Not a few researchers consider the possibility that chemical evolutions had proceeded in space. Building blocks of life such as amino acid and sugar could have been carried by asteroids and comets to the Earth. This possibility has been evidenced by spectroscopic analyses of molecular clouds, experimental studies, and direct analyses of comets and meteorites (e.g., Alexander et al., 2007; Llorca, 2005; Sugahara and Mimura, 2014; Furukawa et al., 2019). The representative is Murchison meteorite (Figure 3.1), which fell near the small town "Murchison" in Victoria, Australia in 1969. Murchison meteorite is the least altered (not heated) carbonaceous chondrite (CM2), which was delivered from undifferentiated parental celestial body subjected to moderate aqueous alteration. Organic matter in Murchison meteorite has been extensively studied, and various types of amino acids were detected, including those not contained in any organisms, in addition to some essential amino acids comprising proteins (e.g., glycine and alanine). There were once some controversies on possibility of contamination of the detected amino acids, which was due to identification

FIGURE 3.1   Murchison meteorite. (Photo courtesy by K. Mimura.)

of some excess of L-type optical isomer over D-type one for some amino acids, as synthetically (namely inorganically) produced amino acids are racemic (the same amount of L- and D-type amino acids) and organisms on the Earth synthesize and utilize L-type of optical isomer of amino acids {also see Genchi (2017) for synthesis and utilization of D-type amino acids by organisms}. Though details are omitted, this controversy was settled, and the indigenousness of the detected amino acids has been fully established. As of 2017, more than 90 amino acids have been identified, and other formation mechanisms such as the formose reaction (formation of sugar from formaldehyde) in addition to the well-known Strecker reaction are proposed (Koga and Naraoka, 2017). Although asteroids could have been promising distributers of raw materials for biopolymers, one problem is that most of the meteoritic amino acids are identified in products of hydrolysis of OM. The complex precursory OM could not have been directly utilized as building blocks for biopolymers, even though they were successfully delivered to the Earth. One more thing to be remembered is that most (~70%) of the OM contained in meteorites is not solvent soluble, called insoluble organic matter (IOM); it is poorly understood how IOM contributed to chemical evolution on the Earth.

Asteroids did not only deliver OM from the space to the Earth, but also could have produced chemical compounds indispensable for building biopolymers, at the time of collision to the Earth. Nakazawa et al. (2005) and Furukawa et al. (2009) experimentally demonstrated that ammonium (NH_4^+) and amino acid (glycine) were formed from N_2 in the atmosphere during collision, with catalysis of metallic Fe, which was available from iron meteorites. More recently, Takeuchi et al. (2020) also showed that impact-induced reactions with strating materials of metallic Fe and Ni, forsterite (Mg_2SiO_4), H_2O, CO_2, and N_2 could produce alanine in addition to glycine. The reaction was supposed to be initiated from reduction of CO_2, N_2, and H_2O, coupled with oxidation of metallic Fe and Ni. Furthermore, Deamer et al. (2002) implied that amphiphilic molecules with both hydrophilic and hydrophobic properties in meteorites contributed to have produced complex biomolecular layers and eventually membranous vesicles as candidates for the first cells, in terrestrial hydrothermal (hot spring) settings.

3.2.3 Deep-Sea Hydrothermal Vent Systems and Origin of Life

Inspired by the discovery of deep-sea black smoker-type hydrothermal vents and related light-independent ecosystems (*Column*), the hypothesis that life was evolved at ancient deep-sea hydrothermal systems was proposed and has long been popular among scientists (e.g., Corliss et al., 1981). There are some reasons for this as noted below.

First of all, black smoker-type hydrothermal systems from which acidic (pH < 2.0) and hot (>300°C) fluids were emanated gave impressions to scientists that deep ocean floors in the Hadean should have been look-alike (Figure 3.2a). Second, the first evolved life is unlikely equipped with highly evolved metabolism "photoautotrophy (photosynthesis)", while chemolithoautotrophy (chemosynthesis) appears to be reasonable as primary autotrophic metabolism (the Section 3.4). Deep-sea hydrothermal vent is an ideal place for chemolithoautotrophs and indeed deep-sea hydrothermal ecosystems are

FIGURE 3.2 (a) Black smoker at a mid-ocean ridge hydrothermal vent in Atlantic Ocean. Reuse from NOAA Photo Library, as a public domain. (b) Beehive carbonate chimney venting 91°C fluid, the highest temperature recorded at the Lost City Hydrothermal Field (Denny et al., 2015). (Photo courtesy by A.R. Denny.)

supported by chemolithoautotrophic primary production. Hydrothermal fluids are rich in reducing substrates such as CH_4 and hydrogen sulfide (H_2S) indispensable for chemolithoautotrophy. Third, abiotic synthesis of NH_3 and CH_4 under hydrothermal condition is demonstrated experimentally and observationally (e.g., Schoonen and Xu, 2004; McCollom and Seewald, 2007). In addition, metals and metal sulfides that can catalyze formation of OCs are prolific in black smoker-type hydrothermal fluids. Indeed, various hydrocarbons and even amino acids have been experimentally synthesized under hydrothermal conditions (e.g., McCollom and Seewald, 2007; Marshall, 1994). Fourth, clay minerals such as montmorillonite [$(Na,Ca)_{0.33}(Al,Mg)_2(Si_4O_{10})(OH)_2$ nH_2O] formed by hydrothermal alteration of oceanic crusts have an ability of abiotic synthesis, polymerizations of OCs, and protecting them from thermal decomposition (e.g., Williams et al., 2005). Fifth, the deep-sea environment appears to have been convenient for preservation of abiotically synthesized OCs. If hydrothermally synthesized OCs were exposed to expected strong UV-radiation due to the absence of ozone screen in the Hadean, they could be immediately decomposed. Finally, LUCA is generally thought to have been thermophile. Akanuma et al. (2013) experimentally demonstrated that LUCA flourished at a temperature above 75°C, although this does not readily mean that the first life had emerged in a high-temperature environment.

Nonblack smoker-type hydrothermal system has also been paid attention. Representative is "Lost City Hydrothermal Field (LCHF)", discovered from the seafloor mountain "Atlantis Massif" near Mid-Atlantic Ridge (Kelley et al., 2001). This hydrothermal system is located off spreading axis and on a 150 million-year-old crust. Massive white chimneys up to 60 m high characterize this hydrothermal system (Figure 3.2b) and are composed of carbonates such as calcite and aragonite ($CaCO_3$) and magnesium hydroxide such as [brucite; $Mg(OH)_2$]. Lost city hydrothermal fluids are relatively low in temperature (< 40°C–90°C) than black smoker fluids and higher in pH (9.0–11) than ambient seawater. While not so enriched in H_2S (0.04 mmol/Kg), the emanated fluids occasionally contain a significant amount of CH_4 (~0.28 mmol/Kg) and H_2 (~0.43 mmol/Kg). Metal concentrations are generally very low. The fluids are formed by reactions of seawater and tectonically exposed mantle peridotite (ultramafic rock) but not by magmatic force. Olivine [$(Fe, Mg)_2 SiO_4$], a major constituent mineral in ultramafic rocks, reacts with H_2O and CO_2 to produce serpentine [$Mg_3Si_2O_5(OH)_4)$], magnetite (Fe_3O_4), and CH_4. This is called serpentinization. Also, reaction of fayalite (Fe_2SiO_4) (one of the two endmembers of solid solution of olivine) with H_2O produces magnetite and H_2, with dissolved silica (also see Chapter 5).

Assuming that the Hadean seawater was acidic and enriched in metals, Russel and his coworkers proposed a hypothesis of evolution of life at monosulfide (FeS/NiS)-bearing mounds precipitated from alkaline hydrothermal fluids rich in H_2 generated by serpentinization; like LCHF (Martin and Russell, 2003; Martin et al., 2008; Russell et al., 2010). Key assertions of this hypothesis include: (1) vesicles compartmentalized by membraneous monosulfide walls in mounds act as templates for primordial cells and as catalysts for synthesis of OCs, (2) steep proton (H^+) gradient across the membraneous walls was produced by acidic seawater outside and alkaline hydrothermal fluids inside, (3) energy produced by this proton gradient and CO_2 reduction by H_2, producing CH_4, having being production of formate ion (CHOO–) and formyl (–CHO). This process is homolog to the simplest metabolic pathway known as the reductive acetyl-coenzyme A (Acetyl-CoA) pathway generating CH_4 and/or acetate ion (CH_3COO^-) and is implied to have been tapped by the first life. Such the idea gives us the opportunity of different approaches to the origin of life, distinct from the chemical evolution of life hypothesis.

3.2.4 Terrestrial Hydrothermal Systems (Hot Springs): Another Candidates for Birthplace of Life

Terrestrial origin of life was first envisaged by Darwin, as "warm little pond" (Darwin, 1859). Other terrestrial sites such as tidal flat and porous surface of hot volcanic rocks were also proposed as alternative birthplaces of life. This concept "evolution of life on land, fueled by OCs supplied from the anoxic atmosphere" has long been retained between scientific communities, because the terrestrial environment has an advantage for concentrating OCs by evaporation and even polymerization by dehydration condensation of monomers (e.g., from amino acids to peptide) could have occured. As discussed above, the situation had turned, being triggered by the revision of the primitive reducing atmospheric composition to neutral or even mildly oxidizing one and the discovery of deep-sea hydrothermal vent systems. Seemingly, the deep-sea hydrothermal origin of life has been regarded as the most plausible model at least for the last two decades. Recently, however, the terrestrial origin of life has been reshaped and rejuvenated, as "the hot spring (geothermal) hypothesis for an origin of life" (Damer and Deamer, 2020 and reference therein). Hot spring combined with natural nuclear system as a birthplace of life has also been proposed (Maruyama et al., 2019).

Hypothesis for hot spring origin of life proposed by Mulkidjanian et al. (2012), personally attractive to me, is introduced here. Both modern and reconstructed Hadean seawaters are likely depleted in key functional ions such as K^+, Zn^{2+}, Mn^{2+} and phosphate (PO_4^{3-}), and are lower in K^+/Na^+ at up to several orders of magnitudes, compared with cell cytoplasm. Proteins and functional systems with affinity to and requirement for these and Mg^{2+} ions are virtually universal and thus assumingly primordial. The extant organisms and plausibly LUCA maintain intracellular ionic concentrations, which is enabled by energy-consuming system of ion-tight membranes. Such highly evolved membrane must be absent in protocells. In other words, protocells are supposed to have "had similar intracellular ionic concentrations with ambient fluids. Therefore, it is argued that protocells had unlikely evolved in marine setting" (Mulkidjanian et al., 2012). Terrestrial hot spring setting including geyser have advantages in providing higher concentrations of K^+, Zn^{2+}, Mn^{2+}, Mg^{2+}, and PO_4^{3-}. As evidenced by some modern examples, terrestrial hydrothermal fluids are generally characterized by much higher ionic concentrations compared with seawater, due to their hydrothermal extraction from bedrock at low water/rock ratios. High K^+/Na^+ ratios could have been achieved by separation of K^+ from Na^+ through vaporization process at fumaroles. In addition, terrestrial hydrothermal fluids are often rich in boron (B), which is suggested to be essential to chemical evolution of ribose (e.g., Scorei, 2012). Sulfide minerals such as sphalerite (ZnS) and alabandite (MnS) that are assumed to be precipitated from hydrothermal fluids were not oxidized under the anoxic Hadean atmosphere. This means no production of sulfuric acid (H_2SO_4), contrastive to oxidative weathering of sulfides. Therefore, small ponds around hydrothermal vents could not have become strongly acidic that is unsuitable condition for life.

The stability of hydrothermally precipitated ZnS and MnS was significant in the context of supply and retention of building blocks of life, because these sulfides have ability to catalyze photosynthesis of OCs and adsorb UV radiation. At terrestrial hydrothermal systems, multiple sources for building blocks of life were available, including hydrothermal, atmospheric, extraterrestrial, and photosynthetic. If stored in small ponds around hot-spring (geyser) vents, these multiple-sourced materials would have been condensed and polymerized by evaporation. This process could have eventually produced membranous compartments of protocells.

3.3 CLASSIFICATION OF LIFE ON THE EARTH

Organisms have been classified in many ways. Representative is a classical system with the hierarchical seven taxonomic ranks including kingdom, phylum, class, order, family, genus, and species. Naming with a combination of genus and species is known as binominal nomenclature, which was proposed by Carl von Linné. The five-kingdoms system, a classification scheme based largely on differences in style of nutrient-uptake, mobility, unicellularity or multicellularity, and eukaryotic or prokaryotic, was proposed by Whittaker (1969) and has long been widely utilized. This system includes Animalia, Plantae, Fungi, Protista and Monera kingdoms. Kingdom Animalia comprises heterotrophic eukaryotic organisms with mobility at least one stage of life cycle. They take foods as nutrient by feeding, represented by vertebrates. Kingdom Fungi represents heterotrophs without mobility but take nutrients through cell membrane, represented by mushrooms. Kingdom Plantae comprises oxygen-producing photoautotrophs with some significant cell and organ differentiation represented by vascular bundle, including green plants and red and brown algae. Kingdom Monera is for unicellular prokaryotic organisms. Kingdom Protista includes relatively diverse taxonomic groups composed mainly of eukaryotic unicellular microorganisms.

C.R. Woese and colloborators had proposed a new taxonomic rank "domain" (Woese et al., 1990). The proposed three-domain system includes Eukarya, Archaea, and Bacteria (Figure 3.3a). Both Archaea and Bacteria are unicellular prokaryotes and have long been classified into a single kingdom as Monera (see above). This new classification introduced by Woese was based on the variation of base sequences of ribonucleic acids in ribosome (rRNA) that is commonly contained in all types of

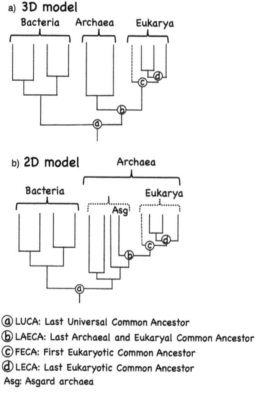

ⓐ LUCA: Last Universal Common Ancestor
ⓑ LAECA: Last Archaeal and Eukaryal Common Ancestor
ⓒ FECA: First Eukaryotic Common Ancestor
ⓓ LECA: Last Eukaryotic Common Ancestor
Asg: Asgard archaea

FIGURE 3.3 Universal phylogenies: three (a) and two (b) domain models. Drawn based on Figure 3.1 in Doolittle (2020).

cellular organisms. Interestingly, rRNA genomic data imply that phylogenic association between Eukarya and Archaea is closer than that between Archaea and Bacteria (Figure 3.3a). Recently based on analyses of more than 3,000 eukaryotic and archaeal gene families, Williams et al. (2019) demonstrated that eukaryotes had been evolved from Archaea specifically close to Asgard archaea. The tree of life composed of the only two domains, Bacteria and Archaea is even proposed (Doolittle, 2020 and references therein) (Figure 3.3b).

In addition to the above-described taxonomic framework, we have used the other different schemes including unicellular vs. multicellular and prokaryotes vs. eukaryotes. Relationships between such the different biological classification schemes and cell structures of prokaryote and eukaryote are shown in Figure 3.4. Prokaryotic cells typically range from 0.1 to 5 µm across and mostly smaller than 10 µm, although some species can be large up to >500 µm (e.g., Schulz and Jørgensen, 2001). Lack of a true nucleus characterizes prokaryotic cells, which contain DNA in an irregularly shaped region called nucleoid. Eukaryotic cells generally range from 10 to 100 µm across and contain a true membrane-bound nucleus. Various organelles with specialized functions are contained, including, e.g., ribosome (contained also in prokaryotic cells), lysosome, mitochondrion, endoplasmic reticulum, chloroplast, and golgi

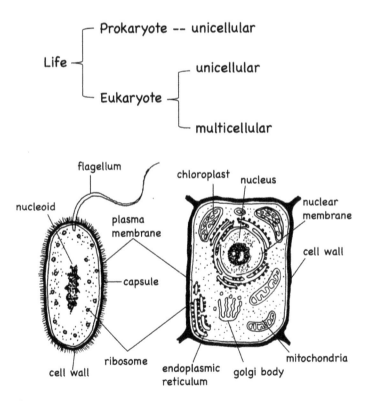

FIGURE 3.4 Basic classification scheme of life and schematic views of eukaryotic and prokaryotic cells. Note that some prokaryotic cyanobacteria exhibit multicellularity in wide sense.

body. The reason why prokaryotic cells is generally small is simple, as well known as "a surface-area-to-volume ratio" (e.g., Harris and Theriot, 2018). Approximating prokaryotic cells as a sphere (r = radius), their volume and surface area can be calculated as $4\pi r^3$ and $4\pi r^2$, respectively. Thus, the ratio of surface area to volume decreases with an increase of cell size. If the cell size becomes to be too large, the surface area of the plasma membrane becomes to be insufficient to support the rate of material (nutrient) diffusion into cell interior. Similarly, wastes could not be diffused out through the membrane quickly. Enlargement of eukaryotic cells and development of specialized organelles are thought to had been closely related to the evolution of transportation system mediated by cytoskeleton (also see Marshall et al., 2012).

3.4 DIVERSITY IN METABOLISMS

3.4.1 AUTOTROPHY AND HETEROTROPHY

In addition to classification based on morphological and functional diversities, organisms can be classified based on how to obtain energy and C as a major framework element of biopolymers. Autotrophs refer to organisms that utilize energy from light or inorganic reducing chemicals to produce OCs and adenosine triphosphate (ATP), which is often referred to as "molecular unit of currency".

Light-dependent autotroph and chemical-dependent one are called photolithoautotroph and chemolithoautotroph, respectively. In either case, autotrophs do not depend on external biosynthesized OCs as a carbon source. Instead, they utilize inorganic C such as CO_2 and bicarbonate (HCO_3^-). Autotrophically synthesized and stored biopolymers, particularly carbohydrates and lipids, are also catabolized to produce ATP.

Heterotrophs are organisms that depend on external OCs as carbon sources. Depending on the energy source, three subtypes are identified, including chemoorganoheterotroph, photoorganoheterotroph, and chemolithoheterotroph. Chemoorganoheterotroph utilizes external OCs as the energy source, whereas photoorganoheterotroph utilizes light. Chemolithoheterotroph refers to heterotrophic organism that utilizes inorganic chemical substrates as the energy source. These five trophic types are schematically illustrated in Figure 3.5. Eukaryotes involve photolithoautotrophs and chemoorganoheterotrophs, whereas prokaryotes do all the seven trophic types. Metabolic diversity in prokaryotes is incomparably higher than eukaryotes. It may be reminded that in addition to these seven specific types, not a few organisms employ mixotrophy, as represented by carnivorous plants. Mixotrophs are known also for some prokaryotes.

3.4.2 CHEMISTRY OF AUTOTROPHY

Oxygenic photosynthesis is a representative metabolism operated by photolithoautotrophs utilizing inorganic matter both as electron donor and acceptor, as shown by the following chemical reaction formula (Equation 3.1),

$$6CO_2 + 6H_2O(+\text{light}) \rightarrow C_6H_{12}O_6 + 6O_2 \tag{3.1}$$

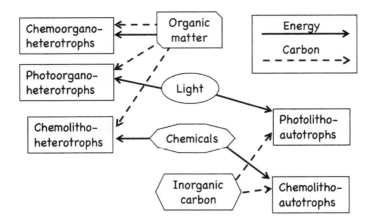

FIGURE 3.5 Classification of organisms based on sources of carbon to construct biomolecules and energy source to produce ATP or to produce biomolecules. Note that abbreviations (e.g., photolithoautotrophs → photoautotrophs, chemolithoautotrophs → chemoautotrophs) are often used.

where H_2O (as electron donor) gives an electron to CO_2. Namely, H_2O is oxidized to be O_2, whereas CO_2 (as electron acceptor) is reduced to be glucose ($C_6H_{12}O_6$). This reaction, oxygenic photosynthesis, proceeds with light energy (not shown in the equation), being operated by cyanobacteria (prokaryote) and green plants and algae (eukaryotes). Anoxygenic photosynthesis is a metabolism performed exclusively by prokaryotic photolithoautotrophs. Chemolithoautotrophs (exclusively prokaryotes) utilize inorganic matter both as electron donor and acceptor to produce OCs using chemical energy. Chemical energy is obtained by oxidation of reduced chemical compounds (as electron donor), including e.g., H_2S, H_2, Fe^{2+}, NH_4^+, Mn^{2+}, and H_2O. Representative electron acceptors are O_2 and NO_3^-, although various combinations of electron donor and acceptor are possible, depending on different redox potentials. Equation 3.2 represents reactions producing chemical energy employed by sulfur oxidizing bacteria (Fenchel et al., 2012). More detailed information about chemolithoautotrophs related to global cycles of S, C, and N will be given in the next chapter.

$$5H_2O + 2SO + 3CO_2 \rightarrow 3(CH_2O) + 2SO_4^{2-} + 4H \qquad (3.2)$$

3.4.3 CHEMISTRY OF HETEROTROPHY

Chemoorganoheterotrophs utilize OCs as energy and carbon sources, including protists, fungi and animals (eukaryotes), and various prokaryotes. The chemical reaction formula of chemoorganoheterotrophy employing O_2 as an electron acceptor is shown below (Equation 3.3), which is a reverse reaction of photolithoautotroph (Equation 3.1).

$$C_6H_{12}O_6 + 6O_2 \rightarrow 6CO_2 + 6H_2O \qquad (3.3)$$

All the eukaryotic chemoorganoheterotrophs are basically O_2-dependent (aerobic), although under O_2-deficient environment, some species such as yeasts could metabolize OCs without O_2. In addition to O_2, prokaryotic chemoorganoheterotroph could use various chemical compounds such as NO_3^-, ferric iron (Fe^{3+}), and sulfate (SO_4^{2-}) as electron acceptor, which are basically species-specific. Examples of chemical formula for the anaerobic respirations are shown below. Equations 3.4 and 3.5 (Fenchel et al., 2012) represent metabolisms of denitrifying and sulfate reducing bacteria, respectively. More detailed information about chemoorganoheterotrophic processes related to global cycles of C, S and N will be given in the next chapter.

$$5(CH_2O) + 5H_2O + 4NO_3^- + 4H^+ \rightarrow 5CO_2 + 2N_2 + 12\ H_2O \qquad (3.4)$$

$$CH_3COO^- + SO_4^{2-} + 2H^+ \rightarrow 2CO_2 + 2H_2O + HS^- \qquad (3.5)$$

It should be emphasized again that modern prokaryotes are much more diverse in metabolic strategy than eukaryotes, notably including species that do not rely on O_2. Therefore, metabolisms of modern prokaryotes should be key to understand ecosystems and its evolution in the Archean, plausibly without O_2 at least in its earlier period.

3.5   ECOSYSTEM: COMPLEX SYSTEM OF LIFE AND ENVIRONMENT

Ecosystem is defined as a semi-closed sustainable (to some extent) system composed of a biological community (a sum of populations of different species) and nonbiological environmental components such as light, water, gases, soils, nutrients, and energies. Nutrients and energies are circulated among biological community and surrounding environment. Organisms comprising biological community can be classified into producer, consumer, detritivore, scavengers, and decomposer, based on their roles in the ecosystem. The term "producer" refers to autotrophs. Representative autotrophs in the terrestrial and marine ecosystems are plants and algae, O_2-producing photoautotrophs. Consumers refer to organisms that obtain nutrients by eating other living organisms, though not equivalent to heterotrophs. Hierarchical classification can be made for consumers such as primary, secondary, and tertiary consumers etc. Primary consumers depend directly on OCs of the primary producers, whereas the secondary consumers eat the primary consumers. Obviously, such relationship is not linear. Therefore, nowadays the term "food web" is preferred to "food chain". Food chain mainly comprised of primary producers and consumers is called grazing food chain.

Detritivores eat remains and excrements of other organisms (detritus) using mouth. Representative detritivores are earthworms and dung beetles. Scavengers are defined for animals that eat dead bodies, as represented by hyenas and vultures. Many arthropods are also classified as scavengers. Decomposers are microorganisms, which are represented by heterotrophic prokaryotes and fungi. These microorganisms without mouths secrete enzymes to the environment, which degrade their foods to be small molecules and absorb them through cell membranes. These three groups comprise detritus food chain. Particularly decomposers are key players in this chain and even in the whole ecosystem, because they play an important role for final mineralization of biomolecules to inorganic nutrients such as NH_4^+, PO_4^{3-}, NO_3^-, and SO_4^{2-}. The mineralization is indispensable for reuse of these essential nutrients by producers.

Archean ecosystems were obviously quite different from modern ones. The earliest ecosystem was likely simply composed of primary producers and decomposers, both of which should have been prokaryotes. Furthermore, the first primary producers were most likely chemolithoautotrophs. Their role in ecosystems was later replaced by anoxigenic photoautotrophs and then oxgenic photoautotrophs, although the timings of emergence of these two types of photoautotrophs have been controversial (Chapter 6). It may be generally considered that unicellular protozoans such as amoeba and ciliate eating other microbes were the first consumers. Their emergence is though to have been back to Neoproterozoic (0.54–1.0 Ga ago) (e.g., Porter and Knoll, 2000). In the wider sense, however, the emergence of consumer-bearing ecosystem can be much older. Eukaryotic cells were likely evolved from some archaea who lacked rigid cell walls. Lack of rigid walls had given archaeal organisms the ability of phagocytosis, which was indispensable for so-called "symbiogenesis", the most plausible hypotheses for the evolution of eukaryotic cells (Yutin et al., 2009; Sagan, 1967, also see *Column* in Chapter 13). Widely accepted eukaryotic fossils are ca. 1.8 Ga-old, as described later. Therefore, the consumer-bearing ecosystems, though again in wider sense, might have evolved already in the Paleoproterozoic or even the Archean.

COLUMN: DISCOVERY OF DEEP-SEA HYDROTHERMAL SYSTEMS

Discovery of deep-sea hydrothermal vent system dates back to 1977. Here I would like to introduce the story of discovery. The Acoustically Navigated Geological Underwater Survey (ANGUS) dived into 2,500 m in depth along the Galapagos Rift, off the South American coast, successfully had monitored temperature spike, indication of emanation of hot waters from hydrothermal vent. This discovery was immediately followed by dive of the manned diving boat "ALVIN" to the same site, which made further discoveries including dense community of e.g., giant white clams, brown mussels, white crabs, and white-stalked tubeworms. Two years later, ALVIN dived into the East Pacific Rise. During this quest, the crews discovered chimney-like rock about six feet in height from which black fluids spouted out. The fluid temperature was 350°C. Obviously, the discovery of black smoker chimneys astonished scientists worldwide (https://www.whoi.edu/oceanus/feature/the-discovery-of-hydrothermal-vents/, https://www.whoi.edu/feature/history-hydrothermal-vents/discovery/1977.html, https://www.whoi.edu/feature/history-hydrothermal-vents/discovery/1979-2.html).

What happened here? Below mid-ocean ridges, hot basaltic magma (~1,000°C) is produced and ascends. Due to high pressure at deep sea, cold seawater penetrates into the oceanic crust through cracks, reacting with basaltic-gabbroic rocks heated by ascending magma. Hydrothermal reactions drastically change physicochemical conditions of seawater. Temperature rises up to 400°C. pH can be down to 1.0. Sulfate (SO_4^{2-}), one of the major dissolved ions, in original seawater is reduced to be H_2S (e.g., Seyfried and Mottl, 1982). Metal ions such as Fe, Mn, Zn, Cu, and Pb in oceanic crust are dissolved into fluids. Silicon and Ca are also dissolved into fluids as silicic acid and Ca^{2+}, respectively, whereas Mg is removed from seawater to form hydrous minerals such as saponite. The formation of hydrous minerals lowers the pH of seawater. Heated fluids with high buoyancy ascend to eventually spout out from the seafloor. Due to mixing with cold and alkaline seawater, sulfide minerals represented by sphalerite (ZnS) and pyrite (FeS_2) precipitate to construct chimneys.

Biological communities flourished at hydrothermal vent systems are distinct from those in the photic zone of ocean and on land. Particulate OM derived from the ocean surface is insufficient to support biomasses around hydrothermal vents. Primary producers in hydrothermal ecosystems are chemolithoautotrophic microorganisms. These prokaryotes do not only comprise biomats around vents, but also live symbiotically with some mollusca (gastropods and bivalves) and annelid (e.g., tubeworms). Discovery of ecosystems independent from the solar energy was impulsive and had influenced significantly long-standing arguments on the origins of life. Namely, the hypothesis that life emerged at deep-sea hydrothermal vent systems was put forward (Corliss et al., 1981), charming a certain number of scientists up to now.

REFERENCES

Akanuma, S., Nakajima, Y., Yokobori, S., Kimura, M., Nemoto, N., Mase, T., Miyazono, K., Tanokura, M., Yamagishi, A. 2013. Experimental evidence for the thermophilicity of ancestral life. *Proceedings of the National Academy of Sciences of the United States of America* 110, 11067–11072.

Alexander, C.M.O'D., Fogel, M., Yabuta, H., Cody, G.D. 2007. The origin and evolution of chondrites recorded in the elemental and isotopic compositions of their macromolecular organic matter. *Geochimica et Cosmochimica Acta* 71, 4380–4403.

Catling, D.C., Zahnle, K.J. 2020. The Archean atmosphere. *Science Advances* 6, eaax1420.

Cleaves, H.J., Chalmers, J.H., Lazcano, A., Miller, S.L., Bada, J.L. 2008. A reassessment of prebiotic organic synthesis in neutral planetary atmospheres. *Origins of Life and Evolution of Biospheres* 38, 105–115.

Corliss, J.B., Baross, J.A., Hoffman, S.E. 1981. An hypothesis concerning the relationship between submarine hot springs and the origin of life on Earth. *Proceedings 26th International Geological Congress. Oceanologica Acta* 4 (Suppl), 59–69.

Damer, B., Deamer, D., 2020. The hot spring hypothesis for an origin of life. *Astrobiology* 20, 429–451. doi:10.1089/ast.2019.2045.

Darwin, C. 1859. *On the Origin of Species: By Means of Natural Selection, or the Preservation of Favoured Races in the Struggle for Life.* John Murray, London.

Deamer, D., Dworkin, J.P., Sandford, S.A., Bernstein, M.P., Allamandola, L.J. 2002. The first cell membranes. *Astrobiology* 2, 371–381.

Denny, A.R., Kelly, D.S., Früh-Green, G.L. 2015. Geologic evolution of the Lost City Hydrothermal Field. *Geochemistry, Geophysics, Geosystems* 17, 375–394.

Doolittle, W.F. 2020. Evolution: Two domains of life or three? *Current Biology* 30, R177–179.

Fenchel, T., Blackburn, H., King, G. 2012. *Bacterial Biogeochemistry. The Ecophysiology of Mineral Cycling.* Elsevier, Amsterdam. 307 p.

Furukawa, Y., Chikaraishi, Y., Ohkouchi, N., Ogawa, N.O., Glavin, D.P., Dworkin, J.P., Abe, C., Nakamura, T. 2019. Extraterrestrial ribose and other sugars in primitive meteorites. *Proceedings of National Academy of Sciences* 116, 24440–24445.

Furukawa, Y., Sekine, T., Oba, M., Kakegawa, T., Nakazawa, H. 2009. Biomolecule formation by oceanic impacts on early Earth. *Nature Geoscience* 2, 62–66.

Genchi, G. 2017. An overview on D-amino acids. *Amino Acids* 49, 1521–1533.

Harris, L.K., Theriot, J.A. 2018. Surface area to volume ratio: A natural variable for bacterial morphogenesis. *Trends in Microbiology* 26, 815–832.

Hashimoto, G.L., Abe, Y., Sugita, S. 2007. The chemical composition of the early terrestrial atmosphere: Formation of a reducing atmosphere from CI-like material. *Journal of Geophysical Research* 112, E05010. doi:10.1029/2006JE002844.

Kasting, J.F. 1993. Earth's early atmosphere. *Science* 259, 920–926.

Kelley, D.S., Karson, J.A., Blackman, D.K., Früh-Green, G.L., Butterfield, D.A., Lilley, M.D., Olson, E.J., Schrenk, M.O., Roe, K.K., Lebon, G.T., Rivizzigno P., the AT3–60 Shipboard Party. 2001. An off-axis hydrothermal vent field near the Mid-Atlantic Ridge at 30°N. *Nature* 412, 145–149.

Kirschvink, J.L., Weiss, B.P. 2002. Mars, Panspermia, and the origin of life: Where did it all begin? *Palaeontologia Electronica* 4, 8–15.

Koga, T., Naraoka, H. 2017. A new family of extraterrestrial amino acids in the Murchison meteorite. *Scientific Reports* 7, 636. doi:10.1038/s41598-017-00693-9.

Kuwahara, H., Eto, M., Kawamoto, Y., Kurihara, H., Kaneko, T., Obayashi, Y., Kobayashi, K. 2012. The use of ascorbate as an oxidation inhibitor in prebiotic amino acid synthesis: A cautionary note. *Origins of Life and Evolution of Biospheres* 42, 533–541.

Larsen, B.B., Miller, E.C., Rhodes, M.K., Wiens, J.J. 2017. Inordinate fondness multiplied and redistributed: The number of species on Earth and the new pie of life. *The Quarterly Review of Biology* 92, 229–265. doi:10.1086/693564.

Llorca, J. 2005. Organic matter in comets and cometary dust. *International Microbiology* 8, 5–12.

Marshall, W.F., Young, K.D., Swaffer, M., Wood, E., Nurse, P., Kimura, A., Frankel, J., Wallingford, J., Walbot, V., Qu, X., Roeder, A.H.K. 2012. What determines cell size? *BMC Biology* 10, 101.

Marshall, W.L. 1994. Hydrothermal synthesis of amino acids. *Geochmica et Cosmochimica Acta* 58, 2099–2106.

Martin, W., Baross, J., Kelley, D., Russell, M.J. 2008. Hydrothermal vents and the origin of life. *Nature Reviews Microbiology* 6, 805–814.

Martin, W., Russell, M.J. 2003. On the origins of cells: A hypothesis for the evolutionary transitions from abiotic geochemistry to chemoautotrophic prokaryotes, and from pro-karyotes to nucleated cells. *Philosophical Transactions of the Royal Society London B* 358, 59–85.

Maruyama, S., Kurokawa, K., Ebisuzaki, T., Sawaki, Y., Suda, K., Santosh, M., 2019. Nine requirements for the origin of Earth's life: Not at the hydrothermal vent, but in a nuclear geyser system. *Geoscience Frontiers* 10, 1337–1357.

McCollom, T.M., Seewald, J.S. 2007. Abiotic synthesis of organic compounds in deep-sea hydrothermal environments. *Chemical Reviews* 107, 382–401.

Miller, S.L. 1953. A production of amino acids under possible primitive Earth conditions. *Science* 117, 528–529.

Miller, S.L. 1955. Production of some organic compounds under possible primitive Earth conditions. *Journal of the American Chemical Society* 77, 2351–2361.

Miyakawa, S., Yamanashi, H., Kobayashi, K., Cleaves, H.J., Miller S.L. 2002. Prebiotic synthesis from CO atmospheres: Implications for the origins of life. *Proceedings of National Academy of Sciences of the United States of America* 99, 14628–14631.

Mulkidjanian, A.Y., Bychkov, A.Y., Dibrova, D.V., Galperin, M.Y., Koonin, E.V. 2012. Origin of first cells at terrestrial, anoxic geothermal fields. *Proceedings of the National Academy of Sciences of the United Steates of America* 109, 5156–5157.

Nakazawa, H., Sekine, T., Kakegawa, T., Nakazawa, S. 2005. High yield shock synthesis of ammonia from iron, water and nitrogen available on the early Earth. *Earth Planetary Science Letters* 235, 356–360.

Porter, S.M., Knoll, A.H. 2000, Testate amoebae in the Neoproterozoic Era: Evidence from vase-shaped microfossils in the Chuar Group, Grand Canyon. *Paleobiology* 26, 360–385.

Russell, M.J., Hall, A.J., Martin, W. 2010. Serpentinization as a source of energy at the origin of life. *Geobiology* 8, 355–371.

Sagan, L. 1967. On the origin of mitosing cells. *Journal of Theoretical Biology* 14, 255–274.

Schulz, H.N., Jørgensen, B.B. 2001. Big bacteria. *Annual Review of Microbiology* 55, 105–137.

Schoonen, M.A.A., Xu, Y. 2004. Nitrogen reduction under hydrothermal vent conditions: Implications for the prebiotic synthesis of C-H-O-N compounds. *Astrobiology* 1, 133–142.

Scorei, R. 2012. Is boron a prebiotic element? A mini-review of the essentiality of boron for the appearance of life on Earth. *Origins of Life and Evolution of Biospheres* 42, 3–17.

Seyfried Jr., W.E., Mottl, M.J. 1982. Hydrothermal alteration of basalt by seawater under seawater-dominated conditions. *Geochimca et Cosmochimica Acta* 46, 985–1002.

Sugahara, H., Mimura, K. 2014. Glycine oligomerization up to triglycine by shock experiments simulating comet impacts. *Geochemical Journal* 48, 51–62.

Takeuchi, Y., Furukawa, Y., Kobayashi, T., Sekine, T., Terada, N., Kakegawa, T. 2020. Impact-induced amino acid formation on Hadean Earth and Noachian Mars. *Scientific Reports* 10, 9220. doi:10.1038/s41598-020-66112-8.

Urey, H.C. 1952. On the early chemical history of the Earth and the origin of life. *Proceedings of the National Academy of Sciences of the United States of America* 38, 351–363.

Whittaker, R.H. 1969. New concepts of kingdoms of organisms. *Science* 163, 150–160.

Williams, L.B., Canfield, B., Voglesonger, K.M., Holloway, J.R. 2005. Organic molecules formed in a "primordial womb". *Geology* 33, 913–916.

Williams, T.A., Cox, C.J., Foster, P.G., Szöllősi, G.J., Embley, T.M. 2019. Phylogenomics provides robust support for a two-domains tree of life. *Nature Ecology and Evolution* 4, 138–147.

Woese, C.R., Kandler, O., Wheelis, M.L. 1990. Towards a natural system of organisms: Proposal for the domains Archaea, Bacteria, and Eucarya. *Proceedings of the National Academy of Sciences of the United States of America* 87, 4576–4579.

Yutin, N., Wolf, M.Y., Wolf, Y.I., Koonon, E.V. 2009. The origins of phagocytosis and eukaryogenesis. *Biology Direct* 4, 9.

4 Life on the Earth 2

4.1 INTRODUCTION

Nutrients refer to substances indispensable for vital activities. Several definitions of nutrients have been proposed: here, an element-based definition (i.e., essential elements) is described. Nutrients can be classified into macronutrients and micronutrients, depending on required quantities. Macronutrients include C, H, N, O, P, S, Ca, Na, K, Mg, and Cl. Silicon (Si) is also a macronutrient for some of unicellular eukaryotes such as radiolaria and diatom, which produce opaline ($SiO_2 \cdot nH_2O$) exoskeleton. Among these, C, H, N, O, P, and S are particularly important, as they comprise biopolymers such as protein, lipid, hydrocarbon, nucleic acid and adenosine polyphosphates (ATP and ADP). Representative micronutrients for a wide variety of organisms are V, Cr, Mo, Mn, Fe, Ni, Co, Cu, and Zn, all of which have a function as cofactor of enzymes. The functions of biopolymers are summarized in Table 4.1. In this chapter, flows and transformation of selected macronutrients are described, first focusing on modern states and then some implications for Archean ones.

4.2 CARBON

Carbon (C) is the most important essential element, because this element comprises frameworks of biopolymers. Global-scale carbon flows comprise fluxes between multiple reservoirs. In the context of the global warming issue, the "surface" reservoirs have been considered, including (1) atmosphere, (2) vegetation (biomass on land), (3) soil, (4) surface ocean, (5) biomass in ocean, (6) deep ocean, (7) oceanic sediment, and (8) fossil fuels. In the geobiological context, deeper reservoirs such as lithosphere, mantle, and even core are involved in the global carbon cycle. Key paths and reactions involved in the surface carbon flows are (1) biological carbon fixation (photo- and

---

TABLE 4.1

Representative Biomolecules and Their Functions

Proteins: C, H, N, S, metals
  • Metabolisms, construction of body, signal transduction and others
Lipids: C, H, O (N, P)
  • Storage of energy, protection of tissues, construction of cell membrane
Carbohydrate: C, H, O
  • Stroage of energy, construction of body
Nucleic acids: C, H, N, O, P
  • Accumulation, preservation, and transmission of genetic information
Adenosine phosphate: C, H, N, O, P
  • Storage of energy

DOI: 10.1201/9780367855208-4

chemosynthesis), (2) respiration of organic matter (OM), (3) dissolution to oceans, (4) precipitation and dissolution of carbonates, (5) deposition, burial and diagenesis of OM. Key paths and reactions involved in deep carbon cycle include (1) degassing related to metamorphism, (2) deep burial related to subduction, (3) subduction-related metamorphism, and (4) flows related to mantle dynamics. Weathering of surface rocks, burial of OM in sediments, degassing related to volcanisms, and alteration of oceanic crusts can be considered as the paths that connect the shallow and the deep carbon cycles. In the following, these carbon cycles are briefly described.

4.2.1 Deep Carbon Cycle

Deep carbon cycle depicts the largest scale carbon cycle within the crustal and mantle reservoirs, which is connected to atmospheric, hydrospheric, and biospheric reservoirs (e.g., Dasgupta and Hirschmann, 2010; Wong et al., 2019). Deeply stored C is released to the surface through volcanisms and orogenic metamorphisms. The deepest carbon reservoir is thought to be placed at the core-mantle boundary. Such deep C is transported to the surface via mantle plume, an upwelling of hot mantle materials from the boundary, occasionally called "hot plume". Hot spot magmatisms at, e.g., Hawaii Island and formation of large igneous provinces (LIPs) represented by, e.g., the Mesozoic Deccan Trap in India are thought to have been caused by mantle plumes. Carbon stored and cycled in the atmosphere, biosphere, and hydrosphere would be transported into the deep Earth through the two paths, (1) formation of carbonate minerals in oceanic crusts mainly in off-axis low temperature hydrothermal system (Coogan et al., 2016) and (2) subduction of sedimentary OM and carbonates. Subducted lithosphere called slab is a huge carbon transporter from the Earth's surface to its deep region. Slabs occasionally break-off and are thought to be stuck around 660 km depth (Feng et al., 2021). The stuck slabs, if their amount exceeds the limit, begin to sink to the deeper mantle and eventually reach the mantle-core boundary. It should be noted that not all C once stored in oceanic crusts reach the mantle-core boundary. A part of the crustal C would be recycled mainly as CO_2 to the Earth's surface through subduction-related metamorphisms and volcanisms. The deep carbon cycle is operated in a timescale from millions to billions of years.

4.2.2 Modern Surface Carbon Cycle

Modern surface carbon cycle described above could be integrated into the three subcycles including (1) terrestrial carbon cycle, (2) oceanic carbon cycle, and (3) sediment carbon cycle.

4.2.2.1 Terrestrial Carbon Cycle

Modern terrestrial carbon cycle is dominated by photosynthesis and respiration by land plants. Here, brief descriptions are given, based on compiled data by Kandasamy and Nath (2016). The total amount of annually fixed C by plants is called gross primary production (GPP). Prior to the Industrial Era, GPP is estimated to have been 108.9 Pg C (Pg = 10^{15}g), which was nearly balanced with the amount of back flux to the atmosphere (107.2 Pg C), including direct respiration by plants and respiration through consumers. Stored C in the land vegetation is assumed to be 450–650 Pg, whereas soils and permafrost are assumed to store 1,500–2,400 Pg C and ~1,700 Pg C, respectively. Flux via weathering is much smaller (0.3 Pg/C/year), but compensate volcanic outgassing (0.1 Pg/C/year).

Although the present volcanic carbon flux is small, enhanced volcanic activity could result in an increase in the temperature via greenhouse effect; in the short term, however, an associated increase in aerosol and dusts could decrease temperature due to an increase in albedo. In the long term, volcanically increased atmospheric CO_2 would be eventually consumed by enhanced chemical weathering (Walker et al., 1981). Equation 4.1 describes dissolution of carbonates represented by calcite, whereas Equation 4.2 (Grotzinger and Press, 2007) describes decomposition of K-feldspar. Increase in temperature due to increase in the atmospheric CO_2 accelerates chemical weathering. Therefore, chemical weathering has a negative feedback effect against volcanic and hydrothermal CO_2 emissions. CO_2-related chemical weathering of rock-forming minerals produces HCO_3^-, which is transported eventually to the oceans. Conversely, if the atmospheric temperature decreases due to, for example, decrease in volcanic emission of CO_2, the rate of chemical weathering would be reduced. This results in recovery of the atmospheric CO_2 concentration.

$$CaCO_3 + CO_2 + H_2O \rightarrow Ca^{2+} + 2HCO_3^- \tag{4.1}$$

$$2KAlSi_3O_8 + 2H_2CO_3 + H_2O \rightarrow Al_2Si_2O5(OH)4 + 4SiO_2 + 2K^+ + 2HCO_3^- \tag{4.2}$$

4.2.2.2 Oceanic Carbon Cycle

Carbon cycle in the modern oceans is schematically illustrated in Figure 4.1, which is dominated by biochemical processes, although seafloor weathering may significantly influence oceanic carbon cycle and even atmospheric CO_2 concentrations (Brady and Gíslason, 1997). The largest carbon reservoir in the oceans is intermediate and deep waters, where 37,100 Pg C as inorganic C and 700 Pg C as dissolved organic carbon (DOC) are stored. The surface ocean and the surface sediment contain 900 Pg C and 1,750 Pg C, respectively. Marine biomass is small as marine carbon reservoir and stores only 3 Pg C (Kandasamy and Nath, 2016).

The atmospheric CO_2 dissolves to the ocean water according to Henry's law. The difference from the other major atmospheric gases such as N_2, O_2, and Ar is that dissolved CO_2 are transformed to three species including carbonic acid (H_2CO_3), HCO_3^-, and carbonate (CO_3^{2-}). Equilibrium relationships between these species are given below (Equations 4.3–4.5).

$$CO_2 + H_2O \leftrightarrow H_2CO_3 \tag{4.3}$$

$$H_2CO_3 \leftrightarrow H^+ + HCO_3 \tag{4.4}$$

$$HCO_3 \leftrightarrow H^+ + CO_3^2 \tag{4.5}$$

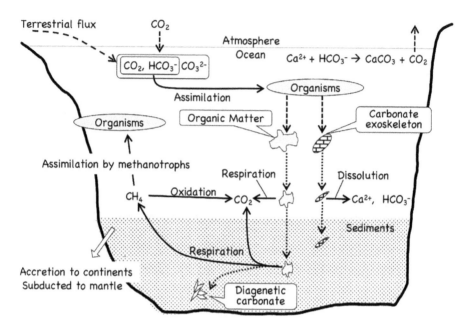

FIGURE 4.1 Schematic model of modern carbon cycle within the Earth's oceans. Ocean floor chemical weathering (e.g., Brady and Gíslason, 1997) is omitted here.

Dissolved CO_2 and occasionally HCO_3^-, some of which are derived from chemical weathering of terrestrial rocks, are taken up by autotrophs represented by algae and cyanobacteria to be converted to biomolecules. Like the terrestrial ecosystem, part of the primary production is lost by self-respiration, death, and prey. Dead bodies of primary producers and consumers and their fragments sink as particulate organic matter (POM). POM is subjected to hydrolysis by extracellular enzymes and respiration by heterotrophs. Dissolved organic matter (DOM) produced by hydrolysis of POM is ideal diet for heterotrophic bacteria.

Consumption of dissolved O_2 by aerobic respiration of OM (POM & DOM), combined with physical processes, results in the formation of oxygen minimum zones, which are assumed to have a size of $100 \times 10^6 km^3$ (Baroni et al., 2020). At the "core" of the oxygen minimum zone, the concentration of dissolved O_2 can be lower than $20 \mu M$ (Paulmier and Ruiz-Pino, 2009). Due to aerobic respiration, nearly 95% of the global marine primary production (~55 Pg/C/year) is assumed to be consumed in the water column. Furthermore, most (~90%) of the OM delivered to the ocean floors is degraded due to microbial respiration, resulting in low global carbon burial (~0.2–0.4 Pg/C/year) (Middelburg, 2019).

In addition to OM, the formation of carbonate minerals is involved in oceanic carbon cycle. In modern ocean environments, precipitation of carbonates is biologically mediated (biomineralization), represented by shells of bivalves and skeletons of corals and tests or scales of unicellular protists (e.g., foraminifera and coccolithophore). Among them, tests and scales of planktonic protists dominate carbonate biomineralization. Localized precipitation of ooids {small egg-like carbonate ($CaCO_3$) granules} at, e.g., Great Bahama Bank was once considered to be totally inorganic, although biological process has been recently recognized to be involved in their formation (Harris et al., 2019). Carbonate precipitation in seawater is described as Equation 4.6.

$$Ca^{2+} + 2HCO_3 \rightarrow CaCO_3 + CO_2 + H_2O \qquad (4.6)$$

Calcareous tests and scales sinking from the ocean surface are subjected to dissolution at depths (reverse reaction of Equation 4.6), due to increase in pressure, and increase in concentration of CO_2 in a consequence of respiration of OM, and decrease in temperature. Calcareous materials that survive dissolution reach the ocean bottom and are subjected to burial and diagenesis. During diagenesis, calcareous tests would experience dissolution and reprecipitation.

4.2.2.3 Sediment Carbon Cycle

Once buried and dissolved O_2 in porewater is consumed due to aerobic respiration, OM would be subjected to different types of respiration. Organic matter would be anaerobically respired by heterotrophic prokaryotes. In anaerobic respiration of OM, nitrate (NO_3^-), tetravalent manganese (Mn^{4+}), ferric iron (Fe^{3+}), and sulfate (SO_4^{2-}) and H_2O are utilized as electron acceptors. These different types of anaerobic respirations generally occur at different depths of sediment. Ideally, below aerobic respiration zone, NO_3^--respiration zone, Mn^{4+}-respiration zone, Fe^{3+}-respiration zone, SO_4^{2-}-respiration zone, and heterotrophic methanogenesis zone occur, with an increase in subsurface depth.

Table 4.2 describes reactions of aerobic respiration and various types of anaerobic respirations with a common starting OM. It should be emphasized here that these reactions are theoretical, and sulfate reducers and heterotrophic methanogenesis can utilize only simple organic compounds such as acetate ion (CH_3COO^-), products of respiration of biopolymers. However, it is valid to show that respirations utilizing O_2,

---

TABLE 4.2
Theoretical Energy Yields When Oxidizing Organic Matter Using Various Electron Acceptors (From Kirchman, 2012)

$Co + 138O_2 \rightarrow 106CO_2 + 16HNO_3 + H_3PO_4 + 122H_2O - 3190$ kj / M glucose

$Co + 94.4HNO_3 \rightarrow 106CO_2 + 55.2N_2 + H_3PO_4 + 177.2H_2O - 3030$ kj / M glucose

$Co + 236MnO_2 + 472H^+ \rightarrow 236Mn^{2+} + 106CO_2 + 8N_2 + H_3PO_4 + 366H_2O - 3090$ kj / M glucose

$Co + 212Fe_2O_3 + 848H^+ \rightarrow 424Fe^{2+} + 106CO_2 + 16NH_3 + H_3PO_4 + 530H_2O - 1410$ kj/M glucose

$Co + 53SO_4^{2-} \rightarrow 106CO_2 + 16NH_3 + 53S^{2-} + H_3PO_4 + 106H_2O - 380$ kj/M glucose

$Co \rightarrow 53CO_2 + 53CH_4 + 16NH_3 + H_3PO_4 - 350$ kj/M glucose

$Co = (CH_2O)_{106} (NH_3)_{16} (H_3PO_4)$

NO_3^-, and MnO_2 (Mn^{4+}) are energetically much efficient compared with the others. These early diagenetic processes mediated by microbes are followed by thermal degradation due to sediment deeper burial, producing kerogen and bitumen. As noted in Chapter 2, kerogen is a complex OM that is not soluble to organic solvents, whereas bitumen can be soluble. Kerogen is a major constituent of OM preserved in Archean sedimentary rocks and occasionally comprises cellularly preserved microfossils.

4.2.3 ARCHEAN CARBON CYCLE

Archean carbon cycle in the oceans is depicted in Figure 4.2. When considering carbon cycle on the Archean Earth, several prerequisites should be given. First, due to assumed much higher mantle temperature than today, carbon flux to the oceans and the atmosphere via volcanic degassing should have been much larger than today. Although it is controversial when the plate tectonics had started, the deep carbon cycle was likely driven at much shorter time scale due to higher plate production rate. Considering presumable smaller continental mass, carbonatization and weathering of oceanic crusts may have played an important role for carbon cycle as a sink for CO_2 in the early Archean (Nakamura and Kato, 2004; Brady and Gíslason, 1997).

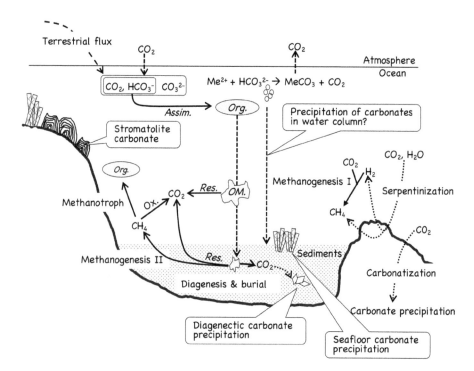

FIGURE 4.2 Schematic model of carbon cycle within the Archean Earth's oceans. Me., Ca^{2+} and potentially Fe^{2+}, Mg^{2+} and Mn^{2+}; Org., Organisms; OM., Organic matter; Res., Respiration including heterotrophic methanogenesis; Assim., Assimilation; Ox., Oxidation. Methanogenesis I and Methanogenesis II represent autotrophic and heterotrophic productions of CH_4, respectively

Second, during the Archean era spanning nearly 1.5 Ga, concentrations of major carbon gases such as CO_2 and CH_4 had changed in a magnitude or more (Catling and Zahnle., 2020). Although it is widely accepted that the atmospheric concentrations of CO_2 and CH_4 were much higher than today throughout the Archean and had compensated the lower solar luminosity to have prevented the Earth from entire freezing, the estimated ranges are highly variable between researchers. For example, based on analyses of even nearly contemporaneous 2.7 Ga-old fossilized soils (paleosols), the upper atmospheric CO_2 concentration ranges widely from 0.015 to 0.75 bar, and the lower one ranges widely from 0.003 to 0.03 bar (Driese et al., 2011; Kanzaki and Murakami, 2015). Assumed higher CH_4 concentrations are attributed to microbial methanogenesis and abiogenic methane formation by serpentinization. In the earlier section, methanogenesis was described as one of the heterotrophic metabolisms. Additionally, methanogenesis is performed autotrophically by methanogens (Equation 4.7: Fenchel et al., 2012).

$$4H_2 + CO_2 \rightarrow CH_4 + 2H_2O \qquad (4.7)$$

Foods for methanogens, CO_2 and H_2, were likely abundant in the Archean, as H_2 in addition to CH_4 were produced by serpentinization. Thus, the problem is when the methanogens had evolved. Recently, Wolfe and Fournier (2018) suggested based on phylogenetic study that the evolution of methanogens can be back to >3.5 Ga ago {also see Roger and Susko (2018) for criticism}. This is consistent with strongly ^{12}C-depleted CH_4 in fluid inclusions in 3.5 Ga-old hydrothermal chert veins (Ueno et al., 2006) ($\delta^{13}C_{PDB} < -50$ ‰: PDB refers to the standard material, Pee Dee Belemnite, although Vienna Pee Dee Belemnite is also used) (Equation 4.8) that can be attributed to isotopic fractionation (*Column*) during autotrophic methanogenesis. Russell et al. (2010) also suggested that methanogenesis was very early in origin. As a final notion in the topic of CH_4, the recent study by Bižić et al. (2020) should be mentioned to; the authors demonstrated that cyanobacteria living in various habitats and environments could produce CH_4. As discussed later, the evolution of cyanobacteria could date back to 3.0 Ga ago and even earlier; CH_4-related Archean carbon cycle, which had traditionally been depicted in the context of Archaeal methanogenesis and serpentinization, may need to be revised, involving contribution of cyanobacteria.

$$\delta^{13}C_{sample} = \left({}^{13}C/{}^{12}C_{sample} / {}^{13}C/{}^{12}C_{standard} - 1 \right) \times 1000 \ (‰) \qquad (4.8)$$

Third, in the absence of organisms producing calcareous skeleton in the Archean, carbonate precipitation appears to have been dominated by inorganic processes (Sumner and Grotzinger, 1996). Biological processes involved in Archean carbonate precipitation were likely restricted to the formation of carbonate stromatolites, although they were in an indirect way (Chapter 7).

Finally, we need to remember that the terrestrial ecosystems were absent in the Archean or, if present, should be much smaller than today and were probably restricted to hot spring systems (Djokic et al., 2017) or rivers (Homann et al., 2018). Terrestrial carbon cycle was dominated by volcanic degassing and chemical weathering.

4.3 SULFUR

4.3.1 Modern Sulfur Cycle

Sulfur (S) cycle on the modern Earth is depicted in Figure 4.3. Sulfur comprises cysteine and methionine, amino acids essential to some plants and animals. Also, it comprises Fe-S cluster, reaction center in metalloproteins such as ferredoxin and hydrogenase. Major sulfide minerals are pyrite (FeS_2), sphalerite (ZnS), and galena (PbS), whereas major sulfate minerals are gypsum ($CaSO_4 \cdot nH_2O$), anhydrite ($CaSO_4$), and barite ($BaSO_4$). These minerals are both volcanogenic and sedimentary. Being exposed to oxidizing atmosphere or water, sulfide minerals could be oxidized to produce SO_4^{2-} (Equation 4.9: Holland, 1984). This reaction can proceed as chemical reaction, and microbial oxidation of sulfide minerals has been well demonstrated (e.g., Geelhoed et al., 2009).

$$FeS_2 + 15/4O_2 + 2H_2O \rightarrow 1/2Fe_2O_3 + 2H_2SO_4 \tag{4.9}$$

Major sulfur containing compounds in volcanic gases are hydrogen sulfide (H_2S), and sulfur dioxide (SO_2). Around volcanic fumaroles where temperatures of spouting gases drastically decrease, native sulfur (S^0) could precipitate from H_2S and SO_2. Gaseous H_2S an SO_2 would be photochemically oxidized with atmospheric O_2 and eventually converted to SO_4^{2-}. Although not involved in Figure 4.3, another source for the atmospheric SO_4^{2-} is dimethyl sulfide (DMS: CH_3SCH_3) from the oceans.

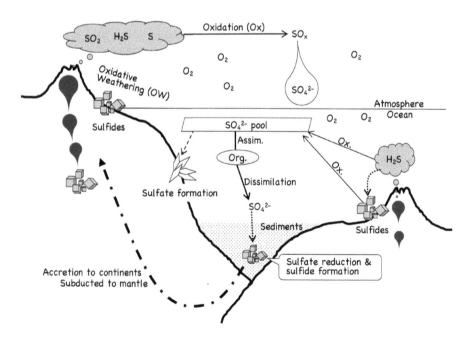

FIGURE 4.3 Schematic model for modern sulfur cycle within the Earth's surface systems. Cycles involving dimethyl sulfide (DMS) described in the text are omitted here.

DMS is microbially converted from dimethylsulfoniopropionate (DMSP) that is utilized as osmotic pressure regulator (e.g., Cui et al., 2020). DMS released from the ocean surface into the atmosphere can be more than 10^7 tons/year (Alcolombri et al., 2015 and references therein). DMS is oxidized finally into SO_4^{2-} via some intermediate substances.

Autotrophs in the terrestrial ecosystems utilize dissolved SO_4^{2-} in ground waters and soil moistures. Sulfate ions taken into living bodies are reduced to H_2S and then assimilated to amino acids such as cysteine, which is finally converted to proteins. In the food webs of ecosystems, sulfur-containing biomolecules are utilized sequentially by consumers, detritivores and scavengers, and finally mineralized to H_2S and SO_4^{2-} by decomposers. Sulfate ions in pore waters of soils is transported to the oceans and is utilized by marine autotrophs and involved in the food webs. The concentration of SO_4^{2-} in the ocean water is thought to be basically constant (2.7 g/kg) through the geologic time after the Cambrian (541–484.5 million years) (Turchyn and Depaelo, 2019). This consistency is maintained by removal of SO_4^{2-} as sulfide minerals such as pyrite (FeS_2) in anaerobic environments and sulfates such as barite ($BaSO_4$) and gypsum ($CaSO_4 \cdot nH_2O$). The former reaction is operated by chemoorganoheterotrophic prokaryotes called sulfate-reducing bacteria (SRB), living in anaerobic environment. Sulfate-reducing bacteria utilizes SO_4^{2-} to oxidize relatively simple organic molecules such as acetate ion (CH_3COO^-), respiration products of other microbes. This process is coupled with reduction of SO_4^{2-} to H_2S, which could react with ferrous iron (Fe^{2+}) to precipitate pyrite (FeS_2), in a peculiar cluster composed of micron-sized pyrite crystals called framboidal pyrite. Ferrous iron is also byproduct of metabolisms of iron-reducing heterotrophic microbes (Table 4.2).

Another important biogeochemical cycle of S is involved in the metabolism of prokaryotic sulfur bacteria. Chemolithoautotrophic sulfur bacteria utilize H_2S and S as an electron donor to reduce CO_2 to produce OM and S or SO_4^{2-} as byproducts (Equation 3.2). Sulfur bacteria are important primary producers for deep-sea hydrothermal vent ecosystems. They comprise bacterial mats around the vent or live symbiotically with other organisms represented by tubeworm (e.g., Wilmot and Vetter, 1990). Free-living sulfur bacteria inhabiting organic-rich anaerobic sediments are also known, including *Thiomargarita* and *Beggiatoa* (e.g., Schulz and Schulz, 2005). In addition, disproportionations of sulfur species are known to be one of the important bacterial metabolisms (e.g., Finster, 2008) (Equation 4.10).

$$4S^0 + 4H_2O \rightarrow SO_4^{2-} + 3HS^- + 5H^+ \qquad (4.10)$$

4.3.2 Archean Sulfur Cycle

Key points in Archean sulfur cycle are (1) assumed much larger volcanic flux of sulfur species such as SO_2 and S than today, (2) photochemical reaction of volcanic SO_2, (3) limited oxidative weathering of subaerially exposed sulfide minerals, and (4) evolution of microbial sulfur metabolisms (Figure 4.4).

In the absence or scarcity of O_2 and thus the absence of ozone (O_3) screen, ultraviolet rays (UV) likely penetrated deeply into the Archean atmosphere. Being reacted

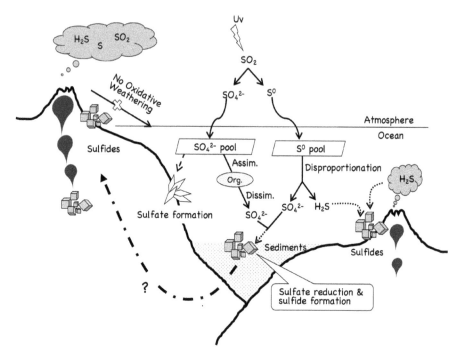

FIGURE 4.4 Schematic model Archean (before the rise of sufficient O_2) sulfur cycle within the Earth's surface systems. (Note that sulfate concentration in Archean seawater was much smaller than today. Also, sulfate reduction was likely not restricted within sediments.)

with UV, SO_2 in the atmosphere derived from volcanoes was dissociated and eventually transformed to SO_4^{2-} and S_8^0 through various and still controversial pathways. This model was initially introduced by Farquhar et al. (2000), to interpret unusual isotopic signatures of S in the Archean sulfide- and sulfate-minerals, known as mass-independent isotopic fractionation of S (MIF-S) (Chapter 6). Rather, identification of MIF-S in the Archean sulfur bearing minerals and its disappearance around 2.5 Ga ago had let many, but if not all, of the researchers to accept the substantially O_2-free ($O_2 < 10^{-5}$ppm) Archean atmosphere. As described in the modern sulfur cycle, representative microbial metabolisms involved in the cycle are reduction of SO_4^{2-} and oxidation of reduced sulfur species. Oxidation of reduced sulfur species is coupled with transport of electron to the acceptor such as O_2 or NO_3^-. As available O_2 is assumed to be insufficient or deficient in the Archean oceans, NO_3^- might have been an important electron acceptor until O_2 had become available. Based on isotopic compositions ($\delta^{33}S_{CDT}$, $\delta^{34}S_{CDT}$, $\delta^{36}S_{CDT}$) {CDT is standard material (Canyon Diablo Troilite), although Vienne CDT is also used} (Equations 4.11–4.13), traces of microbial oxidation of pyrite, and microfossils, researchers have suggested that microbial reduction of SO_4^{2-}, oxidation of sulfides, and disproportionation of S^0 had evolved in the Archean (e.g., Shen et al., 2001; Philippot et al., 2007; Wacey et al., 2011, Czaja et al., 2016, Schopf et al., 2017).

$$\delta^{33}S_{sample} = \left({}^{33}S / {}^{32}S_{sample} / {}^{33}S / {}^{32}S_{standard} - 1 \right) \times 1000 \ (\text{‰}) \qquad (4.11)$$

$$\delta^{34}S_{sample} = \left({}^{34}S / {}^{32}S_{sample} / {}^{34}S / {}^{32}S_{standard} - 1 \right) \times 1000 \ (\text{‰}) \qquad (4.12)$$

$$\delta^{36}S_{sample} = \left({}^{36}S / {}^{32}S_{sample} / {}^{36}S / {}^{32}S_{standard} - 1 \right) \times 1000 \ (\text{‰}) \qquad (4.13)$$

The Archean atmosphere has possibly been substantially free from O_2, except for an assumed temporal oxygenation known as a whiff of oxygen (e.g., Anbar et al., 2007). Oxidative weathering of sulfide minerals can be as early as 3.0 Ga, being operated mainly by terrestrial benthic microbial communities on, e.g., land and riverbeds (Lalonde and Konhauser, 2015; Nabhan et al., 2020), but was not active throughout the Archean. Unlike today, $SO_4{}^{2-}$ was likely not abundant in Archean seawater (e.g., Crowe et al., 2014). This may have been coupled with deficiency of other micronutrients of transition metals such as Ni, Zn, and Mo in the ocean, because these metals are mainly hosted by sulfide minerals.

Whatever seawater $SO_4{}^{2-}$ concentration in the Archean oceans was, some sulfate minerals have been identified as diagenetic, evaporitic, hydrothermal, and possible biogenic products from Archean volcanic-sedimentary successions, including gypsum ($CaSO_4 \cdot 2H_2O$), anhydrite ($CaSO_4$), and barite ($BaSO_4$) (e.g., Baumgartner et al. 2020; Nabhan et al., 2016; Sugitani et al., 2007. Among them, barite ($BaSO_4$) mineralization appears to be a widespread phenomenon during the Archean. In the Paleoarchean (ca. 3.48 Ga-old) Dresser Formation in the Pilbara Craton, extensive barite mineralization had occurred. Boxwork chert-barite dykes are developed within altered basal basalts and are thought to have been hydrothermal in origin. The basal basalt is overlain by sedimentary units composed largely of bedded chert-barite, which had been precipitated from exhalated hydrothermal fluids (Van Kranendonk, 2006). Though much smaller and mostly silicified, barite precipitation has been identified from the slightly younger the ca. 3.4 Ga-old Strelley Pool Formation and the ca. 3.0 Ga-old Farrel Quartzite in the Pilbara Craton, Western Australia, as described later (Sugitani et al., 2003 and unpublished data). Also, in the Barberton greenstone belt, South Africa, sedimentary barite deposit had been described from several horizons of the Barberton (Swaziland) Supergroup (Reimer, 1980). Recently, Lowe et al. (2019) described one of them (the Barite Valley barite deposit), from the ca. 3.2 Ga-old Fig Tree Group, in detail. The barite deposits of various morphologies are hosted by sedimentary rocks assumed to be deposited in shallow water to subaerial environments such as intertidal and fan delta settings, being associated with active faults and fractures in the underlying sequences. The process of barite deposition is interpreted as follows: (1) $SO_4{}^{2-}$ was formed mainly by photolysis of sulfur species in the atmosphere, (2) rainwater carried $SO_4{}^{2-}$ to land surface, (3) meteoric $SO_4{}^{2-}$ was reacted with subsurface Ba-bearing waters vented through fractures and faults into small pools to have been precipitated as barite. Important implication of this study and the formation model of the Dresser Formation is that the Archean barite deposits do not always represent open marine precipitates. As suggested by Lowe et al. (2019), Archean barite deposits may unfortunately place little constraint on Archean seawater composition.

4.4 NITROGEN

4.4.1 Modern Nitrogen Cycle

Modern nitrogen cycle is depicted in Figure 4.5. Nitrogen is a building block of many of organic molecules such as amino acids (proteins), nucleic acids, and chlorophyll. While dinitrogen (N_2) is the dominant gas in the atmosphere (~78%), most organisms cannot directly use this molecule as a nitrogen source, because N_2 is one of the most stable natural chemical compounds. Bioavailable nitrogen-bearing compounds are NH_4^+ and NO_3^-, although nitrite (NO_2^-) can be utilized by a minority such as acidophilic unicellular red alga *Cyanidium caldarium* (Fuggi, 1993). Such bioavailable nitrogen species are produced from atmospheric N_2 through several ways, which can be classified into abiological and biological nitrogen fixations. Representative abiological nitrogen fixation is lightning discharge in the atmosphere, which oxidizes N_2 to NOx (Drapcho et al., 1983). Nitric oxide (NO), if dissolved into oxygen-containing aqueous solution, is shortly oxidized to NO_2^- (e.g., Ignarro et al., 1993 and references therein), which would be microbially oxidized to be NO_3^-. Biological nitrogen fixation is operated by microorganisms called diazotrophs. All the diazotrophs are prokaryotic and are widespread in several phyla of Bacteria and Archaea, which can be interpreted in the context of both vertical and horizontal gene transfers (e.g., Bolhuis et al., 2010). Diazotrophs are either free-living or symbiotic. Symbiosis of bacterial *Rhizobium* with roots of leguminous plants is well known. With this symbiosis, leguminous plants have advantages as pioneer plants or inhabitants in nutrient-poor

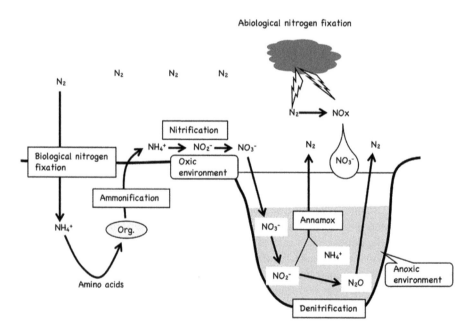

FIGURE 4.5 Schematic model of modern nitrogen cycle within the Earth's surface systems. (Note that biological nitrogen fixation occurs not only on land but also in oceans.)

environments. Symbiosis with termite (intestine) is also well known. It is noteworthy that some filamentous cyanobacteria such as *Anabaena* could produce differentiated thick-walled cells called heterocysts in which nitrogen fixation is operated.

Biological nitrogen fixation is performed using enzymes called nitrogenases, which can be active under the anaerobic condition, whereas irreversibly deactivated by O_2. The enzymes have FeMo-cofactor, although FeV-cofactor or FeFe-cofactor could express in cases. Their evolution is thought to have occurred presumably in the early Archean, probably in the consequence of deficiency of bioavailable nitrogen species produced by abiogenic nitrogen fixation (Glass et al., 2009 and references therein). Nitrogenases could reduce other substrates such as, e.g., acetylene (C_2H_2) and CO_2, in addition to N_2 (Seefeldt et al., 2020 and references therein). Namely, the substrate specificity of this enzyme is low. As can be seen in Equation 4.14 (Kirchman, 2012), 16 ATP molecules are required to fix one molecule of N_2, which implies how important assimilation of N for organisms is.

$$N_2 + 8H^+ + 8e^- + 16ATP \rightarrow 2NH_3 + H_2 + 16ADP + 16Pi \quad (4.14)$$

Nitrogen-bearing biomolecules and their degradation products (proteins and urea) released from organisms into environments would be dissimilated to NH_4^+ by, e.g., proteases and ureases excreted from heterotrophic microbes. Some of NH_4^+ produced within sediments would be incorporated into clay minerals by substituting K^+. Under the aerobic environment, NH_4^+ would be subjected to step-by-step microbial oxidation to NO_3^- ($NH_4^+ \rightarrow NO_2^- \rightarrow NO_3^-$), which is called "nitrification". Oxidation of NH_4^+ to NO_2^- (and NO_3^-) is performed by three groups of microbes including bacteria and archaea {ammonia (ammonium) oxidizing bacteria such as *Nitrosomonas*, ammonia (ammonium) oxidizing archaea such as *Nitrosopumilus*, and comammox bacteria (complete oxidation of NH_4^+ to NO_3^-) such as *Nitrospira*} (Lehtovirta-Morley, 2018 and references therein), whereas oxidation of NO_2^- to NO_3^- by several genera of bacteria including, e.g., *Nitrobacter, Nitrospira, and Nitrospina* (e.g., Spieck and Lipski, 2011). Microbes involved in nitrification are chemolithoautotrophs, which produce biomolecules from CO_2 and H_2O, using chemical energy released via. oxidation of reduced nitrogen-species.

Under the anaerobic environment, heterotrophic microbes operate "denitrification" (reduction of NO_3^- to N_2); final product (N_2) is released to the air (Table 4.2). In 1990s, scientists discovered microbial anaerobic ammonia (ammonium) − oxidation (anammox) (Equation 4.15) (Kuenen, 2008 and references therein), which also contributes to denitrification. Anammox bacteria use NO_2^- not only as an electron acceptor for ammonia (ammonium) − oxidation, but also as an electron donor, which reduces inorganic carbon to OM.

$$NH_4^+ + 1.32NO_2^- + 0.066HCO_3^- + 0.13H^+ \rightarrow 1.02N_2 + 0.26NO_3^- + 2.03H_2O$$

$$+ 0.066CH_2O_{0.5}N_{0.15} \quad (4.15)$$

In summary, nitrogen cycle in the modern Earth is operated in a combination of nitrogen fixation, nitrification, and denitrification. Biological nitrogen fixation requires an anaerobic environment, whereas nitrification favors aerobic environment. Nitrification is a process of oxidation from NH_4^+, products of hydrolysis of proteins, to NO_3^-, being operated by autotrophic nitrifying bacteria, including ammonia (ammonium) oxidation and nitrite oxidation bacteria. Denitrification (dissimilatory nitrate reduction) is a process of stepwise reduction from NO_3^- to N_2, performed by heterotrophic denitrifying bacteria under anaerobic environment. In addition, anaerobic oxidation of NH_4^+, known as annamox, comprises a part of denitrification. Modern nitrogen cycle is closely related to redox conditions and is operated exclusively by microorganisms.

4.4.2 Archean Nitrogen Cycle

The global nitrogen cycle has evolved in response to the change of redox condition of the Earth's surface, which has been closely related to evolution of oxygenic photosynthesis and accumulation of O_2 in the oceans and the atmosphere. Nitrogen cycle on anaerobic Archean Earth is depicted in Figure 4.6, in which supposed bioavailable N species and their sources are also shown, including NOx formation from N_2 by photochemistry and by lightning discharge, and hydrothermal formation of NH_4^+ (Thomazo and Papineau, 2013 and reference therein). It may be also noted that NH_3 and even amino acid formation is induced by meteorite impact (Takeuchi et al., 2020; Shimamura et al., 2016). To elucidate evolution of the global scale nitrogen cycle, isotopic composition of nitrogen ($\delta^{15}N_{air}$: air refers to the standard) of sedimentary

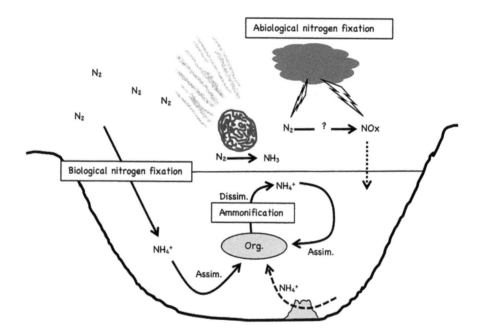

FIGURE 4.6 Schematic model of Archean (before rise of significant O_2) nitrogen cycle within the Earth's surface systems.

OM has been used (Beaumont and Robert, 1999; Pinti et al., 2001; Thomazo et al., 2011; Stüeken et al., 2015a, 2016) (Equation 4.16, Figure 4.7).

$$\delta^{15}N_{sample} = \left({}^{15}N/{}^{14}N_{sample} \big/ {}^{15}N/{}^{14}N_{standard} - 1 \right) \times 1000 \ (\%o) \qquad (4.16)$$

Organic matter from the early Archean (3.6–3.2 Ga ago) sedimentary rocks have negative $\delta^{15}N$ values from $-7\%o$ to $0\%o$. This has been interpreted in the context of anaerobic Archean oceans in which microorganisms consumed N_2 (nitrogen fixation) and NH_4^+ (Beaumont and Robert, 1999). An alternative explanation is that such OM is derived from chemosynthetic bacteria consuming ^{15}N-depleted N_2 and NH_4^+ derived from hydrothermal vents (Pinti et al., 2001; Shen et al., 2006). Organic matter from the late Archean (2.8–2.5 Ga ago) sedimentary rocks is, on the other hand, characterized by the presence of positive $\delta^{15}N$ values up to ca. $+5\ \%o$ (Beaumont and Robert, 1999; Garvin et al., 2009; Godfrey and Falkowski, 2009). This has been interpreted to reflect an increase in oxidized nitrogen species represented by NO_3^-, which is of course attributed to increase in O_2 (Anbar et al., 2007; Reinhard et al., 2009; Stüeken et al., 2015b). A key process for production of OM with positive $\delta^{15}N$ at that time is thought to be denitrification, that is associated with preferential uptake of $^{14}NO_3^-$ relative to $^{15}NO_3^-$, with leaving ^{15}N-enriched NO_3^- in the reservoir. Organic matter with positive $\delta^{15}N$ is likely derived from microorganisms utilizing such NO_3^-.

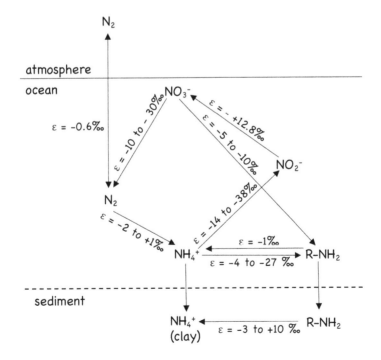

FIGURE 4.7 Illustrated marine nitrogen cycle with isotopic fractionation data ($\varepsilon = \delta^{15}N$ products - $\delta^{15}N$ substrates). R-NH_2 refers to amine group compounds. (Modified from Stüeken et al. (2016).)

Organic matter in the middle Archean (3.2–2.8 Ga ago) has a narrow range of $\delta^{15}N$ values (0 ± 2‰), which is close to the nitrogen isotopic composition of the present-day atmosphere (0‰) (Stüeken et al., 2015a). Stüeken et al. (2015a) attributed this trend to the dominance of N_2 fixing microorganisms. On the other hand, recent geochemical evidence suggests that oxygenic photosynthesis had evolved at least as early as 3.0 Ga (Lyons et al., 2014; Planavsky et al., 2014). If this is the case, how can we explain the narrow range of $\delta^{15}N$ values around 0‰ of OM at that time? One of the possibilities is that despite evolution of oxygenic photosynthesis and an increase in free O_2 in a shallow water, the amount of reduced N species still overwhelmed NO_3^-. Another possibility is that smaller-scale isotopic heterogeneities were masked by measurement on bulk samples. This issue was challenged by Delarue et al. (2018), who analyzed N isotope composition of "individual" microfossils of the Middle Archean (ca. 3.0 Ga-old) Farrel Quartzite (FQ) in the Pilbara Craton, Western Australia (e.g., Sugitani et al., 2007). As described in the later section, microfossils of the Farrel Quartzite (Chapter 9) are morphologically diverse and can be extracted by acid (HF-HCl) maceration of host cherts. This enabled us to analyze "individual" microfossils of different morphologies, using nanoscale secondary ion mass spectrometry (NanoSIMS) at le Muséum national d'histoire naturelle, Paris, France.

The results were surprising. Nitrogen isotopic compositions measured in isolated microfossils of three morphotypes (film, spheroid, lens) and amorphous OM show a large $\delta^{15}N$ isotopic heterogeneity (-21.6 ‰ $< \delta^{15}N < +30.7$ ‰). $\delta^{15}N$ for amorphous OM (n = 8) is from -21.6 to 0 ‰. Values of film-like (possibly fragmented biofilms) (n = 9) and spheroid-like microfossils (n = 6) range from -17.2 to 1.6 ‰, whereas those of lenses (n = 4) range from 0.5 up to 30.7 ‰. These variations cannot be explained by modifications of primary isotopic compositions during early diagenesis and metamorphisms (Lehmann et al., 2002). Thus, the variations could alternatively be interpreted in the context of microbial metabolisms in the water column.

While representative DINs are NH_4^+ and NO_3^- in modern aquatic systems, the concentration of NO_3^- is thought to have been negligible in the Archean oceans (Falkowski and Godfrey, 2008). Our proposed model took this precondition and the Black Sea environment as a modern analogue for the FQ habitat. The Black Sea is characterized by redox stratification of water column and by related large $\delta^{15}N$ variations of OM ranging from negative to positive value (Coban-Yildiz et al., 2006). The negative $\delta^{15}N$ values of the FQ film-like and spheroidal microfossils were explained by their preferential assimilation of $^{14}NH_4^+$ in the anoxic deep zone of the water column, implying benthic lifestyle of these microfossils. This may had resulted in accumulation of $^{15}NH_4^+$ in the deeper zone, due to the large nitrogen isotopic fractionation during ammonium assimilation ($\varepsilon = -4$‰ to -27‰, see Figure 4.7). However, because ammonification of OM of benthic microbes, which does cause little isotopic fractionation, had simultaneously occurred in the deep zone, ^{15}N-enrichment of this reservoir was plausibly ephemeral and weakened. Considering planktonic lifestyle of lenticular microfossils as discussed in Chapter 12, the positive $\delta^{15}N$ values of lenticular microfossils cannot be explained solely by ammonium assimilation. We proposed that aerobic ammonium oxidation led ^{15}N

enrichment of NH_4^+ upward in the water column. Such an enrichment was simultaneously recorded in OM through ammonium assimilation, leaving lenticular microfossils enriched in ^{15}N. Aerobic oxidation of NH_4^+ to NO_2^- requires O_2, suggesting that positive $\delta^{15}N$ values of lenticular microfossils may reflect redox stratification in the water column, which was tightly related to O_2 production by oxygenic photosynthesizers.

4.5 PHOSPHOROUS

4.5.1 MODERN PHOSPHOROUS CYCLE

Phosphorous (P) comprise nucleic acids (DNA and RNA), ATP, and phospholipid. Vertebrates, in addition, require P in order to build their skeletons containing Ca-phosphate as a major component. Major mineral containing phosphorous in rocks is apatite with an ideal chemical formula of $Ca_5(PO_4)_3(F, Cl, OH)$, which occurs in igneous, sedimentary and metamorphic rocks. The abundance of P in the Earth's crust is assumed to be around 0.1 wt%. In addition to its scarcity, low solubility of apatite to waters, and well-expected high recycle rate in terrestrial ecosystems, PO_4^{3-} (representative bioavailable form of P) has affinity with ferric hydroxides $[Fe(OH)_3]$ and aluminum hydroxides $[Al(OH)_3]$ (e.g., Zao et al., 2005; de Vincente et al., 2008), ubiquitous inorganic compounds produced by weathering of rock-forming minerals. Therefore, background concentrations of dissolved phosphate in river waters and thus flux to oceans are generally very low. Scavenging of phosphate by ferric hydroxides also could occur in seawater columns and in hydrothermal plumes and ridge flanks (e.g., Wheat et al., 1996). The surface seawater is highly depleted in PO_4^{3-}, reflecting uptake by primary producers such as algae, whereas deeper waters tend to be relatively enriched, reflecting mineralization of organic P. Such vertical profiles are also known for other nutrients such as dissolved Fe and Si. High productivity zones in oceans are often associated with upwelling of deeper water containing dissolved PO_4^{3-} and other nutrients (Figure 4.8).

Decomposers represented by heterotrophic microbes play an important role of mineralization of P from biopolymers, using enzymes such as alkaline phosphatase and phospholipase. Additionally, microbes are involved in various stages of the global phosphorous cycle. First, removal of P from the seawater column is initiated by burial of phosphorous-bearing OM, which would be decomposed biologically and then thermally to release phosphate. Ruttenberg and Berner (1993) also demonstrated that carbonate fluorapatite $[Ca_5(PO_4)_3F]$ could form authigenically during the early diagenesis closely associated with decomposition of OM. Sulfur-oxidizing bacteria, such as *Beggiatoa* and *Thiomargarita* are also involved in apatite formation in marine sediments (e.g., Brock and Schulz-Vogt, 2011, Schulz and Schulz, 2005). These large vacuolated prokaryotes have high capacity of storage of P in the form of polyphosphate (e.g., $P_2O_7^{4-}$), in addition to S and NO_3^-. Release of phosphate (PO_4^{3-}) from stored polyphosphate to sediment produces local environment saturated with apatite. This is one of key processes of the modern phosphorite formation.

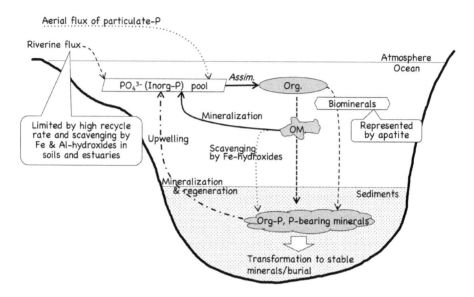

FIGURE 4.8 Schematic model for modern phosphorous cycle within the Earth's surface systems. Phosphorous removal in hydrothermal systems is omitted here.

4.5.2 ARCHEAN PHOSPHOROUS CYCLE

It is not hard to image that the global phosphorous cycle in the Archean was quite different from the modern one (Figure 4.9). Under the anoxic atmosphere, Fe^{2+} liberated from minerals by chemical weathering could not have been oxidized to be Fe^{3+}: precipitation of ferric hydroxide did not occur. Thus, one of the major bottlenecks for phosphorous delivery to oceans, is thought to have been negligible. On the other hand, chemical weathering in the Archean was likely more extensive than later periods (e.g., Sugitani et al., 1996), which potentially means massive formation of aluminum hydroxide in weathering crust. If this is the case, adsorption by aluminum hydroxide may have compensated the deficiency of ferric hydroxide. Thus, the substantial lack of terrestrial ecosystems in the Archean may not always mean that most of the P liberated from the continental crust by weathering and alteration would have been transported into the ocean. In either case, weathering-derived P should have been limited until significant increase in the continental mass.

As briefly mentioned above, in modern ridge-axis hydrothermal systems, seawater phosphate is largely removed by coprecipitation with or adsorption onto ferric hydroxides, which occur both within altered basaltic crusts and hydrothermal plumes (e.g., Wheat et al., 1996). Although it is not well-known how Archean hydrothermal systems worked in phosphorous cycle, silicification that represents hydrothermal alteration of oceanic crusts in the Archean does not appear to largely affect phosphorous concentrations in original rocks (Duchac and Hanor, 1987; Rouchon and Orberger, 2008). Another type of hydrothermal alteration of Archean oceanic crusts is carbonatization that had likely proceeded under CO_2-rich condition. According to data presented by Nakamura and Kato (2004), carbonatization does not largely affect

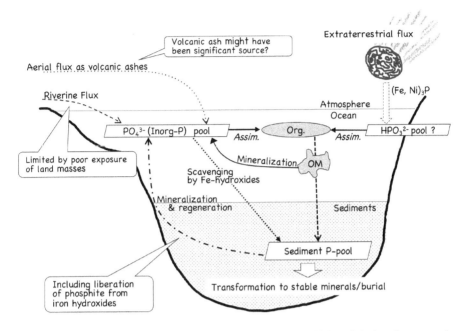

FIGURE 4.9 Schematic model Archean (before the rise of sufficient O_2) phosphorous cycle within the Earth's surface systems. Volcanic- and lightning strike-related phosphorous deliveries are omitted here.

phosphorous concentrations of oceanic crusts, although the recent experimental study implies that this type of alteration may have supplied phosphorus to Archean oceans (Murakami et al., 2019).

Despite the ambiguities in the Archean phosphorous cycle described above, it is likely that deposition of banded iron formation (BIF) and ferruginous cherts (jaspilite) throughout the Archean suggests continuous removal by dissolved PO_4^{3-} from water column by ferric hydroxide, which may had been produced by anoxygenic photosynthesis (Chapter 5). This had likely resulted in low productivity in the Archean oceans, which is known as "an Archean phosphorous crisis" proposed by Bjerrum and Canfield (2002) and strengthened by Jones et al. (2015).

Such the general view of Archean phosphorous cycles has been changed by studies related to extraterrestrial source of P. Pasek and Lauretta (2008) and references therein suggested that schreibersite [$(Fe, Ni)_3P$], one of phosphides, ubiquitous trace mineral in meteorites was a potential source for phosphorous on the early Earth. Schreibersite reacts with water to release phosphite (HPO_3^{2-}) and hypophosphite ($H_2PO_2^-$) that are relatively reduced forms of PO_4^{3-}. Indeed, HPO_3^{2-} was detected from the 3.5 Ga-old carbonates of the Coucal Formation of the Coonterunah Supergroup in the Pilbara Craton, Western Australia (Pasek et al., 2013 and references therein). Herschy et al. (2018) also demonstrated that reduction of PO_4^{3-} to HPO_3^{2-} by Fe^{2+} could occur at relatively low temperature ($<200°C$) diagenetic condition. Namely, P could have been diagenetically liberated as HPO_3^{2-} from ferric hydroxide. This transformation is suggested to have widely occurred and thus contributed to accumulation of HPO_3^{2-} as a major nutrient in the Archean oceans. Other potential sources for phosphorous in the Archean have also been proposed, including, e.g., volatilization of phosphorous oxide (P_4O_{10})(from volcanoes), which violently reacts with water to be phosphates (Yamagata et al., 1991), lightning-induced formation of HPO_3^{2-} in volcanic cloud (Schwartz, 2006), and terrestrial phosphide formation including schreibersite induced by lightning strikes (Hess et al., 2021).

COLUMN: ISOTOPE AND ISOTOPE FRACTIONATION

Isotope refers to the same element with different mass numbers, which are attributed to difference in neutron numbers. Isotopes of the same element have the same proton number. Two types of isotopes are known including radioactive isotopes and stable isotopes. Radioactive isotopes decay to be different nuclides, with constant half-lives. This has long been utilized to determine ages of rocks. Stable isotopes are not radioactive. Almost all the elements have stable isotopes. Isotopes have identical chemical features and thus behave in the same manner. However, due to difference in mass number, isotopes of the same elements could behave differently in quantity (isotopic fractionation). For example, $H_2^{16}O$ vaporizes preferentially relative to $H_2^{18}O$. Thus, if glaciers are evolving, the seawater $^{18}O/^{16}O$ value is getting larger, and vice versa. This change of isotope ratio in seawater is recorded in calcium carbonate skeletons of organisms such as foraminifera. Therefore, we can decipher the past climate change by analyzing $^{18}O/^{16}O$ values of foraminifera in marine sediment cores. Isotopic fractionation occurs in many biochemical reactions. In this chapter, carbon, sulfur, and nitrogen isotopic fractionations are referred to. It is well known that autotrophic organisms and thus heterotrophic ones is enriched in ^{12}C to ^{13}C. Thus, light isotopic compositions of organic matter in geologic samples have often been used for their biogenic origin. Sulfate reducing bacteria prefers ^{32}S than heavier isotopes. Therefore, pyrite with significantly light $\delta^{34}S$ values could be taken as evidence for biological sulfate reduction and thus biosignature.

REFERENCES

Alcolombri, U., Ben-Dor, S., Feldmesser, E., Levin, Y., Tawfik, D.S., Vardi, A. 2015. Identification of the algal dimethyl sulfide-releasing enzyme: A missing link in the marine sulfur cycle. *Science* 348, 1466–1469.

Anbar, A.D., Duan, Y., Lyons, T.W., Arnold, G.L., Kendall, B., Creaser, R.A., Kaufman, A.J., Gordon, G.W., Scott, C., Garvin, J., Buick, R. 2007. A whiff of oxygen before the Great Oxidation Event? *Science* 317, 1903–1906.

Baroni, I.R., Palastanga, V., Slomp, C.P. 2020. Enhanced organic carbon burial in sediments of oxygen minimum zones upon ocean deoxygenation. *Frontiers in Marine Science* 6, 839. doi:10.3389/fmars.2019.00839.

Baumgartner, R.J., Van Kranendonk, M.J., Fiorentini, M.L., Pagès, A., Wacey, D., Kong, C., Saunders, M., Ryan, C. 2020. Formation of micro-spherulitic barite in association with organic matter within sulfidized stromatolites of the 3.48 billion-year-old Dresser Formation, Pilbara Craton. *Geobiology* 18, 415–425.

Beaumont, V., Robert, F. 1999. Nitrogen isotope ratios of kerogens in Precambrian cherts: A record of the evolution of atmosphere chemistry? *Precambrian Research* 96, 63–82.

Bižić, M., Klintzsch, T., Ionescu, D., Hindiyeh, M.Y., Günthel, M., Muro-Pastor, A.M., Eckert, W., Urich, T., Keppler, F., Grossart, H.-P. 2020. Aquatic and terrestrial cyanobacteria produce methane. *Science Advances* 6, eaax5343. doi:10.1126/sciadv.aax5343.

Bjerrum, C.J., Canfield, D.E. 2002. Ocean productivity before about 1.9 Gyr ago limited by phosphorous adsorption onto iron oxides. *Nature* 417, 159–162.

Bolhuis, H., Severin, I., Confuruis-Guns, V., Wollenzien, U.I.A., Stal, L.J. 2010. Horizontal transfer of the nitrogen fixation gene cluster in the cyanobacterium *Microcoleus chthonoplastes*. *The ISME Journal* 4, 121–130.

Brady, P.V., Gíslason, S.R. 1997. Seafloor weathering controls on atmospheric CO_2 and global climate. *Geochimica et Cosmochimica Acta* 61, 965–973.

Brock, J., Schulz-Vogt, H.N. 2011. Sulfide induces phosphate release from polyphosphate in cultures of a marine *Beggiatoa* strain. *The ISME Journal* 5, 497–506.

Catling, D.C., Zahnle, K.J. 2020. The Archean atmosphere. *Science Advances* 6, eaax1420.

Coban-Yildiz, Y., Altabet, M.A., Yilmaz, A., Tugrul, S. 2006. Carbon and nitrogen isotopic ratios of suspended particulate organic matter (SPOM) in the Black Sea water column. *Deep-Sea Research Part II* 53, 1875–1892.

Coogan, L.A., Parrish, R.R., Roberts, N.M.W. 2016. Early hydrothermal carbon uptake by the upper oceanic crust: Insight from in situ U-Pb dating. *Geology* 44, 147–150.

Crowe, S.A., Paris, G., Katsev, S., Jones, C., Kim, S.-T., Zerkle, A.L., Nomosatryo, S., Fowle, D.A., Adkins, J.F., Sessions, A.L., Farquhar, J., Canfield, D.E. 2014. Sulfate was a trace constituent of Archean seawater. *Science* 346, 735–739.

Cui, Y., Wong, S.-K., Kaneko, R., Mouri, A., Tada, Y., Nagao, I., Chun, S.-J., Lee, H.-G., Ahn, C.-Y., Oh, H.-M., Sato-Takabe, Y., Suzuki, K., Fukuda, H., Nagata, T., Kogure, K., Hamasaki, K. 2020. Distribution of dimethylsulfoniopropionate degradation genes reflects strong water current dependencies in the Sanriku coastal region in Japan: From mesocosm to field study. *Frontiers in Microbiology.* doi:10.3389/fmicb.2020.01372.

Czaja, A.D., Beukes, N.J., Osterhout, J.T. 2016. Sulfur-oxidizing bacteria prior to the Great Oxidation Event from the 2.52 Ga Gamohaan Formation of South Africa. *Geology* 44, 983–986.

Dasgupta, R., Hirschmann, M.M. 2010. The deep carbon cycle and melting in Earth's interior. *Earth and Planetary Science Letters* 298, 1–13.

Delarue, F., Robert, F., Sugitani, K., Tartèse, R., Duhamel, R., Derenne, S. 2018. Nitrogen isotope signatures of microfossils suggest aerobic metabolism 3.0 Gyr ago. *Geochemical Perspectives Letters* 7, 32–36.

de Vicente, I., Huang, P., Andersen, F.O., Jensen, H.S. 2008. Phosphate adsorption by fresh and aged aluminum hydroxide. Consequences for lake restoration. *Environmental Science & Technology* 42, 6650–6655.

Djokic, T., Van Kranendonk, M.J., Campbell, K.A., Walter, M.R., Ward, C.R. 2017. Earliest signs of life on land preserved in ca. 3.5 Ga hot spring deposits. *Nature Communications* 8, 1–9. 15263.

Drapcho, D.L., Sisterson, D., Kumar, R. 1983. Nitrogen fixation by lightning activity in a thunderstorm. *Atmospheric Environment* 17, 729–734.

Driese, S.G., Jirsa, M.A., Ren, M., Brantley, S.L., Sheldon, N.D., Parker, D., Schmitz, M. 2011. Neoarchean paleoweathering of tonalite and metabasalt: Implications for reconstructions of 2.69 Ga early terrestrial ecosystems and paleoatmospheric chemistry. *Precambrian Research* 189, 1–17.

Duchac, K.C., Hanor, J.S. 1987. Origin and timing of the metasomatic silicification of an early archean komatiite sequence, Barberton Mountain Land, South Africa. *Precambrian Research* 37, 125–146.

Falkowski, P.G., Godfrey, L.V. 2008. Electrons, life and the evolution of Earth's oxygen cycle. *Philosophical Transactions of Royal Society B* 363, 2705–2716.

Farquhar, J., Bao, H.M., Thiemens, M. 2000. Atmospheric influence of Earth's earliest sulfur cycle. *Science* 289, 756–758.

Fenchel, T., King, G.M., Blackburn, T.H. 2012. *Bacterial Biogeochemistry: The Ecophysiology of Mineral Cycling. Elsevier, Amsterdam*, 248pp.

Feng, J., Yao, H., Wang, Y., Poli, P., Mao, Z. 2021. Segregated oceanic crust trapped at the bottom mantle transition zone revealed from ambient noise interferometry. *Nature Communications* 12, 2531.

Finster, K. 2008. Microbiological disproportionation of inorganic sulfur compounds. *Journal of Sulfur Chemistry* 29, 281–292.

Fuggi, A. 1993. Uptake and assimilation of nitrite in the acidophilic red alga *Cyanidium caldarium* Geitler. *New Phytologist* 125, 351–360.

Garvin, J., Buick, R., Anbar, A.D., Arnold, G.L., Kaufman A.J. 2009. Isotopic evidence for an aerobic nitrogen cycle in the latest Archean. *Science* 323, 1045–1048.

Geelhoed, J.S., Sorokin, D.Y., Epping, E., Tourova, T.P., Banciu, H.L., Muyzer, G., Stams, A.J.M., Van Loosdrecht, M.C.M. 2009. Microbial sulfide oxidation in the oxic-anoxic transition zone of freshwater sediment: involvement of lithoautotrophic Magnetospirillum strain J10. *FEMS Microbiology Ecology* 70, 54–65.

Glass, J.B., Wolfe-Simon, F., Anbar, A.D. 2009. Coevolution of metal availability and nitrogen assimilation in cyanobacteria and algae. *Geobiology* 7, 100–123.

Godfrey, L.V., Falkowski, P.G. 2009. The cycling and redox state of nitrogen in the Archaean ocean. *Nature Geoscience* 2, 725–729.

Grotzinger, J., Press, S. 2007. *Understanding Earth*. W. H. Freeman and Company, New York. 579 p.

Harris, P.M., Diaz, M.R., Eberli, G.P. 2019. The formation and distribution of modern ooids on Great Bahama Bank. *Annual Review of Marine Science* 11, 491–516.

Herschy, B., Chang, S.J., Blake, R., Lepland, A., Abbott-Lyon, H., Sampson. J., Atlas, Z., Kee, T.P., Pasek, M.A. 2018. Archean phosphorus liberation induced by iron redox geochemistry. *Nature Communications* 9, 1346.

Hess, B.L., Piazolo, S., Harvey, J. 2021. Lightning strikes as a major facilitator of prebiotic phosphorus reduction on early Earth. *Nature Communications* 12, 1535.

Homann, M., Sansjofre, P., Van Zuilen, M., Heubeck, C., Gong, J., Killingworth, B., Foster, I.S., Airo, A., Van Kranendonk, M.J., Ader, M., Lalonde, S.V. 2018. Microbial life and biogeochemical cycling on land 3,220 million years ago. *Nature Geoscience* 11, 665–671.

Ignarro, L.J., Fukto, J.M., Griscavage, J.M., Rogers, N.E., Byrns, R.E. 1993. Oxidation of nitric oxide in aqueous solution to nitrite but not nitrate: Comparison with enzymatically formed nitric oxide from L-arginine. *Proceedings of the National Academy of Sciences of the United States of America* 90, 8103–8107.

Jones, C., Nomosatryo, S., Crowe, S.A., Bjerrum, C.J., Canfield, D.E. 2015. Iron oxides, divalent cations, silica, and the early earth phosphorous crisis. *Geology* 43, 135–138. doi:10.1130/G36044.1.

Kandasamy, S., Nath, B.N. 2016. Perspectives on the terrestrial organic matter transport and burial along the land-deep sea continuum: Caveats in our understanding of biogeochemical processes and future needs. *Frontiers in Marine Science* 3, 259. doi:10.3389/fmars.2016.00259.

Kanzaki, Y., Murakami, T. 2015. Estimates of atmospheric CO_2 in the Neoarchean–Paleoproterozoic from paleosols. *Geochimica et Cosmochimica Acta* 159, 190–219.

Kirchman, D.L. 2012. *Processes in Microbial Ecology*. Oxford University Press, Oxford, 318p.

Kuenen, J.G. 2008. Anammox bacteria: from discovery to application. *Nature Reviews Microbiology* 6, 320–326.

Lalonde, S.V., Konhauser, K.O. 2015. Benthic perspectives on Earth's oldest evidence for oxygenic photosynthesis. *Proceedings of the National Academy of Sciences of the United States of America* 112, 995–1000.

Lehmann, M.F., Bernasconi, S.M., Barbieri, A., McKenzie, J.A. 2002. Preservation of organic matter and alteration of its carbon and nitrogen isotope composition during simulated and in situ early sedimentary diagenesis. *Geochimica et Cosmochimica Acta* 66, 3573–3584.

Lehtovirta-Morley, L.E. 2018. Ammonia oxidation: Ecology, physiology, biochemistry and why they must all come together. *FEMS Microbiology Letters* 365, fny058. doi:10.1093/femsle/fny058.

Lowe, D.R., Drabon, N., Byerly, G.R. 2019. Crustal fracturing, unconformities, and barite deposition, 3.26–3.23 Ga, Barberton Greenstone Belt, South Africa. *Precambrian Research* 327, 34–46.

Lyons, T.W., Reinhard, C.T., Planavsky, N.J. 2014. The rise of oxygen in Earth's early ocean and atmosphere. *Nature* 506, 307–315.

Middelburg, J.J. 2019. Marine carbon biogeochemistry. A primer for earth system scientists. *Springer Open*, 118p.

Murakami, A., Sawaki, Y., Ueda, H., Orihashi, Y., Machida, S., Komiya, T. 2019. Behavior of phosphorus during hydrothermal alteration of basalt under CO_2-rich condition. *JPGU 2019*, BCG07-P02.

Nabhan, S., Luber, T., Scheffler, F., Heubeck, C. 2016. Climatic and geochemical implications of Archean pedogenic gypsum in the Moodies Group (~3.2 Ga), Barberton Greenstone Belt, South Africa. *Precambrian Research* 275, 119–134.

Nabhan, S., Marin-Carbonne, J., Mason, P.R.D., Heubeck, C. 2020. In situ S-isotope compositions of sulfate and sulfide from the 3.2 Ga Moodies Group, South Africa: A record of oxidative sulfur cycling. *Geobiology* 18, 426–444.

Nakamura, K., Kato, Y. 2004. Carbonatization of oceanic crust by the seafloor hydrothermal activity and its significance as a CO_2 sink in the Early Archean. *Geochimica et Cosmochimica Acta* 68, 4595–4618.

Pasek, M., Lauretta, D. 2008. Extraterrestrial flux of potentially prebiotic C, N, and P to the early Earth. *Origins of Life and Evolution of Biospheres* 38, 5–21.

Pasek, M.A., Harnmeijer, J.P., Buick, R., Gull, M., Atlas, Z. 2013. Evidence for reactive reduced phosphorus species in the early Archean ocean. *Proceedings of National Academy of Sciences of the United States of America* 110, 10089–10094.

Paulmier, A., Ruiz-Pino, D. 2009. Oxygen minimum zones (OMZs) in the modern ocean. *Progress in Oceanography* 80, 113–128. doi:10.1016/j.pocean.2008.08.001.

Philippot, P., Van Zuilen, M., Lepot, K., Thomazo, C., Farquhar, J., Van Kranendonk, M.J. 2007. Early Archaean microorganisms preferred elemental sulfur, not sulfate. *Science* 317, 1534–1537.

Pinti, D.L., Hashizume, K., Matsuda, J. 2001. Nitrogen and argon signatures in 3.8 to 2.8 Ga metasediments: Clues on the chemical state of the Archean ocean and the deep biosphere. *Geochimica et Cosmochimica Acta* 65, 2301–2315.

Planavsky, N.J., Asael, D., Hofmann, A., Reinhard, C.T., Lalonde, S.V., Knudsen, A., Wang, X., Ossa, F.O., Pecoits, E., Smith, A.J.B., Beukes, N.J., Bekker, A., Johnson, T.M., Konhauser, K.O., Lyons, T.W., Rouxel, O.J. 2014. Evidence for oxygenic photosynthesis half a billion years before the Great Oxidation Event. *Nature Geoscience* 7, 283–286.

Reimer, T.O. 1980. Archean sedimentary baryte deposits of the Swaziland Supergroup (Barberton Mountain Land, South Africa). *Precambrian Research* 12, 393–410.

Reinhard, C.T., Raiswell, R., Scott, C., Anbar, A.D., Lyons, T.W. 2009. A late Archean sulfidic sea stimulated by early oxidative weathering of the continents. *Science* 326, 713–716.

Roger, A.J., Susko, E. 2018. Molecular clocks provide little information to date methanogenic Archaea. *Nature Ecology & Evolution* 2, 1676–1677.

Rouchon V., Orberger, B. 2008. Origin and mechanisms of K-Si-metasomatism of ca. 3.4–3.3 Ga volcaniclastic deposits and implications for Archean seawater evolution: Examples from cherts of Kittys Gap (Pilbara craton, Australia) and Msauli (Barberton Greenstone Belt, South Africa). *Precambrian Research*, 165, 169–189.

Russell, M.J., Hall, A.J., Martin, W. 2010. Serpentinization as a source of energy at the origin of life. *Geobiology* 8, 355–371.

Ruttenberg, K.C., Berner, R.A. 1993. Authigenic apatite formation and burial in sediments from non-upwelling, continental margin environments. *Geochimica et Cosmochimica Acta* 57, 991–1007.

Schopf, J.W., Kudryavtsev, A.B., Osterhout, J.T., Williford, K.H., Kitajima, K., Valley, J.W., Sugitani, K. 2017. An anaerobic ~ 3400 Ma shallow-water microbial consortium: Presumptive evidence of Earth's Paleoarchean anoxic atmosphere. *Precambrian Research* 299, 309–318.

Schulz, H.N., Schulz, H.D. 2005. Large sulfur bacteria and the formation of phosphorite. *Science* 307, 416–418.

Schwartz, A.W. 2006. Phosphorus in prebiotic chemistry. *Philosophical Transactions: Biological Sciences* 361, 1743–1749.

Seefeldt, L.C., Yang, Z.-Y., Lukoyanov, D.A., Harris, D.F., Dean, D.R., Raugei, S., Hoffman, B.M. 2020. Reduction of substrates by nitrogenases. *Chemical Reviews* 120, 5082–5106.

Shen, Y., Buick, R., Canfield, D.E. 2001. Isotopic evidence for microbial sulphate reduction in the early Archaean era. *Nature* 410, 77–81.

Shen, Y., Pinti, D.L., Hashizume, K. 2006. Biogeochemical cycles of sulfur and nitrogen in the Archean ocean and atmosphere. *Archean Geodynamics and Environments, Geophysical Monograph Series* 164, 305–320.

Shimamura, K., Shimojo, F., Nakano, A., Tanaka, S. 2016. Meteorite impact-induced rapid NH_3 production on Early Earth: Ab Initio molecular dynamics simulation. *Scientific Reports* 6, 38953. doi:10.1038/srep38953.

Spieck, E., Lipski, A. 2011. Cultivation, growth physiology, and chemotaxonomy of nitrite-oxidizing bacteria. *Methods in Enzymology* 486, 109–130.

Stüeken, E.E., Buick, R., Guy, B.M., Koehler, M.C. 2015a. Isotopic evidence for biological nitrogen fixation by molybdenum-nitrogenase from 3.2 Gyr. *Nature* 520, 666–669.

Stüeken, E.E., Buick, R., Anbar, A.D. 2015b. Selenium isotopes support free O_2 in the latest Archean. *Geology* 43, 259–262.

Stüeken, E.E., Kipp, M.A., Koehler, M.C., Buick, R. 2016. The evolution of Earth's biogeochemical nitrogen cycle. *Earth-Science Reviews* 160, 220–239.

Sugitani, K., Grey, K., Allwood, A., Nagaoka, T., Mimura, K., Minami, M., Marshall, C.P., Van Kranendonk, M.J., Walter, M.R. 2007. Diverse microstructures from Archaean chert from the Mount Goldsworthy–Mount Grant area, Pilbara Craton, Western Australia: Microfossils, dubiofossils, or pseudofossils? *Precambrian Research* 158, 228–262.

Sugitani, K., Horiuchi, Y., Adachi, M., Sugisaki, R. 1996. Anomalously low Al_2O_3/TiO_2 values for Archean cherts from the Pilbara Block, Western Australia––possible evidence for extensive chemical weathering on the early earth. *Precambrian Research* 80, 49–76.

Sugitani, K., Mimura, K., Suzuki, K., Nagamine, K., Sugisaki, R. 2003. Stratigraphy and sedimentary petrology of an Archean volcanic-sedimentary succession at Mt. Goldsworthy in the Pilbara Block, Western Australia: implications of evaporite (nahcolite) and barite deposition. *Precambrian Research* 120, 55–79.

Sumner, D.Y., Grotzinger, J.P. 1996. Were kinetics of Archean calcium carbonate precipitation related to oxygen concentration? *Geology* 24, 119–122.

Takeuchi, Y., Furukawa, Y., Kobayashi, T., Sekine, T., Terada, N., Kakegawa, T. 2020. Impact-induced amino acid formation on Hadean Earth and Noachian Mars. *Scientific Reports* 10, 9220.

Thomazo, C., Ader, M., Philippot, P. 2011. Extreme ^{15}N-enrichments in 2.72-Gyr-old sediments: evidence for a turning point in the nitrogen cycle. *Geobiology* 9, 107–120.

Thomazo, C., Papineau, D. 2013. Biogeochemical cycling of nitrogen on the early Earth. *Elements* 9, 345–351.

Turchyn, A.V., DePaolo, D.J. 2019. Seawater chemistry through Phanerozoic time. *Annual Review of Earth and Planetary Sciences* 47, 197–224.

Ueno, Y., Yamada, K., Yoshida, N., Maruyama, S., Isozaki, Y. 2006. Evidence from fluid inclusions for microbial methanogenesis in the early Archaean era. *Nature* 440, 516–519.

Van Kranendonk, M. J. 2006. Volcanic degassing, hydrothermal circulation and the flourishing of early life on Earth: A review of the evidence from c. 3490–3240 Ma rocks of the Pilbara Supergroup, Pilbara Craton, Western Australia. *Earth-Science Reviews* 74, 197–240.

Wacey, D., Saunders, M., Brasier, M.D., Kilburn, M.R. 2011. Earliest microbially mediated pyrite oxidation in ~3.4 billion-year-old sediments. *Earth and Planetary Science Letters* 301, 393–402.

Walker, J.C.G., Hays, P.B., Kasting, J.F. 1981. A negative feedback mechanism for the long-term stabilization of Earth's surface temperature. *Journal of Geophysical Research* 86, 9776–9782.

Wheat, C.G., Feely, R.A., Mottl, M.J. 1996. Phosphate removal by oceanic hydrothermal processes: An update of the phosphorus budget in the oceans. *Geochimica et Cosmochimica Acta* 60, 3593–3608.

Wilmot Jr., D.B., Vetter, R.D. 1990. The bacterial symbiont from the hydrothermal vent tubeworm Riftia pachyptila is a sulfide specialist. *Marine Biology* 106, 273–283.

Wolfe, J.M., Fournier, G.P. 2018. Horizontal gene transfer constrains the timing of methanogen evolution. *Nature Ecology & Evolution* 2, 897–903.

Wong, K., Mason, E., Brune, S., East, M., Edmonds, M., Zahirovic, S. 2019. Deep carbon cycling over the past 200 million years: A review of fluxes in different tectonic settings. *Frontiers in Earth Science* 7, article 263. doi:10.3389/feart.2019.00263.

Yamagata, Y., Watanabe, H., Saitoh, M., Namba, T. 1991. Volcanic production of polyphosphates and its relevance to prebiotic evolution. *Nature* 352, 516–519.

Zao, B., Zang, Y., Yuan, H., Yang, M.Y. 2015. Granular ferric hydroxide adsorbent for phosphate removal: Demonstration preparation and field study. *Water Science and Technology* 72, 2179–2186.

5 Topics of the Early Precambrian Earth 1

5.1 INTRODUCTION

The Archean era (4.0–2.5Ga) occupies over 30% of the Earth's history (Figure 1.1). The Archean Earth was quite different from the Phanerozoic and even the Proterozoic Earth, although the whole picture is poorly known. The Earth's environment and life had significantly changed during the Archean. In this chapter and next, events and aspects basically (not exclusively) unique to the Archean Earth are focused, including e.g., (1) process of Archean cratonization, (2) active volcanism related to high mantle temperature, (3) frequent large asteroid impacts and their implications on early evolution of life, (4) formation of Archean iron formations (IFs) before the rise of oxygen, (5) Great Oxidation Event and earlier oxygenation, and (6) physicochemical features of early oceans and atmosphere.

5.2 ARCHEAN CRATONS

5.2.1 DISTRIBUTIONS AND COMPOSITIONS OF ARCHEAN CRATONS

Archean cratons composed largely of rocks older than 2.5 Ga are distributed widely in 35 regions in the world (Kusky and Polat, 1999) (Figure 5.1). They comprise parts of Proterozoic shields, where Precambrian basement rocks including granitic rocks and high-grade metamorphic rocks are exposed. Note that the Archean cratons shown in Figure 5.1 do not always contain rocks of all Archean ages (4.0–2.5 Ga).

Archean cratons have roots of cratonic lithospheric mantle characterized by strong and buoyant properties, which is distinct from their Proterozoic successors (Hansen, 2015). Archean cratons are also unique in their structures, represented by thin greenstone belts sandwiched by ovoid to elongated bodies of granitoids (Figure 5.2). Greenstone belts are composed largely of mafic to ultramafic volcanic rocks, with various minor components including intermediate volcanic rocks, volcaniclastic rocks, and sedimentary rocks (chert, IF, carbonate, and siliciclastic rock). These packages are generally steeply dipping and deformed by granitoid bodies to be often keel-shaped (Figure 5.2a). Granitoids are characterized by potassium (K)-poor in composition. This type of granitic rocks is called tonalite-trondhjemite-granodiorite series (TTG).

Many of the Archean cratons had been subjected to high grade metamorphism. Obviously, sedimentary rocks subjected to high temperature and high pressure metamorphism, are not suitable for search for life and study on early environments. In this context, the Pilbara Craton, Western Australia, and the Kaapvaal Craton,

DOI: 10.1201/9780367855208-5

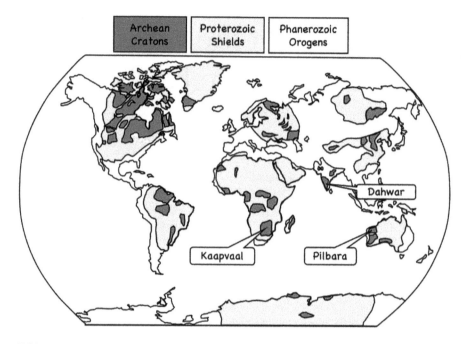

FIGURE 5.1   Distribution of Archean cratons and Proterozoic shields. (After Furnes et al. (2015)). Precambrian greenstone sequences represent different ophiolite types.)

South Africa (Figure 5.1) are the best targets. These two cratons have sedimentary rocks ranging widely from 3.6 to 2.7 Ga (Paleoarchean to Neoarchean), many of which are of low metamorphic grade up to greenschist facies, which means that the rocks had been heated less than ca. 400°C (also see Figure 2.10). In the south of the Pilbara Craton, the Hamersley basin in which the Neoarchean Fortescue Group and the Neoarchean to Paleoproterozoic Hamersley Group occur (Figure 5.2b). The Kaapvaal Craton also contains Neoarchean (~2.67 to ~2.46 Ga) Transvaal Supergroup dominated by terrigenous clastic rocks, carbonates, and IFs. Cheney (1996) proposed that these two cratons comprised a supercontinent "Vaalbara". This supercontinent could be back to 3.1 Ga, and possibly back to 3.6 Ga ago (Zegers et al., 1998), although Evans and Muxworthy (2018) presented paleomagnetic data that are inconsistent with the existence of a single supercontinent between ~2.87 and ~2.71 Ga ago.

Other cratons and provinces containing rocks older than 3.0 Ga include the Superior Craton and the Nain province of the North Atlantic Craton, and the Dharwar Craton. From the Nuvvuagittuq greenstone belt of the Superior Craton and the Saglek block of the Nain province, Canada, possible oldest traces of life have been reported, although their reliabilities have not yet been established (Chapter 8). The Isua supracrustal (or greenstone) belt (ISB), West Greenland, also belongs to the Nain province of North Atlantic Craton and has been extensively studied for search for early life, and several important discoveries were made (Chapter 8). The Dharwar

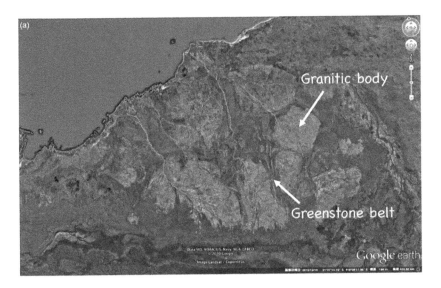

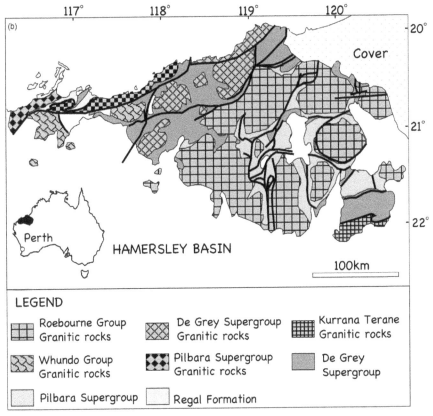

FIGURE 5.2 (a) Google image of the Pilbara Craton , which is nearly identical in scale in the figure in (b). Examples of granitic body and greenstone belt are shown. (b) Simplified geology of the northern Pilbara Craton. (Modified from Van Kranendonk et al. (2006).)

Craton, India contains 3.4–2.5 Ga-old rocks (e.g., Sunder Raju et al., 2013), although only a few reports on the trace of Neoarchean life had appeared (Sharma and Shukla, 2004). To my knowledge, from the other Archean cratons, no remarkable discoveries on early life have been made. This is mainly due to the scarcity of sedimentary rocks of low metamorphic grade.

5.2.2 Origins of Archean Cratons

Origins of Archean craton have been controversial; the situation is different between terranes comprising cratons and between cratons. For example, the East Pilbara Terrane (EPT) and the West Pilbara Super Terrane (WPST) comprising the northern Pilbara Craton formed in different fashions (Figure 5.2). Some researchers once tried to interpret the origin of EPT in the context of accretionary tectonics, that is, accretion of oceanic crusts (greenstone belt) onto continents (granitoid bodies) (e.g., Isozaki et al., 1997). Now it seems well established that EPT represents oceanic plateau, formed initially by cyclic eruptions and accumulation of basaltic lavas. Within the oceanic plateau, TTG was formed and uprose in a diapiric fashion. On the other hand, the WPST is thought to have been formed in a tectonic setting similar to Phanerozoic subduction zones (e.g., Van Kranendonk et al., 2007). The origin of the Kaapvaal Craton, South Africa, more specifically, the Barberton greenstone belt, has been still controversial. A geotectonic setting for mafic to ultramafic lavas of this belt is diversely interpreted, including continental arc or back-arc, and juvenile oceanic (mid-ocean ridge) settings (Grosch and Slama, 2017 and reference therein). Finally, a unique idea for the formation of Archean cratons proposed by Hansen (2015) is introduced. The author proposed in an analogy with quasi-circular crustal plateaus on Venus that the formation of Archean cratons was triggered by large asteroid impacts. This seems to be fascinating hypothesis, considering accumulating evidence for repeated large asteroid impacts during the early Archean as described later.

5.3 EARLY CONTINENTAL GROWTH AND ITS IMPLICATIONS

5.3.1 Models of Continental Growth

Formation of the evolved crust might have been back to 4.4 Ga ago in the Haeden (e.g., Wilde et al., 2001, Iizuka et al., 2015), but at that time, its volume is assumed to have been small and the Earth's crust was dominated by mafic to ultramafic components. As can be seen in Figure 5.3, various models of continental growth have been proposed and its growth history has long been controversial (e.g., gradual vs. stepwise). As summarized by Hawkesworth et al. (2019), the variation of the continental growth model depends on methodologies. For example, the slowest growth model represented by the curves 8 and 9 in Figure 5.3 is based on the distribution of presently preserved rocks of different ages on the Earth. Intermediate growth models represented by the curves 6 and 7 are drawn based on the present distribution of rocks of different ages; in this case, however, the employed ages are so-called "model ages", which are estimated when their precursory source materials were

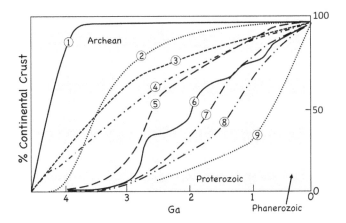

FIGURE 5.3 Models of growth of continental crust. (Modified from Hawkesworth et al. (2019). (1) Armstrong (1981), (2) Pujol et al. (2013), (3) Dhuime et al. (2012), (4) Belousova et al. (2010), (5) Taylor and McLennan (1985), (6) Condie and Aster (2010), (7) Allègre & Rousseau (1984), (8) Hurley and Rand (1969), and (9) Goodwin (1996).)

differentiated from the mantle. The other models (the curves from 1 to 5) intend to give constrains on the crustal volumes at various ages, employing isotopic ratios of certain radiogenic elements or trace elements as proxies. Due to accumulation of precise data of trace elements and isotopes in zircon ($ZrSiO_4$, see *Column*), one of the most resistant minerals, it seems now widely accepted that ~70% of the present volume of the continental crust was produced by 3.0 Ga ago, with a higher production rate than post 3.0 Ga (e.g., Dhuime et al., 2015).

5.3.2 PLATE TECTONICS AND CONTINENTAL GROWTH AND IMPLICATIONS

Growth of continental crust is closely related to plate tectonics because subduction-related magmatism could produce felsic magma effectively and accretion of sediments and volcanic islands substantially increases mass of continent (Chapter 2). Although late onset of plate tectonics in the Neoproterozoic was proposed by Stern (2008), many researchers appear to have come to believe that plate tectonics had started no later than the Neoarchean (2.8–2.5 Ga ago). However, it should be noted that it spans ~2 Ga from the Hadean to the Neoarchean. It is thus fundamentally different in significance for the Earth's history whether the plate tectonics had started in the Hadean or in the Neoarchean. Obviously, there is no consensus on this issue. Turner et al. (2020), for example, estimated SiO_2 contents and Th/Nb ratios in melts (magma), from which 4.3 to 3.3 Ga-old zircon grains had been crystallized: these two signatures were used for identification of derivation of this highly resistant mineral (e.g., intraplate magmatism vs. subduction-related magmatism). They concluded that the melts from which zircon were crystallized were commonly andesitic, like melts representatively formed in modern subduction zones, and implied that the Earth's early crustal compositions and tectonic regime had little changed during this period (from the Hadean to the Mesoarchean). In other words, modern-style plate tectonics had started already in the Hadean. Similar implications have been drawn from isotopic proxies of Si and Mo (Deng et al., 2019; McCoy-West et al., 2019), although a study using B isotopes of TTG suggests that before 2.8 Ga, modern style subduction could not play an important role for formation of juvenile continental crust (Smit et al., 2019).

The other view, probably more widely accepted, emphasizes that the onset of modern-style plate tectonics had occurred around 3.0 Ga ago (e.g., O'Neill et al., 2020 and references therein). Cawood et al. (2018) also suggested that modern-style plate tectonics has commenced 2.5 Ga ago, with a transition phase between 3.2 and 2.5 Ga. Various pieces of evidence have been presented. A few of them are introduced here. In the Barberton greenstone belt, South Africa, components of greenstone successions had drastically changed from the older Onverwacht Group dominated by mafic to ultramafic volcanisms to the younger Fig Tree Group dominated by terrigenous clastic sedimentation. This transition occurred 3.2 Ga ago, which may require accelerated and enhanced orogeny that is potentially linked to starting of modern-style plate tectonics. The start may have been triggered by large asteroid impacts (Lowe et al., 2014). A similar trend has been recognized in the Pilbara Craton, Western Australia, although initiation of major terrigenous clastic sedimentation is recognized as somewhat younger succession of the 2.99 Ga-old De Grey Supergroup. This event was preceded by the well-established subduction-related accretion events that were associated with collision of East and West Pilbara terranes called the Prinsep Orogeny at 3.07 Ga ago (Van Kranendonk et al., 2007).

It may be emphasized again here that the growth of continental crusts was closely related to the evolution of atmosphere, hydrosphere, and biosphere via an increase in the rate of chemical weathering, which had likely reduced the atmospheric CO_2 and increases inflow of rock-derived nutrients such as Ca, K, Mg, and P into the ocean (Figure 5.4).

5.4   KOMATIITE VOLCANISM AND ITS SIGNIFICANCE

5.4.1   KOMATIITES AND ITS ORIGIN

The Archean, particularly the early Archean, was characterized by active volcanisms, distinctive from the Phanerozoic (ca. 5.4 Ga ago to the present) and even the Proterozoic (2.5–0.54 Ga). This is indicated by rock records (e.g., Byerly, 2015; Hickman, 2012). For example, the ca. 3.5 to 3.3 Ga-old Onverwacht Group of the Barberton greenstone belt, South Africa, is ~12 km in thickness and is dominated by mafic to ultramafic volcanic and volcaniclastic rocks. The ca. 3.5- to 3.0 Ga-old Pilbara Supergroup of the Pilbara Craton, Western Australia, is up to 15–20 km thick, and more than its 90% is composed of mafic to ultramafic volcanic and volcaniclastic rocks.

Komatiite is ultramafic volcanic rock that was lithified from highly magnesian magma with MgO >18 wt%. This type of ultramafic rock is abundant in the Archean but is rare in the Proterozoic and the Phanerozoic. Komatiite typically has

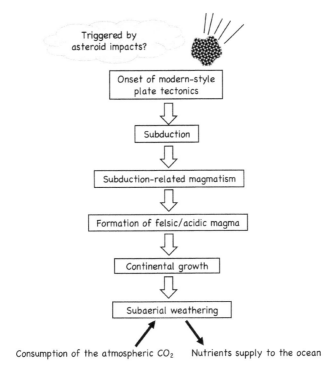

FIGURE 5.4   Onset of the plate tectonics and potentially to likely induced various events. (Note that the reality was not so simple.)

spinifex texture, which is comprised of millimeter to decimeter-sized plate-like crystals of olivine [$(Mg, Fe)_2SiO_4$] and pyroxene [$(Mg, Fe) SiO_3$] in a fine-grained matrix (e.g., Shore and Fowler, 1999) (Figure 5.5). Large-scale melting of mantle at high temperature is required for generation for komatiite, because liquid produced by partial melting tends to be relatively depleted in MgO and enriched in SiO_2 than parental mantle (Chapter 2). Indeed, estimated eruption temperatures of Archean komatiites (n = 16) range from 1,400°C to 1,640°C, except an outlying one (1,240°C) (Barnes and Arndt, 2019), significantly higher than currently erupting magmas (900°C–1,200°C). It is quite reasonable to assume that the Archean mantle was much hotter than today (Berry et al., 2008), considering higher abundance of radioactive elements and heat stored during accretion of protoplanets (planetary accretion).

5.4.2   SERPENTINIZATION OF KOMATIITE AND ITS IMPLICATIONS

While the origin and formation process of komatiite have long been hot research topics, alteration of komatiite and other ultramafic rocks, so-called "serpentinization", has recently been paid special attention in the context of the origin of life and early metabolisms (e.g., Holm et al., 2015). Serpentinization refers to the formation

FIGURE 5.5   Spinifex texture of komatiite, in the North Pole region, Pilbara Craton, Western Australia.

of hydrous mineral such as serpentine $[Mg_3Si_2O_5(OH)_4]$ from olivine (Equation 5.1) (Russel et al., 2010). By-products of this reaction are various, depending on composition of olivine (Fe/Mg ratio) and species of reactants. For example, serpentinite and brucite $[Mg(OH)_2]$ are produced by reaction of forsterite (Mg_2SiO_4: Mg-rich endmember of olivine) and H_2O. On the other hand, H_2 is produced by reaction of fayalite (Fe_2SiO_4: Fe-rich endmember of olivine) and H_2O (Equation 5.2) (Sleep et al., 2004). In the presence of CO_2, magnetite (Fe_3O_4) and methane (CH_4) could be produced. It is also known that Ni-Fe alloys are produced during serpentinization (Sleep et al., 2004).

In the presence of H_2, carbon compounds such as CO, CO_2, and HCO_3^- could be reduced to CH_4 and other OCs (e.g., Lazar et al., 2015; Holm and Charlou, 2001; Shock and Canovas, 2010). This reaction is accelerated in the presence of catalysts such as Ni-Fe alloys and magnetite, which is known as Fischer-Tropsch-type (FTT) synthesis. In fact, Sforna et al. (2018) recently described the occurrences of the three distinct types of poorly structured condensed carbonaceous matter from serpentinite of the Northern Apennine ophiolites, Italy. The three types of carbonaceous matter are associated with different mineral phases formed at temperatures ($\leq 200°C$) lower than that expected for FTT.

$$18Mg_2SiO_4 + 6Fe_2SiO_4 + 26H_2O + CO_2 \rightarrow 12Mg_3Si_2O5(OH)_4 + 4Fe_3O_4 + CH_4 \quad (5.1)$$

$$3Fe_2SiO_4 + 2H_2O \rightarrow 2Fe_3O_4 + 3SiO_2 + 2H_2 \quad (5.2)$$

5.5 LARGE ASTEROID IMPACT AND ITS IMPLICATIONS

5.5.1 PALEO- AND MESOARCHEAN RECORDS OF ASTEROID IMPACTS

Sedimentary records in the Archean suggest that large asteroid impacts occurred more frequently than later periods (e.g., Byerly et al., 2002; Glikson et al., 2016; Lowe et al., 2014). D.R. Lowe and his colloborators had performed extensive field works in the Barberton greenstone belt, South Africa, and have recognized at least eight large (20–70 km across) asteroid impacts during the relatively short period (ca. 200 million years from 3.47 to 3.22 Ga) corresponding to the Onverwacht Group and the overlying Fig Tree Group. The identification had been mainly based on spherules that were interpreted to have been produced by vaporization of impactor and target and subsequent condensation. On the other hand, to date only two well-established impact layers have been reported from the contemporaneous Warrawoona Group of the Pilbara Craton, Western Australia (Glikson et al., 2016; Byerly et al., 2002). The other two putatives have been identified recently by the author (one from the ca. 3.4 Ga-old Strelley Pool Formation and the other from the ca. 3.0 Ga-old Farrel Quartzite), as described later.

5.5.2 IDENTIFICATION OF LARGE ASTEROID IMPACTS

By the way, how can we identify spherulitic and related structures (spherule, tear-shaped structure, dumbbell-shaped structure) in sedimentary rocks as the impact origin? Such structures are not rare; they can be formed by other processes including, e.g., authigenic carbonate precipitation (oolite), rapid cooling of splashes of magma (e.g., Pele's tears), and lightning in volcanic clouds (e.g., Genareau et al., 2015). Also, what other lines of evidence of large asteroid impacts could be obtained? Direct evidence for large impacts is terrestrial crater. However, well-identified impact craters are mostly younger than 2.5 Ga ago (e.g., Jourdan et al., 2012). Thus, we have no choice but to seek other evidence for Archean impact events.

Krull-Davatzes et al. (2019) complied criteria for recognizing distal impact ejecta units, referring works by, e.g., Simonson et al. (2015) and Smith et al. (2016). The criteria include those specific for impact spherules and those for beds containing extraterrestrial components. Impact spherules (Figure 5.6) formed by condensation of vaporized impact and/or target materials have (1) fibroradial and barred relict mineral textures, (2) oxide-rich rim, (3) aerodynamic shapes such as dumbbells and teardrops and internal bubbles, and (4) association with Ni-rich chromite. Other criteria applicable for beds concentrating extraterrestrial materials are associations with e.g., (1) shocked minerals (TiO_2-II, shocked quartz and shocked zircon), (2) chondrite-level high concentrations of Ni, (3) chondritic ratios of Ni/Co and Ni/Cr, (4) high concentrations of platinum group elements (PGE: Ru, Rh, Pd, Os, Ir, and Pt), and (5) isotope anomalies of Cr. Also, isotopic compositions of Os, distinct basically between extraterrestrial and terrestrial materials, have been used for identifying younger impact event (e.g., Onoue et al., 2016).

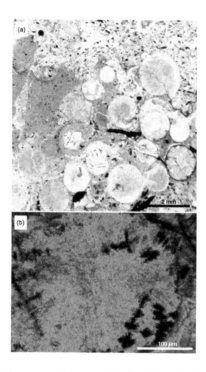

FIGURE 5.6 Examples for ancient impact spherules, from the Barberton greenstone belt, South Africa. (a) Morphologically and texturally diverse spherules from the Barberton greenstone belt. Transmitted light. (b) Oxide-rimmed spherule with skeletal spinel from the Barberton greenstone belt. Plane-polarized light. (Photomicrograph courtesy by Krull-Davatzes et al. (2019).)

Another clue is identification of sediments deposited under a sudden high-energy environment. Impacts of large asteroids onto oceans and even lands could cause tsunamis, which could rework previously deposited sediments on ocean floors and beaches. Reworked sediments and rocks could be transported far away from their original site of deposition and redeposited. Thus, "tsunami deposit" is identifiable as an unusual layer containing coarse clasts within sedimentary succession representing a quiet and low-energy depositional environment (Figure 5.7). Although "tsunami" could alternatively be caused by earthquakes, volcanic eruptions, and rarely landslides, identification of such high-enery deposits in the field provides opportunities to find ancient impact events.

5.5.3 Implications of Large Asteroid Impacts

It is not hard to imagine that large asteroid impacts had influenced the Earth's environment. Environmental disturbance includes crustal deformation, terrain amalgamation, induction of volcanisms, and evaporation of ocean water, and even triggering

FIGURE 5.7 (a) Large rip-up clast (the arrow) associated with spherule bed at the Barberton greenstone belt, South Africa. Photo courtesy of A.K. Davatzes (Krull-Davatzes et al., 2019). (b) Sandstone with large rip-up clasts at the Goldsworthy greenstone belt, Western Australia—possibly produced by impact-induced tsunami.

modern-like plate tectonic regime as suggested earlier (e.g., Lowe et al., 2014; Lowe and Byerly, 2015). More recently, O'Neill et al. (2020) numerically simulated that short-lived subduction can be initiated by impact of asteroid of moderate size (Ø ~70 km). Additionally, it is reasonable to assume that the early life was influenced significantly by asteroid impacts. Impact-induced high-pressure and temperature severely damaged ecosystems close to the impact centers. Destruction of ecosystems away from the impact center, on the other hand, may have provided new niches including habitats and nutrients with surviving life. Possibly, impact-induced environmental agitation was closely related to the evolution of specific organisms, which would be discussed later (Chapter 13).

5.6   ARCHEAN SEAWATER COMPOSITIONS AND PRODUCTS 1: IRON FORMATIONS

5.6.1   Clues to Archean Seawater Compositions

Physicochemical conditions and the secular changes of Archean oceans were closely related to the evolution of early life. Although Archean oceans had gone, traces of them can be preserved as fluid inclusions. Fluid inclusions refer to micron-scale tiny inclusions composed of liquid and/or vapor phase occasionally with solid phases. While they may occur in many types of minerals, those occurring within hydrothermal megaquartz veins have often been subjected to paleoenvironmental studies. Such fluid inclusions, however, cannot be regarded as having pristine seawater compositions. They had been hydrothermally altered to various degrees. The "fossilized" hydrothermally altered "Archean seawater" has been subjected to a wide variety of analyses such as microthermometry, ion-chromatography, and mass spectrometry, providing information about salinity, ionic compositions, and microbial metabolisms in Archean oceans and even abiotic formation of organic compounds under hydrothermal conditions (e.g., Screiber et al., 2017; Marty et al., 2018; Burgess et al., 2020 and references therein). Although details of the fluid inclusion studies are not referred to here, they are expected to provide some important constraints on Archean seawater compositions and evolutions, with careful considerations of some ambiguities that arise from, e.g., extent of hydrothermal alterations and coexistence of fluid inclusions of multiple generations (e.g., Farber et al., 2015).

5.6.2   What Are Iron Formations?

In Figure 5.8, distributions of IFs through the Earth's history are shown. IFs are iron- and silica-rich sedimentary rocks that characterize the Precambrian era. Concentration of SiO_2 varies from less than 10 to ~70 wt%, whereas $_{total}Fe$ varies from less than 10 to ~60 wt% (Holland, 1984), although Fe enrichment may reflect secondary enrichment (supergene enrichment). Representative unweathered, primary IFs contain Fe of 30–45 wt%. In the context of Archean seawater compositions, Figure 5.8 may give readers a somewhat erroneous impression, because the oldest-known IF has been described from >3.75 Ga-old Nuvvuagittuq Supracrustal Belt in northern Québec, Canada (Mloszewska et al., 2012), and in the durations when no IFs appear to be produced in this figure (e.g., 3.5–3.2 Ga), small-scale IFs, and other iron-rich chemical sediments called jaspilite or ferruginous cherts occur.

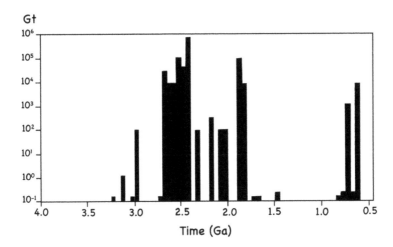

FIGURE 5.8   Abundance of iron formations through time. (Modified from Bekker et al. (2014).)

The youngest IF is thought to have been formed closely related to the global glaciation (the Snowball Earth), in the Cryogenian (7.20–6.35 Ga) of the Neoproterozoic (Ilyin, 2009).

Representative iron-bearing minerals in IFs are hematite (Fe_2O_3) and magnetite (Fe_3O_4). Other iron-bearing minerals are siderite ($FeCO_3$), ankerite [$Ca(Fe, Mg, Mn)(CO_3)_2$], and iron-bearing silicates such as stilpnomelane. An Fe-poor layer is composed dominantly of quartz (SiO_2). It should be noted that iron-bearing minerals are not always primary. They are products of diagenesis, metamorphism, and alteration. Sun and Li (2017) performed detailed electron microscopic, X-ray diffraction, and Mössabuer spectroscopic studies on 3.8 to 2.2 Ga-old IFs and suggested that only three types of Fe-bearing minerals were originated from primary precipitates from seawaters. They include compositionally and texturally homogeneous euhedral magnetite, aggregates of hematite nanocrystals, and submicrometer-sized euhedral hematite (Figure 5.9a). Siderite in IFs has once been utilized to constrain the partial pressure of the Archean atmospheric CO_2. However, this proposition now seems to be not always valid: siderite could not be precipitated directly from seawater. Carbon isotopic studies also suggest that many of siderite in IFs were produced microbially through dissimilatory iron reduction {anaerobic respiration using Fe^{3+} as electron acceptor} (e.g., Johnson et al., 2013).

Iron formations have been classified into banded iron formation (BIF) and granular iron formation (GIF), based on textures. Banded iron formation is characterized by alternation of iron-rich and iron-poor layers at various thickness and results in banded to laminated occurrences (Figure 5.9b). Sand-sized granules characterize GIF. The former dominates Archean IFs. Such textural variations are related to difference in depositional settings. Namely, BIFs are likely deposited under a low

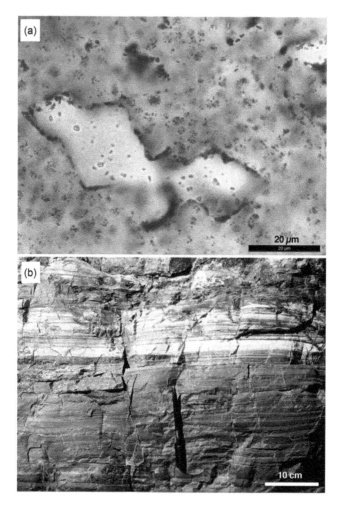

FIGURE 5.9 (a) Photomicrograph tiny hematite supposed to be primary precipitates from ca. 3.0 Ga-old banded iron formation at the Goldsworthy greenstone belt in the Pilbara Craton, Western Australia. Clusters correspond to the matrix individual quartz grains. Rhombic voids represent dissolved carbonates of secondary origin. Transmitted light. (b) Photograph of an outcrop of banded iron formation from which the specimen for (a) was collected.

energy environment, probably below wave base, although mechanisms generating bands and laminae, very rhymical in some cases, have been variously interpreted, including microbial processes, hydrothermal activity, seasonal forcing, climate control, and others (e.g., Lantink et al., 2019 and references therein). On the other hand, it has long been interpreted that GIF represents high-energy depositional environment at shallow marine setting, where waves reworked iron-rich sediments into granules, although a recent study suggests involvement of microbial processes in the formation of granules (e.g., Dodd et al., 2018).

5.6.3   Iron-Rich and Anoxic Deep Seawaters?

Modern seawaters are strongly depleted in dissolved Fe. This is simply because of full oxygenation of the modern atmosphere and the shallow oceans. Iron takes three oxidation numbers, including 0, 2, and 3. Ferrous iron (Fe^{2+}) is much more soluble than ferric iron (Fe^{3+}) at neutral to alkaline pH. Under oxic environment, Fe^{2+} could be readily oxidized to Fe^{3+}, which would be precipitated as ferric hydroxides [$Fe(OH)_3$]. Thus, fluxes of dissolved Fe via hydrothermal vents and chemical weathering of rock-forming minerals to open oceans are limited. Furthermore, as Fe is an essential nutrient for organisms, it is quickly absorbed by primary producers such as phytoplankton. In some places called "High nutrient, low-chlorophyll (HNLC) regions" in the present oceans, the primary production is suppressed by deficiency of available Fe, despite other essential elements such as N and P being abundant (Martin and Fitzwater, 1988).

Returning to the Early Precambrian, worldwide deposition of IFs and other ferruginous chemical sediments can be taken direct evidence for prolonged iron-enrichment in seawater, more specifically deep water. In other words, the ocean deep water was a huge reservoir of dissolved Fe^{2+}. Although Fe^{3+} can also be dissolved under strongly acidic condition (pH < 2), such low pH is unlikely except for the earliest stage of ocean formation. As discussed later, many of researchers consider that the pH of Archean seawater was around neutral or weakly acidic (e.g., Halevy and Bachan, 2017). Thus anoxic condition is required as a huge reservoir of dissolved Fe^{2+}, although the surface environments should have changed and fluctuated through the Archean. In extension, this may suggest poverty of hydrogen sulfide (H_2S) in deep oceans, because H_2S would react with Fe^{2+} to be precipitated as such as pyrites (FeS_2), which potentially reduces the budget of dissolved Fe.

While the deep ocean water was likely the reservoir of dissolved Fe^{2+}, what sources for that were available? It seems widely accepted that the major source of dissolved Fe in the Archean oceans was high temperature (~ >400°C) hydrothermal fluid like that venting from the present-day mid-ocean ridge (MOR) spreading centers. Such modern deep-sea high-temperature hydrothermal fluids are strongly acidic (pH < 2) and are enriched in metals represented by Fe, Mn, and Si. The fluids are known to exhibit characteristic shale-normalized rare-earth elements (REEs) pattern such as an enrichment of $_{63}Eu$ relative to neighboring $_{62}Sm$ and $_{64}Gd$ (Figure 5.10, also see *Column* in Chapter 8 for REEs systematics). This is called positive Eu-anomaly and is attributed to hydrothermal alteration of the oceanic crust (e.g., Nakada et al., 2017, and references therein) and retention in anoxic hydrothermal fluids. Similar positive Eu-anomaly has been reported from many of the Precambrian IFs in the worlds, which has often been taken as evidence for that Fe of IFs is derived from high-temperature hydrothermal fluids.

It is however possible that Archean high-temperature hydrothermal fluids were distinct from modern ones, as suggested by Shibuya et al. (2007), who conducted petrological analyses of hydrothermally altered Archean basalts and thermodynamic calculations, suggesting that high-temperature (~350°C) hydrothermal fluids in the Archean could have been highly alkaline, silica-rich, and iron-poor. Also noteworthy

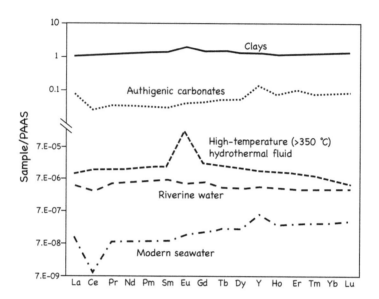

FIGURE 5.10 PAAS (Post Archean Australian Shale Composite) (McLennan, 1989)-normalized patterns of various natural fluids and sediments. Modified from Tostevin et al. (2016). Note that the patterns of riverine water and clays show that their REE sources are basically equivalent to the average shale (the upper continental crust).

is that positive Eu-anomaly can be produced also by reaction at low temperatures less than 100°C (Nakada et al., 2017 and references therein). As discussed in Section 5.7.3, off-axis low-temperature hydrothermal activities were likely widespread in the Archean, particularly before the onset of modern-style plate tectonics. Though metal concentrations of such low-temperature hydrothermal fluids should have been much lower than high temperature ones, they could have produced iron- and silica-rich deposits as exemplified by modern analogous at Loihi Seamount, where iron-oxide deposits, with variously enriched in amorphous silica, precipitated from low-temperature iron-rich and H_2S-depleted fluids related to off-axis hydrothermal activities (Rouxel et al., 2018).

This low temperature hydrothermal model for the source of Fe and Si in the Archean IFs and related ferruginous sediments is fascinating, but this model could not explain all the IFs. Indeed, Frei and Polat (2007) argued that ~3.7 Ga-old BIFs at the ISB, West Greenland, were sourced from hydrothermally fertilized bottom waters for Fe and surface seawaters for Si, which was assumed to have been derived from mafic land masses older than 4.0 Ga. Also considering that the Archean Eon had continued more than 1.5 Ga from 4.0 to 2.5 Ga and that the Earth's surface environment had drastically changed during this period, including the onset of modern style plate-tectonics, accelerated growth of continents, and decrease in the upper mantle temperature and related decline of volcanic activities, sources for Fe and Si in IFs could have changed through time. For further discussion for origin of IFs, Bekker et al. (2014) and references therein are recommended.

5.6.4 Origins of Early to Mesoarchean Iron Formations

As shown in Figure 5.8, the major production period of IFs started about 2.7 Ga ago and continued to 1.8 Ga with intermittent relatively calm periods. This major production is widely believed to have been caused by oxygenation of the Earth, as a consequence of flourish of cyanobacteria, oxygen-producing photoautotrophs (also see Chapter 6). Namely, Fe^{2+} was oxidized by O_2 to be precipitated as ferric hydroxides $[Fe(OH)_3]$ (Equation 5.3) (Holland, 1984).

$$4Fe^{2+} + O_2 + 10H_2O \rightarrow 4Fe(OH)_3 + 8H^+ \tag{5.3}$$

While the oxidation of Fe^{2+} could proceeded as a purely chemical process, it seems more likely that microbial metabolism (chemolithoautotrophy) was involved in this reaction (e.g., Holm, 1989). Although a few researchers consider that the evolution of cyanobacteria can be back to Paleoarchean, such a view is minority and therefore other mechanisms of Fe-oxidation at that time is required, including oxidation of Fe^{2+} directly by ultraviolet radiation and by hydrogen peroxide (H_2O_2) that was produced by photochemical reaction of H_2O: see Pecoits et al. (2015) for review. A more plausible and widely accepted idea is that the formation of IFs before the evolution of cyanobacteria was attributed to anoxygenic photosynthesis using Fe^{2+} as electron donor (e.g., Posth et al., 2013) (Equation 5.4). This metabolism is operated by photoferroautotrophs such as purple bacteria and green sulfur bacteria as extant species (Widdel et al., 1993). Photoferroautotroph, anoxygenic photoautotroph, is generally considered to have evolved earlier than oxygenic photoautotrophs. They had likely been major primary producers until the establishment of ecosystem based on primary production by oxygenic photosynthesis, probably around the Great Oxidation Event (GOE: 2.4–2.0 Ga) (e.g., Planavsky et al., 2021 and references therein).

$$4Fe^{2+} + CO_2 + 11H_2O + hv \rightarrow CH_2O + 4Fe(OH)_3 + 8H^+ \tag{5.4}$$

This idea, however, appears to be not consistent with the following observations. According to Thompson et al. (2019) and references therein, while laboratory experiments reveal that photoferroautotrophic cells are tightly associated with Fe^{3+} hydroxides as their own metabolic by-products, organic carbon contents in IFs are much lower than expected from stoichiometry of reaction as represented by the Equation 5.4. One explanation for this is respiratory consumption of organic matter (OM) associated with reduction of Fe^{3+} as ferric hydroxide to Fe^{2+}. Considering poverty of OM in IFs, it can be assumed that most of OM was consumed by this process. If this is the case, secondary minerals containing Fe^{2+} such as siderite and magnetite should be much more abundant. However, dominant mineral in IFs is hematite, diagenetic product of ferric hydroxides. This inconsistency has been challenged by Thompson et al. (2019), who experimentally showed that in the presence of abundant dissolved silica, likely in the Archean seawater as discussed below, cells of photoferrotrophs and ferric hydroxides as their metabolic by-products do not deposit together, due to cell surface repulsion. As a result, ferric hydroxide deposited as IFs and OM did as shales. Although repulsion between cell surface and ferric hydroxides is well demonstrated, this phenomenon does not always mean separate deposition of IFs and OM. The enigma is still not unresolved.

5.7 ARCHEAN SEAWATER COMPOSITIONS AND PRODUCTS 2: CHERTS

5.7.1 SILICA AND CHERT

In the simplest and widest sense, chert refers to rock composed predominantly of microcrystalline quartz, whatever its origin is. More specifically, the term "chert" refers to sedimentary rock composed originally of chemically precipitated silica or siliceous organic remains. Silicon (Si) is essential to wide range of phyla and in the most case is used as opaline silica ($SiO_2 \cdot nH_2O$) including plant opal, diatom frustule, radiolarian test, and sponge spicule. The global silicon cycle is simple, compared with other nutrients. Silicon is liberated as silicic acids (e.g., H_4SiO_4) from rock-forming silicate minerals by weathering. While a part of weathering-derived dissolved silicic acids (DSi) is utilized by terrestrial biota including aquatic species, most of DSi is transported to oceans. Thus, weathering of continental crust is one of the major sources of DSi in oceans. Other sources for DSi include weathering and hydrothermal alterations of oceanic crusts. Concentrations of DSi in open ocean waters display a distinctive vertical profile characterized by depletion in the surface and increase to the depth. This trend reflects a biological uptake by diatom and/or radiolaria in the photic zone and dissolution of sinking opaline exoskeletons at depths. Opaline skeletons that escape dissolution deposit on ocean floors. Ocean floor sediments rich in opaline skeletons (>30%) are called siliceous ooze. Mesozoic radiolarian bedded cherts (radiolarites) (Figure 2.8) and Cenozoic diatomite (though not called chert) represent lithified siliceous ooze. Due to biological uptake and dissolution, concentration of DSi in seawaters ranges widely (e.g., 0.8–4.0 mg/kg as silicon; Nonose et al., 2014), being undersaturated with respect to silica. In addition to biogenic origin, cherts could be formed in other various ways, including e.g., chemical precipitation of hydrated silica (gel/ooze) from hydrothermal fluids, diagenesis of silicic clastic sediments such as acidic tuff, and replacement of other sediments and rocks. For further details of modern oceanic silica cycle, the readers may read Tréguer and De La Rocha (2013).

5.7.2 ARCHEAN PRIMARY CHERTS, INDICATIVE OF SILICA-RICH OCEAN

In the Archean era, on the contrary, cherts are exclusively of abiogenic origin, as silica-secreting microorganisms had evolved much later. Although molecular clock studies suggest that the evolution of sponge can be back to the Precambrian, the record of spicules in sediments became abundant in the early Cambrian era (Muscente et al., 2015). Due to the absence of silica-requiring and -secreting organisms, DSi had been stagnated in the Precambrian oceans. In addition, stronger hydrothermal activities, particularly in the Archean and the Paleoproterozoic, likely delivered a significant amount of DSi to the oceans. Absence of plants and thus no production of plant opal mean that most of DSi produced by continental weathering would have been transported to the oceans. The growth of continents had increased delivery of DSi to the oceans.

Silica enrichment in the Archean seawater is widely accepted (e.g., Maliva et al., 2005), and many previous studies have suggested that Archean cherts, if not all, were originated from primary precipitated amorphous silica. However, it has been equivocal whether the seawater was supersaturated with respect to amorphous silica. Stefurak et al. (2014) gave on answer to this issue. They performed detailed sedimentological, petrographic, and geochemical studies on the early Archean cherts in the Barberton greenstone belt, South Africa and identified unique structures denoted as silica granules from fine to coarse sand in size (200–600 μm as equivalent sphere diameter). The silica granules occur in pure chert layers (white chert beds) alternating black and ferruginous cherts. The granules are compositionally homogeneous, range from nearly spherical to oblate spheroidal in shape and were likely relatively soft and easily compacted, and lack structures and inclusions suggestive of their secondary origins. The authors interpreted that the granules were formed by aggregation of opal ($SiO_2 \cdot nH_2O$)-A microsphere (~2 μm), which was produced by aggregation of colloidal silica nanospheres (5 nm). This works provides direct evidence for the formation of primary chert in the Archean ocean. Opal-A is diagenetically converted to less soluble opal-CT and eventually to microquartz, although this process involves multiple stages of dissolution and precipitation (e.g., Knauth, 1994).

5.7.3 Formation Processes of Secondary Cherts

Secondary cherts originated from nonsiliceous sediments and rocks are also common in the Archean. Several processes have been proposed for such the secondary chert formation. Here hydrothermal silicification and syndepositional silicification and later pervasive silicification are reviewed.

5.7.3.1 Hydrothermal Alteration of Oceanic Crusts

Hofmann and Harris (2008) identified that volcanic units of the Barberton greenstone belt have an upper "silica alteration zone" with thickness of several tens of meters, with a transition down to unaltered volcanic rocks. The "silica alteration zone", which contains secondary chert, is characterized by chert dykes/veins and capped by sedimentary units, either chemical and biogenic sediments (laminated to banded chert, massive chert and BIF) or silicified clastic rocks originated from shales, sandstones, and volcaniclastic rocks. In other words, the chert dykes/veins appear to be terminated at the base of the cap units. It was postulated that the cap units had played a key role in the increase in hydrothermal fluid pressure within the underlying volcanic units. When the pressure exceeded a critical point, the volcanic unit was hydraulically fractured and the complex dykes/vein systems were formed (e.g., Hofmann and Bolhar, 2007). The term "hydrothermal system" tends to remind us black smokers at oceanic spreading centers, from which high temperature (~400°C) and metal- and sulfide-rich fluids emanate. However, hydrothermal fluids responsible for silicification of Archean volcanic units may have been distinct from the black smoker-type fluids, which can simply be deduced from the lack of massive sulfide deposits. Temperature of hydrothermal fluid has also been esti-mated based on fractionation of oxygen isotopes (^{18}O vs.^{16}O) of quartz, proposing a relatively low temperature condition (100°C–150°C) (Hofman and Harris, 2008).

5.7.3.2 Syndepositional Silicification

As to silicification of Archean clastic sedimentary rocks, Rouchon and Orberger (2008) and Ledevin et al. (2014) proposed a unique model, which involves sorption of dissolved silica onto sinking detrital particles. It is suggested that fine-grained micaceous particles, with a large reactive surface and with long resident time in water column, effectively adsorbed silica (Ledevin, 2019). Noteworthy, this sorption did not always require that the seawater was saturated with respect to silica. Sinking detrital particles acted as "scavenger" of dissolved silica, eventually forming ocean floor sediment composed of mixture of siliciclastic detritus and chemically precipi-tate silica coating detritus. Further amorphous silica precipitation was proceeded by adsorption of porewater silica on a sediment particle surface, increasing silica content of the sediment. Release of silica during devitrification of volcanic ash also occurred. Due to circulation of silica-rich fluids, the sediment was eventually silici-fied to be chert.

5.7.3.3 Pervasive Silicification

Replacement of carbonate by silica and resultant secondary chert formation dur-ing the Archean is evident in the field level. At the Trendall Locality in the Pilbara Craton, Western Australia, we can see spectacular outcrop of ca. 3.4 Ga-old carbon-ate stromatolites (see Chapter 7). At this locality, chert-replaced stromatolites are common. The carbonate and cherty facies are transitional, suggesting that silicifi-cation was pervasive (Figure 5.11a). Interestingly, in some places, silicification of carbonate appears to be blocked by cross cutting vein of chert; namely, the Si-rich fluids could not permeate through the chert vein. (Figure 5.11b). This indicates that Si-rich fluids responsible for silicification of stromatolite invaded after the formation of this vein and thus after the complete lithification of stromatolites. Microscopic-scale evidence for later silicification was obtained from the author's own examination of Archean chert from the Pilbara Craton. One example is described here. Some of finely laminated cherts exhibit alternation of laminae composed of pure chert and those of carbonate, which has been heterogeneously silicified. Silicified portion is composed dominantly of microcrystalline quartz with disseminated tiny dusts of unknown origin and small amount carbonate particles (Figure 5.11c).

5.7.3.4 Chert Formation and Element Remobilization

Secondary chert formation involves significant elemental remobilization. Duchac and Hanor (1987), for example, reported that some komatiites, MgO-rich (~20 wt%) and SiO_2-poor (~45 wt%) ultramafic volcanic rocks, from the Barberton green-stone belt, South Africa had been metasomatically silicified to be cherty rocks ($SiO_2 = 78$–91 wt%), preserving original spinifex texture. Ledevin (2019) summa-rized chemical disturbance during silicification, involving a significant increase of Si and K, which is associated with an enrichment of Rb and Ba. Most of the other

FIGURE 5.11 (a) Photograph of partially (the arrowed portion) silicified carbonate stromatolite of the ca. 3.4 Ga-old Strelley Pool Formation in the Pilbara Craton, Western Australia. (b) Pervasively silicified portion of the same outcrop shown in (a). The dashed line shows the cross-cutting chert vein, which worked as barrier against silicification. Namely, the left of the dashed line is not pervasively silicified. (c) Partial silicification of the carbonate layer in laminated chert of the ca. 3.0-Ga old Cleaverville Formation at the Goldsworthy greenstone belt in the Pilbara Craton. The silicified portion is dusty and contains disseminated carbonate particles. Open nicol.

major elements, particularly Ca and Na, would be depleted during silicification. Depletion has been identified also for many of minor elements such as Pb, Sr, Ni, and Co. Elements immobile during metasomatic silicification are represented by two major elements, Al and Ti. Zirconium, Nb, Hf, and Ta, which are so-called "high-field strength elements (HFSE)" are also immobile and some of transition metals such as Sc, Cr and V as well. Rare earth elements (REEs) are generally immobile, although LREE (light rare earth elements: La-Eu, in wider sense) are less persistent. As described later for microfossil-bearing cherts from the Pilbara Craton, OM can be mobilized during later metasomatic silicification. However, earlier silicification does not appear to remobilize OM. Rather, earlier silicification (silica encapsulation) might have played an important role for preservation of organic molecules against metamorphism (Chapter 11). While element remobilization during metasomatic silicification has been well studied, places to which the elements had been remobilized seem to have rarely been discussed.

5.8 HOW WAS THE EARTH'S ATMOSPHERE OXIDIZED

5.8.1 ARCHEAN ATMOSPHERE

According to the most recent review of the Archean atmosphere (Catling and Zahle, 2020), H_2 is assumed to had been less than 0.01 from 3.0 to 2.7 Ga, whereas CH_4 is assumed to be $> \sim 5,000$ ppmv around 3.5 Ga ago and had declined toward the end of Archean, with a lower limit of 20 ppmv required for production of elemental sulfur (S_8^0) responsible for the mass independent fractionation of sulfur (S-MIF) (Chapter 6). Concentration of O_2 is assumed to have been less than 10^{-6} PAL (present atmospheric level) until 2.4 Ga ago, again required for S-MIF. Such abundant CH_4, together with CO_2, had compensated for low solar luminosity in the Archean and contributed to prevent the Earth from freezing: it has been widely accepted that the solar luminosity 4.0 Ga ago was around 70% of the present level (e.g., Feulner, 2012). This hypothesis has long been taken as prerequisite to place constraints on the Archean atmospheric compositions. Namely, the Archean atmosphere likely contained more greenhouse gases such as CO_2 and CH_4 than present (e.g., Kasting, 1993). In either case, such an anoxic atmosphere had evolved to be oxic through many millions of years.

Needless to say, one of the factors contributing to the oxidizing process of the Earth's atmosphere was the evolution of oxygenic photosynthesis. In addition, a decrease in the atmospheric reducing gases represented by H_2 and CH_4 should be considered. The decrease of these gases had proceeded in several ways, including (1) microbial consumption and production of H_2 and CH_4 and (2) hydrogen escape, in addition to direct oxidation by O_2.

5.8.2 MICROBIAL CONSUMPTION AND PRODUCTION OF H_2 AND CH_4

Dihydrogen (H_2) is utilized by some autotrophic bacteria, called hydrogen-oxidizing bacteria, as electron donor to assimilate CO_2. This metabolism is operated aerobically by using O_2 or anaerobically by using, e.g., Fe^{3+} (Hedrich and Johnson, 2013; Smith et al., 1994). Some bacteria or archaea called methanotrophs have ability to metabolize CH_4 either aerobically or anaerobically. In this metabolism, CH_4 is utilized as electron donor and carbon source. In aerobic methanotrophy, oxygen is used to oxidize CH_4, whereas in anaerobic one, other electron acceptors are required, including, e.g., NO_3^-, Fe^{3+}, and SO_4^{2-} (Stein et al., 2012; Haroon et al., 2013). Contrary to methanotrophs, methanogens refer to methane producing archaea. As mentioned earlier, methanogens are either heterotrophic or autotrophic. Heterotrophic methanogens metabolize simple organic matter such as acetate ion (CH_3COO^-), formate ion ($COOH^-$) and methanol (CH_4O), products of heterotrophic degradation of substrates (Kouzuma et al., 2017). Autotrophic methanogens convert H_2 in combination with CO_2 into CH_4 and H_2O.

A phylogenetic study suggests that methanogenesis had evolved as early as 3.5 Ga ago (Wolfe and Fournier, 2018). Methanotrophs, which require oxidizing substances (see above), likely had evolved later. Indeed, extremely light carbon isotopic values of kerogen ($\delta_{13}C_{PDB} < -50‰$), evidence for methanotrophy (Figure 7.2), have been recorded from the Neoarchean sedimentary successions around 2.7 Ga ago (e.g., Lepot et al., 2019). Therefore, balance of microbial production and consumption of CH_4 and H_2 had likely changed through the Archean. It may be also noted that the decrease in CH_4 and H_2 is coupled with consumption of oxidizing substances. Therefore, as far as production rates of O_2 and other oxidizing substances did not surpass those of H_2 and CH_4 (both biotic and abiotic), the atmosphere could not have substantially been oxygenated.

5.8.3 Hydrogen Escape

Hydrogen escape refers to the loss of H from planetary atmosphere to outer space, one of so-called "atmospheric escape". Atmospheric escape is known to have drastically changed atmospheric compositions of planets and their satellites: it is still occurring now (Gronoff et al., 2020; Catling and Zahnle, 2009). Three categorized processes of atmospheric escape are known. One of them is called impact erosion, which refers to atmospheric escape that occurs when asteroids or comets impact planets. This process is not discussed here. The others include thermal escape and nonthermal escape. Key points for these two processes include (1) weight of particle (atom and molecule), which is related to attained velocity, (2) velocity of particle, which is closely related to temperature, and (3) charge of particle, which is closely related to the magnetic field. For example, H_2, the lightest gas, would be heated by UV at the upper atmosphere and could easily attain velocity that overcomes gravity of the Earth, escaping to the space. Although the charged particle (ion) is tethered to magnetic field, it could escape to the space along with magnetic field lines that are driven away by the solar wind to the space (polar wind).

History of hydrogen escape from the Archean atmosphere has been imposed from secular trend of Xe (xenon) isotopes preserved in fluid inclusions of rocks and minerals of various ages. The degree of fractionation of Xe isotopes is characterized by gradual decrease from the Paleoarchean to around 2.0 Ga ago, close to the present value of the air (Catling and Zahnle, 2020, Zahnle et al., 2019 and references therein). To briefly explain this trend, Xe^+ (ionized Xe by solar-UV radiation at the upper atmosphere) was dragged out by H via polar wind to the space. During this process, isotopic fractionation, due to mass differences of Xe isotopes, had occurred. Loss of Xe required abundant H as "dragger", which was likely formed by UV-photochemical decomposition of CH_4 and/or H_2 at the upper atmosphere. Molecular oxygen as a deionizer of Xe^+ must have been depleted in order that CH_4 and H_2 were stably present at their source regions and Xe^+ was not deionized. Therefore, this hypothetical model points to the fact that the Archean atmosphere was anoxic, which is consistent with S-MIF and the GOE.

5.8.4   Archean Ocean Temperature and pH

5.8.4.1   Temperature: Hot or Temperate?

Temperature and pH in the Archean ocean and their secular variations, particularly temperature, have been topics of great debate. How do we assume temperatures of ancient ocean water? Temperature-dependent isotopic fractionation of oxygen between water and precipitated silica (chert) is a clue to reveal temperature change of seawater through the Earth's history (Equation 5.5) (Knauth and Epstein, 1976).

$$1,000\ln(\Delta^{18}O) = \left(3.09 \times 10^6 T^{-2}\right) - 3.29 \tag{5.5}$$

where $\Delta^{18}O = \delta^{18}O_{silica} - \delta^{18}O_{water}$ and $\delta^{18}O = [(^{18}O/^{16}O)sample/(^{18}O/^{16}O)_{SMOW} - 1] \times 1,000$ (‰). SMOW stands for the present-day Standard Mean Ocean Water isotopic composition. Based on systematic increase in $\delta^{18}O$ in cherts from ~20‰ to ~35‰ during the last 3.5 Ga, it was assumed that the temperature of seawater from which cherts were precipitated has decreased by 50°C–80°C since Archean times (Knauth and Epstein, 1976; Knauth and Lowe, 2003; Robert and Chaussidon, 2006). This hypothesis assumes that the $\delta^{18}O$ value of seawater has not been changed significantly. This assumption appears to be supported indirectly by $\delta^{18}O$ values of ca. 3.8 Ga-old ophiolites (Pope et al., 2012) and of carbonaceous matter indigenous to Precambrian cherts, demonstrating that the oxygen isotopic composition of seawater was nearly constant around 0 ± 5‰ (Tartèse et al., 2017). Such a high temperature Archean oceans model, however, is not consistent with oxygen isotopic composition of phosphate (Blake et al., 2010) and combined oxygen and hydrogen isotopic analyses (Hren et al., 2009). This issue may be the most problematic in the Archean environment.

5.8.4.2   Acidic, Neutral, or Alkaline?

The pH values of Archean oceans have also long been a topic of debate, though much less than temperature. Kempe and Degens (1985) argued that the ancient ocean associated with high volcanic (hydrothermal) activity was similar to modern soda lakes such as Lake Magadi, East Africa and Lake Mono, California, USA, being characterized by high alkalinity, a high pH and low Ca and Mg concentrations. The highly alkaline early "soda" ocean was gradually replaced by "halide" ocean, due to addition of hydrothermally leached chlorine (Cl) and removal of dissolved carbonates by organisms and carbonate-precipitations. Blättler et al. (2017) also considered that the ancient ocean was alkaline but rejected the early soda ocean hypothesis, based on isotope compositions of Ca ($^{44}Ca/^{40}Ca$) of carbonates from 2.7 to 1.9 Ga, which exhibit very limited variability, consistent with a high ratio of Ca to carbonate alkalinity.

A mildly acidic ocean model is also recently proposed by Halevy and Bachan (2017) and Krissansen-Totton et al. (2018). The former employed modeling using pCO_2 and more than 10 parameters governing ocean chemistry including, e.g., HCO_3^-, HS^-, NH_4, Na, and Cl. The latter applied a self-consistent geological carbon cycle model employing two separate boxes representing the atmosphere-ocean system and the sea-floor pore space. Regardless of difference in methodology, both studies suggested that the early Archean ocean was mildly acidic to neutral (pH = 6.5–7.0).

COLUMN: ZIRCON, WINDOWS TO THE HADEAN (4.6–4.0 GA)

Zircon ($ZrSiO_4$) is enriched in radiometric elements such as U and Th and thus known as powerful tool for dating of geologic materials. It occurs as primary accessory mineral in igneous and metamorphic rocks, being common in felsic rocks such as granite, but rarely in mafic rocks. This mineral is one of the most resistant rock-forming minerals and therefore could survive weathering, metamorphism, and even melting, though not always. For example, zircon grains in sedimentary rocks are in many cases recycled ones and thus their ages tend to be older than the age of sedimentation. Zoned texture of zircon, if present, indicates its survival during melting and metamorphism of its parent rock and overgrowth. The age of its rim should be younger than its core. The oldest recorded mineral on the Earth is zircon. To date, the well-established oldest zircon, from the Jack Hills in Western Australia, is dated 4.4 Ga (Wilde et al., 2001). Although the oldest rock from the Acasta Gneiss of the Slave Craton, Canada, is dated 4.03 Ga (Bowring and Williams, 1999), such the Hadean rocks are quite rare. Therefore, information of the Hadean Earth has relied exclusively on zircons. For example, graphite inclusions with light carbon isotopic signature ($\delta^{13}C_{PDB} = -24 \pm 5\permil$) in 4.1 Ga-old zircon has been implied to be the oldest biosignature (Bell et al., 2015). Data of REEs and oxygen isotopes of the above-mentioned oldest zircon were used to recalculate those in melts where this zircon had grown, giving light rare-earth element enrichment with negative Eu-anomaly and $\delta^{18}O$ values of 8.5‰–9.5‰ (Wilde et al., 2001). Based on these data, the authors argued that the parental melt was granitic and that such the high values $\delta^{18}O$ in magma require contribution of supracrustal material interacted with a liquid hydrosphere at low temperature, emphasizing the presence of continental crust and oceans in the Hadean.

REFERENCES

Allègre, C.J., Rousseau, D. 1984. The growth of the continent through geological time studied by Nd isotope analysis of shales. *Earth and Planetary Science Letters* 67, 19–34.

Armstrong, R.L. 1981. Radiogenic isotopes: The case for crustal recycling on a near-steady-state no-continental-growth Earth. *Philosophical Transactions of the Royal Soceity of London, Series A* 301, 443–472.

Barnes, S.J., Arndt, N.T. 2019. Distribution and geochemistry of komatiites and basalts through the Archean. In M.J. van Kranendonk, V.C. Bennett, J.E. Hoffmann eds. *Earth's Oldest Rocks* 2nd edition. pp. 103–132. Elsevier, Amsterdam.

Bekker, A., Planavsky, N.J., Krapež, B., Rasmussen, B., Hofmann, A., Slack, J.F., Rouxel, O.J., Konhauser, K.O. 2014. Iron formations: Their origins and implications for ancient seawater chemistry. In H.D. Holland and K.K. Turekian eds. *Treatise on Geochemistry* 2nd edition 9, 561–628.

Bell, E.A., Boehnke, P., Harrison, T.M., Mao, W.L. 2015. Potentially biogenic carbon preserved in a 4.1 billion-year-old zircon. *Proceedings of National Academy of Sciences of the United States of America* 112, 14518–14521.

Belousova, E.A., Kostitsyn, Y.A., Griffin, W.L., Begg, G.C., O'Reilly, S.Y., Pearson, N.J. 2010. The growth of the continental crust: Constraints from zircon Hf-isotope data. *Lithos* 119, 457–466.

Berry, A.J., Danyushevsky, L.V., O'Neill, H.S.C., Newville, M., Sutton, S.R. 2008. Oxidation state of iron in komatiitic melt inclusions indicates hot Archaean mantle. *Nature* 455, 960–963. doi:10.1038/nature07377.

Blake, R.E., Chang, S.J., Lepland, A. 2010. Phosphate oxygen isotopic evidence for a temperate and biologically active Archaean ocean. *Nature* 464, 1029–1032.

Blättler, C.L., Kump, L.R., Fischer, W.W., Paris, G., Kasbohm, J.J., Higgins, J.A. 2017. Constraints on ocean carbonate chemistry and pCO_2 in the Archaean and Palaeoproterozoic. *Nature Geoscience* 10, 41–45.

Bowring, S.A., Williams, I.S. 1999. Priscoan (4.00–4.03 Ga) orthogneisses from northwestern Canada. *Contributions to Mineralogy and Petrology* 134, 3–16.

Burgess, R., Goldsmith, S.L., Sumino, H., Gilmour, J.D., Marty, B., Pujol, M., Konhauser, K.O. 2020. Archean to Paleoproterozoic seawater halogen ratios recorded by fluid inclusions in chert and hydrothermal quartz. *American Mineralogist* 105, 1317–1325.

Byerly, G.R. 2015. Onverwacht Group. In M. Gargaud, W.M. Irvine eds. *Encyclopedia of Astrobiology*. Springer, Berlin. doi:10.1007/978-3-662-44185-5_5128.

Byerly, G.R., Lowe, D.R., Wooden, J.L., Xie, X. 2002. An Archean impact layer from the Pilbara and Kaapvaal cratons. *Science* 297, 1325–1327.

Catling, D.C., Zahnle, K.J. 2009. The planetary air leak. *Scientific American* 36–43.

Catling, D.C., Zahnle, K.J. 2020. The Archean atmosphere. *Science Advances* 6, eaax1420.

Cawood, P.A., Hawkesworth, C.J., Pisarevsky, S.A., Dhuime, B., Capitanio, F.A., Nebel, O. 2018. Geological archive of the onset of plate tectonics. *Philosophical Transactions of the Royal Society Series A* 376. doi:10.1098/rsta.2017.0405.

Cheney, E.S. 1996. Sequence stratigraphy and plate tectonic significance of the Transvaal succession of southern Africa and its equivalent in Western Australia. *Precambrian Research* 79, 3–24.

Condie, K.C., Aster, R.C. 2010. Episodic zircon age spectra of orogenic granitoids: The supercontinent connection and continental growth. *Precambrian Research* 180, 227–236.

Deng, Z., Chaussidon, M., Guitreau, M., Puchtel, I.S., Dauphas, N., Moynier, F. 2019. An oceanic subduction origin for Archaean granitoids revealed by silicon isotopes. *Nature Geoscience* 12, 774–778.

Dhuime, B., Hawkesworth, C.J., Cawood, P.A., Storey, C.D. 2012. A change in the geodynamics of continental growth 3 billion years ago. *Science* 335, 1334–1336.

Dhuime, B., Wuestefeld, A., Hawkesworth, C.J. 2015. Emergence of modern continental crust about 3 billion years ago. *Nature Geoscience* 8, 552–555.

Dodd, M.S., Papineau, D., She, Z., Fogel, M.L., Nederbragt, S., Pirajno, F. 2018. Organic remains in late Palaeoproterozoic granular iron formations and implications for the origin of granules. *Preacmbrian Research* 310, 133–152.

Duchac, K.C., Hanor, J.S. 1987. Origin and timing of the metasomatic silicification of an early Archean komatiite sequence, Barberton Mountain Land, South Africa. *Precambrian Research* 37, 125–146.

Evans, M.E., Muxworthy, A. 2018. Vaalbara palaeomagnetism. *Canadian Journal of Earth Sciences* 56, 912–916. doi:10.1139/cjes-2018-0081.

Farber, K., Dziggel, A., Meyer, F.M., Prochaska, W., Hofmann, A., Harris, C. 2015. Fluid inclusion analysis of silicified Palaeoarchaean oceanic crust – A record of Archaean seawater? *Precambrian Research* 266, 150–164.

Feulner, G. 2012. The faint young sun problem. *Reviews of Geophysics* 50, RG2006. doi:10.1029/2011RG000375.

Frei, R., Polat, A. 2007. Source heterogeneity for the major components of ~3.7 Ga Banded Iron Formations (Isua Greenstone Belt, Western Greenland): Tracing the nature of interacting water masses in BIF formation. *Earth and Planetary Science Letters* 253, 266–281.

Furnes, H., Dilek, Y., de Wit, M. 2015. Precambrian greenstone sequences represent different ophiolite types. *Gondwana Research* 27, 649–685.

Genareau, K., Wardman, J.B., Wilson, T.M., McNutt, S.R., Izbekov, P. 2015. Lightning-induced volcanic spherules. *Geology* 43, 319–322.

Glikson, A., Hickman, A., Evans, N.J., Kirkland, C.L., Park, J.-W., Rapp, R., Romano, S. 2016. A new ~3.46 Ga asteroid impact ejecta unit at Marble Bar, Pilbara Craton, Western Australia: A petrological, microprobe and laser ablation ICPMS study. *Precambrian Research* 279, 103–122.

Goodwin, A.M. 1996. *Principles of Precambrian Geology*. p. 327. Academic Press, London.

Gronoff, G., Arras, P., Baraka, S., Bell, J.M., Cessateur, G., Cohen, O., Curry, S.M., Drake, J.J., Elrod, M., Erwin, J., Garcia-Sage, K., Garraffo, C., Glocer, A., Heavens, N.G., Lovato, K., Maggiolo, R., Parkinson, C.D., Wedlund, C.S., Weimer, D.R., Moore, W.B., 2020. Atmospheric escape processes and planetary atmosphere evolution. *Journal of Geophysical Research: Space Physics* 125, e2019JA027639.

Grosch, E.G., Slama, J. 2017. Evidence for 3.3-billion-year-old oceanic crust in the Barberton greenstone belt, South Africa. *Geology* 45, 695–698.

Halevy, I., Bachan, A. 2017. The geologic history of seawater pH. *Science* 355, 1069–1071.

Hansen, V.L. 2015. Impact origin of Archean cratons. *Lithosphere* 7, 563–578.

Haroon, M.F., Hu, S., Shi, Y., Imelfort, M., Keller, J., Hugenholtz, P., Yuan, Z., Tyson, G.W. 2013. Anaerobic oxidation of methane coupled to nitrate reduction in a novel archaeal lineage. *Nature* 500, 567–570.

Hawkesworth, C., Cawood, P.A., Dhuime, B. 2019. Rates of generation and growth of the continental crust. *Geoscience Frontiers* 10, 165–173.

Hedrich, S., Johnson, D.B. 2013. Aerobic and anaerobic oxidation of hydrogen by acidophilic bacteria. *FEMS Microbiology Letters*. 349, 40–45.

Hickman, A.H. 2012. Review of the Pilbara Craton and Fortescue basin, Western Australia: Crustal evolution providing environments for early life. *Island Arc* 21, 1–31.

Hofmann, A., Bolhar, R. 2007. Carbonaceous cherts in the Barberton greenstone belt and their significance for the study of early life in the Archean record. *Astrobiology* 7, 355–388.

Hofmann, A., Harris, C. 2008. Silica alteration zones in the Barberton greenstone belt: A window into subseafloor processes 3.5–3.3 Ga ago. *Chemical Geology* 257, 221–239.

Holland, H.D. 1984. *The Chemical Evolution of the Atmosphere and Oceans*. Princeton University Press, Princeton, 582p.

Holm, N.G. 1989. The $^{13}C/^{12}C$ ratios of siderite and organic matter of a modern metalliferous hydrothermal sediment and their implications for banded iron formations. *Chemical Geology* 77, 41–45.

Holm, N.G., Charlou, J.L. 2001. Initial indications of abiotic formation of hydrocarbons in the Rainbow ultramafic hydrothermal system, Mid-Atlantic Ridge. *Earth and Planetary Science Letters* 191, 1–8.

Holm, N.G., Oze, C., Mousis, O., Waite, J.H., Guilbert-Lepoutre, A. 2015. Serpentinization and the formation of H_2 and CH_4 on celestial bodies (planets, moons, comets). *Astrobiology* 15, 587–600.

Hren, M.T., Tice, M.M., Chamberlain, C.P. 2009. Oxygen and hydrogen isotope evidence for a temperate climate 3.42 billion years ago. *Nature* 462, 205–208.

Hurley, P.M., Rand, J.R. 1969. Pre-Drift continental nuclei. *Science* 164, 1229–1242.

Iizuka, T., Yamaguchi, T., Hibiya, Y., Amelin, Y. 2015. Meteorite zircon constraints on the bulk Lu-Hf isotope compositon and early differentiation of the Earth. *Proceedings of National Academy of Sciences of the United States of America* 112, 5331–5336.

Ilyin, A.V. 2009. Neoproterozoic banded iron formations. *Lithology and Mineral Resources* 44, 78–86.

Isozaki, Y., Kabashima, T., Ueno, Y., Kitajima, K., Maruyama, S., Kato, Y., Terabayashi, M. 1997. Early Archean mid-ocean ridge rocks and early life in the Pilbara Craton., W. Australia. *EOS* 78, F399.

Johnson, C.M., Ludois, J.M., Beard, B.L., Beukes, N.J., Heimann, A. 2013. Iron formation carbonates: Paleoceanographic proxy or recorder of microbial diagenseis? *Geology* 41, 1147–1150.

Jourdan, F., Reimold, W.U., Deutsch, A. 2012. Dating terrestrial impact structures. *Elements* 8, 49–53.

Kasting, J.F. 1993. Earth's early atmosphere. *Science* 259, 920–926.

Kempe, S., Degens, E.T. 1985. An early soda ocean? *Chemical Geology* 53, 95–108.

Knauth, L.P. 1994. Petrogenesis of chert. *Reviews in Mineralogy and Geochemistry* 29, 233–258.

Knauth, L.P., Epstein, S. 1976. Hydrogen and oxygen isotope ratios in nodular and bedded cherts. *Geochimica et Cosmochimica Acta* 40, 1095–1108.

Knauth, L.P., Lowe, D.R. 2003. High Archean climatic temperature inferred from oxygen isotope geochemistry of cherts in the 3.5 Ga Swaziland Supergroup, South Africa. *Geological Society of America Bulletin* 115, 566–580.

Kouzuma, A., Tsutsumi, M., Ishii, S., Ueno, Y., Abe, T., Watanabe, K. 2017. Non-autotrophic methanogens dominate in anaerobic digesters. *Scientific Reports* 7, 1510.

Krissansen-Totton, J., Arney, G.N., Catling, D.C. 2018. Constraining the climate and ocean pH of the early Earth with a geological carbon cycle model. *Proceedings of the National Academy of Sicences of the United States of America* 115, 4105–4110.

Krull-Davatzes, A., Goderis, S., Simonson, B.M. 2019. Archean asteroid impacts on Earth: Stratigraphic and isotopic age correlations and environmental consequences. In M.J. van Kranendonk, V.C. Bennett, J.E. Hoffmann eds. *Earth's Oldest Rocks* 2nd edition. pp. 169–182. Elsevier, Amsterdam.

Kusky, T.M., Polat, A. 1999. Growth of granite-greenstone terranes at convergent margins, and stabilization of Archean cratons. *Tectonophysics* 305, 43–73.

Lantink, M.L., Davies, J.H.F.L., Mason, P.R.D., Schaltegger, U., Hilgen, F.J. 2019. Climate control on banded iron formations linked to orbital eccentricity. *Nature Geoscience* 12, 369–374.

Lazar, C., Cody, G.D., Davis, J.M. 2015. A kinetic pressure effect on the experimental abiotic reduction of aqueous CO_2 to methane from 1 to 3.5 kbar at 300°C. *Geochimica et Cosmochimica Acta* 151, 34–48.

Ledevin, M. 2019. Archean cherts: Formation processes and paleoenvironments. In M.J. van Kranendonk, V.C. Bennett, J.E. Hoffmann eds. *Earth's Oldest Rocks* 2nd edition. pp. 913–944. Elsevier, Amsterdam.

Ledevin, M., Arndt, N., Simionovici, A., Jaillard, E., Ulrich, M. 2014. Silica precipitation triggered by clastic sedimentation in the Archean: New petrographic evidence from cherts of the Kromberg type section, South Africa. *Precambrian Research* 255, 316–334.

Lepot, K., Williford, K.H., Philippot, P., Thomazo, C., Ushikubo, T., Kitajima, K., Mostefaoui, S., Valley, J.W. 2019. Extreme ^{13}C-depletions and organic sulfur content argue for S-fueled anaerobic methane oxidation in 2.72 Ga old stromatolites. *Geochimica et Cosmochimica Acta* 244, 522–547.

Lowe, D.R., Byerly, G.R. 2015. Geologic record of partial ocean evaporation triggered by giant asteroid impacts, 3.29–3.23 billion years ago. *Geology* 43, 535–538.

Lowe, D.R., Byerly, G.R., Kyte, F.T. 2014. Recently discovered 3.42–3.23 Ga impact layers, Barberton Belt, South Africa: 3.8 Ga detrital zircons, Archean impact history, and tectonic implications. *Geology* 42, 747–750.

Maliva, R.G., Knoll, A.H., Simonson, B.M. 2005. Secular change in the Precambrian silica cycle: Insights from chert petrology. *Geological Society of America Bulletin* 117, 835–845.

Martin, J.H., Fitzwater, S.E. 1988. Iron deficiency limits phytoplankton growth in the northeast Pacific subarctic. *Nature* 331, 341–343.

Marty, B., Avice, G., Bekaert, D.V., Broadley, M.W. 2018. Salinity of the Archaean oceans form analysis of fluid inclusions in quartz. *Comptes Rendus Geoscience* 350, 154–163.

McCoy-West, A.J., Chowdhury, P., Burton, K.W., Sossi, P., Nowell, G.M., Fitton, J.G., Kerr, A.C., Cawood, P.A., Williams, H.M. 2019. Extensive crustal extraction in Earth's early history inferred from molybdenum isotopes. *Nature Geoscience* 12, 946–951.

McLennan, S.M. 1989. Rare earth elements in sedimentary rocks; influence of provenance and sedimentary processes. *Reviews in Mineralogy and Geochemistry* 21, 169–200.

Mloszewska, A.M., Pecoits, E., Cates, N.L., Mojzsis, S.J., O'Neil, J., Robbins, L.J., Konhauser, K.O. 2012. The composition of Earth's oldest iron formations: The Nuvvuagittuq Supracrustal Belt (Québec, Canada). *Earth and Planetary Science Letters* 317–318, 331–342.

Muscente, A.D., Michel, F.M., Dale, J.G., Xiao, S. 2015. Assessing the veracity of Precambrian 'sponge' fossils using *in situ* nanoscale analytical techniques. *Precambrian Research* 263, 142–156.

Nakada, R., Shibuya, T., Suzuki, K., Takahashi, Y. 2017. Europium anomaly variation under low-temperature water-rock interaction: A new thermometer. *Geochemistry International* 55, 822–832.

Nonose, N., Cheong, C., Ishizawa, Y., Miura, T., Hioki, A. 2014. Precise determination of dissolved silica in seawater by ion-exclusion chromatography isotope dilution inductively coupled plasma mass spectrometry. *Analytica Chimica Acta* 840, 10–19.

O'Neill, C., Marchi, S., Bottke, W., Fu, R. 2020. The role of impacts on Archaean tectonics. *Geology* 48, 174–178.

Onoue, T., Sato, H., Yamashita, D., Ikehara, M., Yasukawa, K., Fujinaga, K., Kato, Y., Matsuoka, A. 2016. Bolide impact triggered the Late Triassic extinction event in equatorial Panthalassa. *Scientific Reports* 6, 29609.

Pecoits, E., Smith, M.L., Catling, D.C., Philippot, P., Kappler, A., Konhauser, K.O. 2015. Atmospheric hydrogen peroxide and Eoarchean iron formations. *Geobiology* 13, 1–14.

Planavsky, N.J., Crowe, S.A., Fakhraee, M., Beaty, B., Reinhard, C.T., Mills, B.J.W., Holstege, C., Konhauser, K.O. 2021. Evolition of the structure and impact of Earth's biosphere. *Nature Reviews Earth & Environment* 2, 123–139.

Pope, E.C., Bird, D.K., Rosing, M.T. 2012. Isotope composition and volume of Earth's early oceans. *Proceedings of the National Academy of Sciences of the United States of America* 109, 4371–4376.

Posth, N.R., Konhauser, K.O., Kappler, A. 2013. Microbiological processes in banded iron formation deposition. *Sedimentology* 60, 1733–1754.

Pujol, M., Marty, B., Burgess, R., Turner, G., Philippot, P. 2013. Argon isotopic composition of Archaean atmosphere probes early Earth geodynamics. *Nature* 498, 87–90.

Robert, F., Chaussidon, M. 2006. A palaeotemperature curve for the Precambrian oceans based on silicon isotopes in cherts. *Nature* 443, 969–972.

Rouchon, V., Orberger, B. 2008. Origin and mechanisms of K-Si-metasomatism of ca. 3.4–3.3 Ga volcaniclastic deposits and implications for Archean seawater evolution: Examples from cherts of Kittys Gap (Pilbara Craton, Australia) and Msauli (Barberton Greenstone Belt, South Africa). *Precambrian Research* 165, 169–189.

Rouxel, O., Toner, B., Germain, Y., Glazer, B. 2018. Geochemical and iron isotopic insights into hydrothermal iron oxyhydroxide deposit formation at loihi seamount. *Geochimica et Cosmochimica Acta* 220, 449–482.

Russel, M.J., Hall, A., Martin, W. 2010. Serpentinization as a source of energy at the origin of life. *Geobiology* 8, 355–371.

Screiber, U., Mayer, C., Schmitz, O.J., Rosendahl, P., Bronja, A., Greule, M., Keppler, F., Mulder, I., Sattler, T., Schöler, H.F. 2017. Organic compounds in fluid inclusions of Archean quartz—Analogues of prebiotic chemistry on early Earth. *PLoS One.* doi:10.1371/journal.pone.0177570.

Sforna, M.C., Brunelli, D., Pisapia, C., Pasini, V., Malferrari, D., Ménez, B. 2018. Abiotic formation of condensed carbonaceous matter in the hydrating oceanic crust. *Nature Communications* 9, 5049. doi:10.1038/s41467-018-07385-6.

Sharma, M., Shukla, M. 2004. A new Archaean stromatolite from the Chitradurga Group, Dharwar Craton, India and its significance. *Palaeobotanist* 53, 5–16.

Shibuya, T., Kitajima, K., Komiya, T., Terabayashi, M., Maruyama, S. 2007. Middle Archean ocean ridge hydrothermal metamorphism and alteration recorded in the Cleaverville area, Pilbara Craton, Western Australia. *Journal of Metamorphic Geology* 25, 751–767.

Shock, E., Canovas, P. 2010. The potential for abiotic organic synthesis and biosynthesis at seafloor hydrothermal systems. *Geofluids* 10, 161–192.

Shore, M., Fowler, A.D. 1999. The origin of spinifex texture in komatiites. *Nature* 397, 691–694.

Simonson, B.M., Goderis, S., Beukes, N.J. 2015. First detection of extraterrestrial material in ca. 2.49 Ga impact spherule layer in Kuruman Iron Formation, South Africa. *Geology* 43, 251–254.

Sleep, N.H., Meibom, A., Fridriksson, Th., Coleman, R.G., Bird, D.K. 2004. H_2-rich fluids from serpentinization: Geochemical and biotic implications. *Proceedings of the National Academy of Sciences of the United States of America* 101, 12818–12823.

Smit, M.A., Scherstén, A., Næraa, T., Emo, R.B., Scherer, E.E., Sprung, P., Bleeker, W., Mezger, K., Maltese, A., Cai, Y., Rasbury, E.T., Whitehouse, M.J. 2019. Formation of Archean continental crust constrained by boron isotopes. *Geochemical Perspectives Letters* 12, 23–26.

Smith, F.C., Glass, B.P., Simonson, B.M., Smith, J.P., Krull-Davatzes, A.E., Booksh, K.S. 2016. Shock-metamorphosed rutile grains containing the high-pressure polymorph TiO_2-II in four Neoarchean spherule layers. *Geology* 44, 775–778.

Smith, R.L., Ceazan, M.L., Brooks, M.H. 1994. Autotrophic, hydrogen-oxidizing, denitrifying bacteria in groundwater, potential agents for bioremediation of nitrate contamination. *Applied and Environmental Microbiology* 60, 1949–1955.

Stefurak, E.J.T., Lowe, D.R., Zenter, D., Fischer, W.W. 2014. Primary silica granules—A new mode of Paleoarchean sedimentation. *Geology* 42, 283–286.

Stein, L.Y., Roy, R, Dunfield, P.F. 2012. Aerobic methanotrophy and nitrification: Processes and connections. *Encyclopedia of Life Sciences.* John Wiley and Sons, Ltd, Chichester. doi:10.1002/9780470015902.a0022213.

Stern, R.J. 2008. Neoproterozoic crustal growth: The solid earth system during a critical episode of Earth history. *Gondwana Research* 14, 33–50.

Sun, S., Li, Y.-L. 2017. Geneses and evolutions of iron-bearing minerals in banded iron formations of >3760 to ca. 2200 million-year-old: Constraints from electron microscopic, X-ray diffraction and Mössbauer spectroscopic investigations. *Precambrian Research* 289, 1–17.

Sunder Raju, P.V., Eriksson, P.G., Catuneanu, O., Sarkar, S., Banerjee, S. 2013. A review of the inferred geodynamic evolution of the Dharwar craton over the ca. 3.5–2.5 Ga period, and possible implications for global tectonics. *Canadian Journal of Earth Sciences* 51, 312–325.

Tartèse, R., Chaussidon, M., Gurenko, A., Delarue, F., Robert, F. 2017. Warm Archean oceans reconstructed from oxygen isotope composition of early-life remnants. *Geochemical Perspectives Letters* 3, 55–65. doi:10.7185/geochemlet.1706.

Taylor, S.R., McLennan, S.M. 1985. *The Continental Crust: Its Composition and Evolution: An Examination of the Geochemical Record Preserved in Sedimentary Rocks.* P.312. Blackwell Scientific Publications, Oxford.

Thompson, K.J., Kenward, P.A., Bauer, K.W., Warchola, T., Gauger, T., Martinez, R., Simister, R.L., Michiels, C.C., Llirós, M., Reinhard, C.T., Kappler, A., Konhauser, K.O., Crowe, S.A. 2019. Photoferrotrophy, deposition of banded iron formations, and methane production in Archean oceans. *Science Advances* 5, eaav2869.

Tostevin, R., Shields, G.A., Tarbuck, G.M., He, T., Clarkson, M.O., Wood, R.A. 2016. Effective use of cerium anomalies as a redox proxy in carbonate-dominated marine settings. *Chemical Geology* 438, 146–162.

Tréguer, P.J., De La Rocha, C.L. 2013. The world ocean silica cycle. *Annual Review of Marine Science* 5, 477–501.

Turner, S., Wilde, S., Wörner, G., Schaefer, B., Lai, Y.-J. 2020. An andesitic source for Jack Hills zircon supports onset of plate tectonics in the Hadean. *Nature Communications* 11, 1241. doi:10.1038/s41467-020-14857-1.

Van Kranendonk, M.J., Hickman, A.H., Smithies, R.H., Williams, I.R., Bagas, L., Farrell, T.R. 2006. Revised lithostratigraphy of Archean supracrustal and intrusive rocks in the northern Pilbara Craton, Western Australia. P.57. *Geological Survey of Western Australia, Record* 2006/15.

Van Kranendonk, M.J., Smithies, R.H., Hickman, A.H., Champion, D.C. 2007. Review: Secular tectonic evolution of Archean continental crust: Interplay between horizontal and vertical processes in the formation of the Pilbara Craton. *Terra Nova* 19, 1–38.

Widdel, F., Schnell, S., Heising, S., Ehrenreich, A., Assmus, B., Schink, B. 1993. Ferrous iron oxidation by anoxygenic phototrophic bacteria. *Nature* 362, 834–836.

Wilde, S.A., Valley, J.W., Peck, W.H., Graham, C.M. 2001. Evidence from detrital zircons for the existence of continental crust and oceans on the Earth 4.4 Gyr ago. *Nature* 409, 175–178.

Wolfe, J.M., Fournier, G.P. 2018. Horizontal gene transfer constrains the timing of methanogen evolution. *Natire Ecology & Evolution* 2, 897–903.

Zahnle, K.J., Gacesa, M., Catling, D.C. 2019. Strange messenger: A new history of hydrogen on Earth, as told by Xenon. *Geochimica et Cosmochimica Acta* 244, 56–85.

Zegers, T.E., de Wit, M.J., Dann, J., White, S.H. 1998. "Vaalbara, Earth's oldest assembled continent? A combined structural, geochronological, and palaeomagnetic test. *Terra Nova* 10, 250–259.

6 Topics of the Early Precambrian Earth 2

6.1 INTRODUCTION

During the Archean era, microbes have made significant innovation in their metabolisms, which was evolution of photosynthesis. Once microbes had acquired an ability of photosynthesis, they could utilize infinite energy of light to synthesize biomolecules from inorganic carbon such as CO_2 and H_2O. This ability might had liberated autotrophic microbes from specific habitats where both reductants and oxidants, indispensable for chemolithoautotrophy, were available. Evolution of photosynthesis may have drastically increased in primary production, and as discussed in Chapter 5, anoxygenic photosynthesis might have contributed to deposition of iron formations (IFs) before the rise of oxygen.

Rise of oxygen has driven away obligate anaerobic microbes to subsurface or deep-water habitats, whereas an energy revolution caused by evolution of aerobic respiration made a way of evolution of eukaryotes and multicellularity. In this chapter, this evolutionary revolution and related events are focused.

6.2 PHOTOSYNTHESIS AND ITS EVOLUTION

6.2.1 ANOXYGENIC AND OXYGENIC PHOTOSYNTHESIS

As described in detail in Chapter 3, autotrophs are classified into chemolihtoautotroph and photolithoautotroph. It is widely accepted that chemolithoautotrophs had evolved earlier than photolithoautotrophs. Metabolism employed by the photolithoautotrophs is photosynthesis. Two distinct types of photosynthesis are identified for the extant organisms, including anoxygenic photosynthesis and oxygenic one. Oxygenic photosynthesis is a function widely recognized for plants, algae, and cyanobacteria. Chloroplasts in plants and algae had originated from cyanobacteria (see *Column* in the Chapter 13). Anoxygenic photosynthesis is known for relatively wide lineages of prokaryotes, including e.g., Proteobacteria, Chloroflexi, Acidobacteria, Chlorobi, Firmicutes, and Gemmatimonadetes (Hug et al., 2016; Ward et al., 2019). Oxygenic photosynthesizers utilize H_2O as electron donor, although some species of cyanobacteria could facultatively utilize hydrogen sulfide (H_2S) and perform anoxygenic photosynthesis (Hamilton et al., 2018). They oxidize H_2O to produce O_2, whereas reduce inorganic carbon such as CO_2 to organic matter (OM). Anoxygenic photosynthesizers do not utilize H_2O as electron donor and thus cannot produce O_2. Instead, they utilize other reductants such as H_2S, elemental sulfur (S^0), and ferrous iron (Fe^{2+}) as electron donor. Anoxygenic photoferroautotrophy have likely been responsible for origin of IFs before the evolution of cyanobacteria as discussed in Chapter 5.

DOI: 10.1201/9780367855208-6

Pigment-protein complex with some cofactors is called reaction center (RC) and plays a fundamental role in both anoxygenic and oxygenic photosynthesis. Eight types of protein subunits in reaction centers are known, including PshA, PscA, PsaA, PsaB, L, M, D1, and D2. Depending on utilization of these protein subunits and related most reactive wavelength, two types of reaction centers (RC1 and RC2) are identified. RC1 with PshA, PscA, PsaA, and PsaB is most reactive at 700 nm, while RC2 with L, M, D1, and D2 at 680 nm (e.g., Sanchez-Baracaldo and Cardona, 2019). Usage of these protein subunits is different between photosynthetic prokaryotic lineages. For example, Proteobacteria utilize L and M, whereas Acidobacteria utilize PscA. Cyanobacteria utilize D1 and D2 in Photosystem II (PSII)(RC2), whereas PsaA and PsaB in Photosystem I (PSI)(RC1). PSI and PSII have strong reducing power and strong oxidizing power, respectively. The former is utilized to reduce CO_2, whereas the latter is utilized to decompose H_2O to release O_2. These two photosystems with distinct functions are combined to complete oxygenic photosynthesis.

Photosystems utilized in anoxygenic photosynthesis are not identical to either PSI or PSII and called PSI- or PSII-like photosystems (e.g., Sanchez-Baracaldo and Cardona, 2019). Subunits L and M utilized for anoxygenic photosynthesis are homologs of subunits D1 and D2 utilized in PSII with RC2, respectively. Subunits PshA and PscA, again utilized for anoxygenic photosynthesis, are known to be very similar in some aspect such as folding motifs to PsaA and PsaB utilized in PSI with RC1. Thus, the evolution of anoxygenic and oxygenic photosynthesis is closely related to the evolution of RC1 and RC2.

As seen in many reviews (e.g., Hohmann-Marriott and Blankenship, 2011; Fischer et al., 2016; Martin et al., 2018), anoxygenic photosynthesis is generally thought to have evolved first. Martin et al. (2018) suggested mainly based on physiological arguments that progenitor (prototype cell) with RC1 had evolved after the evolution of chlorophyll-like pigment absorbing long-wavelength light emitted from deep-sea hydrothermal site (also see Nürnberg et al., 2018), and RC2 had later evolved from RC1 by gene duplication. Through combination of RC1 and RC2 and subsequent evolution of oxygen-evolving (water-splitting) complex with Mn- and Ca-containing metalloenzyme core, oxygenic photosynthesis with PSI and PSII had evolved. T. Cardona and his colloborators challenged such the story of RC1 first and argued from comparative structural biology and phylogenetic analyses that the earliest and thus ancestral reaction centers, from which RC1 and RC2 had evolved, were already equipped with the ability of water oxidation (e.g., Sanchez-Baracaldo and Cardona, 2019).

It is generally accepted that the evolution of oxygenic photosynthesis is synonym of evolution of cyanobacteria. The timing of emergence of cyanobacteria has been extensively studied from various disciplines, including molecular phylogeny, isotope geochemistry, trace element geochemistry, organic chemistry, and paleontology. Obviously, results obtained by different approaches are not always identical to each other. Different results have been obtained even within the same discipline. Phylogenetic and molecular clock analyses, for example, give the timing of the evolution of the ancestral cyanobacteria from 2.4 to 3.6 Ga ago (e.g., Schirrmeister et al., 2013; Wang et al., 2019; Sanchez-Baracaldo and Cardona, 2019; Garcia-Pichel et al.,

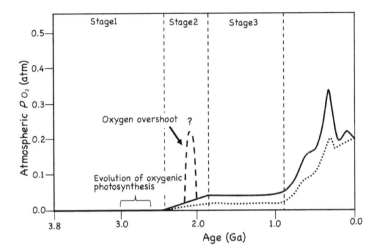

FIGURE 6.1 Estimated evolution of atmospheric pO_2 (Modified from Holland (2006). Solid and lower dashed lines correspond to the upper and the lower estimates, respectively. Stage 1 (3.85–2.45Ga): virtually free from O_2 in the atmosphere and possible local shallow-water oxygen oases. Stage 2 (2.45–1.85Ga): rise of the atmospheric oxygen up to 0.04 atm and mildly oxygenated shallow oceans. Stage 3 (1.85–0.85Ga): boring period without change of atmospheric oxygen levels. Oxygen overshoot: an assumed temporal increase in the atmospheric oxygen, due to extensive increase in primary production and burial of OM (Bekker and Holland. 2012).

2019). In either case, we, if not all, probably have a consensus that the evolution of cyanobacteria had occurred at least 2.7 Ga ago and can be plausibly back to 3.0 Ga ago or even somewhat older, although it needs to be emphasized that the emergence of cyanobacteria does not readily mean contemporaneous oxygenation of the shallow-water and the atmospheric (Figure 6.1).

6.2.2 OXYGENIC PHOTOSYNTHESIS: ENERGETIC AND PHYSIOLOGICAL PERSPECTIVE

Evolution of oxygenic photosynthesis and accumulation of O_2 in environments is one of the most critical events in the Earth's history, in the context of both biological and nonbiological sides of things (Figure 6.2). The presence or absence of O_2 is closely related to the type of respiration. Aerobic respiration utilizes O_2 as electron acceptor, whereas anaerobic respiration utilizes other material such as nitrate (NO_3^-) and sulfate (SO_4^{2-}). Organisms employing aerobic respiration are called aerobes, whereas those doing anaerobic respiration are called anaerobes. More precisely, organisms can be classified into obligate aerobes, obligate anaerobes, facultative aerobes, and facultative anaerobes. Obligate aerobes cannot metabolize substrates without O_2, whereas obligate anaerobes cannot live with that. Molecular oxygen is absolutely toxic to anaerobes. Facultative aerobe metabolizes substrate using O_2 if it is available (aerobic respiration) and could perform anaerobic respiration or fermentation under the oxygen-deficient environment. Facultative anaerobe prefers oxygen-free

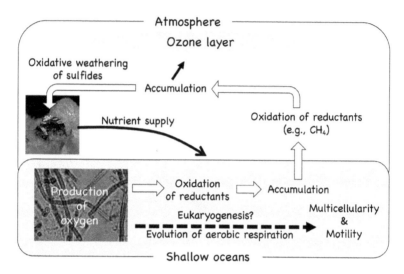

FIGURE 6.2   Schematic view of oxygen production by evolution of oxygenic photosynthesis in shallow oceans and related biological and environmental events.

environment, whereas could survive with O_2. Yeasts are known to be eukaryotic facultative aerobes. They perform aerobic respiration in oxygen-rich environment, whereas perform fermentation in oxygen-deficient environment. Ideally, yeasts could produce 38ATP from one molecule of glucose by aerobic respiration but only 2ATP by alcohol fermentation. Another type is lactic acid fermentation, which is operated by some bacteria and animal cells. Lactic acid fermentation is an important energy source under high exercise intensity such as long sprint running. Namely, human is obligate aerobe, but may be called facultative aerobe at cellular level. Also, as shown in Table 4.2 of Chapter 4, energies obtained by representative anaerobic respirations of 1M glucose, except for those using nitrate and manganese oxides (MnO_2), are less than half of that obtained by aerobic respiration.

The threshold concentration of O_2 for these two different types of respiration is 1% of the present atmospheric level (PAL), which is often called "Pasteur point", after Louis Pasteur. As described above, aerobic respiration could produce much more ATP than anaerobic respiration and fermentation. This is very critical, considering that every vital activity requires energy stored as ATP. Eukaryotes require much more energy than prokaryotes, because they have much larger and complex-structured cells. Obviously, the higher motility of eukaryotic multicellular organisms depends on aerobic respiration.

6.2.3   OXYGENIC PHOTOSYNTHESIS: NUTRITIONAL PERSPECTIVE

Many redox-sensitive metals such as Fe, Mn, Ni, Zn, Mo, and Cu are essential elements as cofactors or metalloenzymes. Cofactors refer to chemicals (except for proteins) required for catalytic activities, whereas metalloenzymes refer to enzymes with metallic ions as active centers. Availability of these metals for organisms depends on not only their abundance in environments but also their forms. Obviously, dissolved forms are much more available than solid forms for organisms. In this context, development of oxidizing environment, which was first established in the shallow oceans and then in the atmosphere, may have been ambivalent in terms of availability of these metals. On one hand, as many of these metals are associated with sulfide minerals such as pyrite (FeS_2) or comprise their own sulfide minerals, their oxidative weathering is essential to release themselves from solid phase as dissolved forms (e.g., Wang et al., 2019). On the other hand, oxygenation of the Earth might have possessed negative "feedback" effects on bioavailability of these metals as discussed below.

Ocean water oxygenation had likely accelerated oxidation of ferrous ion (Fe^{2+}) and subsequent precipitation of ferric hydroxides [$Fe(OH)_3$]. Further oxygenation had caused oxidation of dissolved divalent manganese ion (Mn^{2+}) to trivalent and tetravalent ion (Mn^{3+} and Mn^{4+}), which would result in precipitation of manganese oxides such as birnessite ($MnO_2.nH_2O$). Although anoxygenic photoautotrophs could have participated in precipitation of ferric hydroxides and even manganese oxides before rise of O_2 (Johnson et al., 2013), the precipitation of these compounds was likely enhanced under aerobic condition. These hydroxides (or oxides) might have lowered the bioavailability of themselves and some other nutrients via coprecipitation and adsorption, like phosphates (e.g., Jones et al., 2015), as discussed in Chapter 4.

That aside, it may be worth mentioned that Ni comprises metalloenzymes essential for wide clades including bacteria, archaea, fungi, and plants (Boer et al., 2014). Eight Ni-bearing metalloenzymes are known, including, e.g., superoxide dismutase, NiFe hydrogenase, CO dehydrogenase, and Methyl-CoM reductase that play an important role in prokaryotic metabolisms (Ragsdale, 2009). The roles of these Ni-bearing enzymes are interesting in the context of evolution of Earth's surface environment. For example, Ni superoxide dismutase is involved in intracellular defense system against reactive oxygen species and is suggested to have emerged around 2.0 Ga ago, which is widely known as the time of completion of the first atmospheric oxygenation. Konhauser et al. (2009) compiled Ni/Fe ratios of IFs and demonstrated that marine Ni concentrations profoundly irreversibly dropped between 2.7 and 2.5 Ga ago. This trend was attributed to a decrease in upper-mantle temperature and consequent decrease in eruption of Ni-rich ultramafic magma (komatiite), Based on the requirement of Ni for methanogenesis, it was further suggested that this Ni-famine had declined ecological niches of methanogens, leading to the Earth's oxygenation at the Great Oxidation Event (GOE) (Konhauser et al., 2015). Wang et al. (2019), on the other hand, suggested that despite the expected decrease in Ni content of the continental crust around the end of the Archean, the onset of oxidative weathering of Ni-bearing sulfides (see above) has sustained sufficient methane production.

Response of some other metals such as Mo, V, Cr, and U to the surface oxygenation is assumed to have been different from Fe and Mn. These metals could comprise oxyanion (e.g., MoO_4^{2-}, VO_4^{3-}, CrO_4^{2-}, UO_4^{2-}) in oxidizing aqueous environment. Among these metals, Mo is outstanding in the context of biochemistry and the Earth's oxygenation. Molybdenum is essential for most organisms.

Over 50 Mo-containing enzymes have been identified. Representatives are nitrogenase (Nif with MoFe-cofactor), xanthine oxidase, sulfite oxidase, and aldehyde dehydrogenase. Especially, nitrogenase is of particular importance, because this enzyme catalyzes biological nitrogen fixation, the process of conversion of atmospheric N_2 into ammonia (NH_3), which would be assimilated into glutamate, anion of glutamic acid. Without this function, organisms need to take up abiologically produced bioavailable nitrogen species (Chapter 4). Such abiologically produced nitrogen species had likely become unavailable as early as 3.5 Ga (Kasting and Siefert, 2002) (nitrogen crisis), although a different view on the timing of nitrogen crisis (2.2 Ga) was presented by Navarro-Gonzalez et al. (2001). In either case, evolution of nitrogenase was indispensable for expanding biomass on the Earth. Also, organisms with the ability of nitrogen fixation could produce "alternative nitrogenases" containing V or Fe, which are denoted as Vnf and Anf, respectively. These alternatives are produced and utilized "as an emergency evacuation" in the situation of low availability of Mo. According to the isotopic data of N in sediments and the phylogenic study suggesting that Nif is ancestral to the alternatives (Vnf and Anf), Stüeken et al. (2015) implied that the Mo-dependent nitrogenase (Nif) had already been active at 3.2 Ga. In recent review on the evolution of nitrogenase, on the other hand, Mus et al. (2019) argued that the establishment of Nif was around 2.1 Ga, and before this, nitrogenase (proto-nitrogenase) was Mo-independent and instead may have contained Fe-S cluster. This seems reasonable considering flux of Mo from the continents to the oceans was likely low before the GOE. Molybdenum has a very low abundance of ~1 ppm in the crust, and is hosted mainly by sulfide mineral, molybdenite (MoS_2). Molybdenite occurs in various rock types, including felsic rocks such as granite, rhyolite, and pegmatite, others subjected to hydrothermal alteration, and porphyry copper deposits. Like other sulfide minerals, molybdenite is stable under anaerobic environment. Thus, the rise of the atmospheric oxygen accelerated oxidative weathering of molybdenite, resulting in an increase in flux of Mo into oceans as oxyanion (e.g., MoO_4^{2-}) and its bioavailability (Equation 6.1) (Franscoli and Hudson-Edwards, 2018).

$$2MoS_2 + 9O_2 + 6H_2O \rightarrow MoO_4^{2-} + 4SO_4^{2-} + 12H^+ \qquad (6.1)$$

6.3 GEOCHEMICAL AND MINERALOGICAL RECORDS OF THE GREAT OXIDATION EVENT (GOE) AND EARLIER OXYGENATION

6.3.1 OCCURRRENCES OF REDOX-SENSITIVE MINERALS AND RELATED SEDIMENTS

Some types of sedimentary rocks or ores disappear, whereas other types appear between 2.0 and 2.5 Ga. The changes of geological records were likely related to the rise of atmospheric O_2 (Figure 6.3). In this section, we first have a look of transition of sedimentary record around the GOE period, which is followed by introductions of monumental studies on sulfur isotopes and some relatively new studies suggesting temporal and local oxygenation before the GOE.

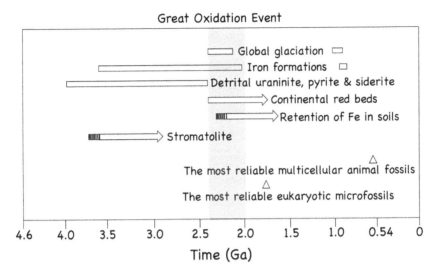

FIGURE 6.3   Main geological and biological events related to the Great Oxidation Event. Modified from Wiese and Reitner (2011).

6.3.1.1   Uraninite, Pyrite, and Siderite

Pyrite (FeS_2) is the most common sulfide mineral and occurs in a wide variety of rocks. Uraninite (UO_2, but U_3O_8 is more common in nature) can occur as an accessory mineral in acidic igneous rocks, although a major primary occurrence is associated with granitic pegmatites. Siderite ($FeCO_3$) occurs as a diagenetic mineral in sedimentary rocks and as a gangue mineral in hydrothermal ores. These minerals are stable under anoxic environment whereas could be rapidly chemically weathered in the presence of O_2 (oxidative weathering) (Equations 6.2 and 6.3) (Holland, 1984; Bachan and Kump, 2015). Also see Equation 4.8 for oxidative weathering of pyrite.

$$UO_2 + 1/2O_2 + 2HCO_3^- \rightarrow UO_2(CO_3)^{2-} + H_2O \qquad (6.2)$$

$$4FeCO_3 + O_2 \rightarrow 2Fe_2O_3 + 4CO_2 \qquad (6.3)$$

Before the rise of atmospheric O_2, these minerals liberated from host rocks due to mechanical weathering and erosion could retain as sand grains, which were transported together with other detrital materials to be deposited elsewhere as terrigenous clastic sediments. Once lithified to be sandstone/conglomerate and isolated from O_2, these detrital minerals could survive for geological time. They have significantly higher specific gravities >4.0 (uraninite; 7.5–9.7, pyrite; ~5.0, siderite; 4.0), contrastive to the major rock-forming minerals such as quartz, felspar and mica with specific gravities less than 3.0. Therefore, these heavy minerals, particularly uraninite and pyrite could be hydraulically separated from the other minerals and concentrated as placer deposits. Such placer model has been proposed for the origin of some representative uraniferous-pyritic-auriferous deposits in South Africa (Witwatersrand) and Canada (Elliot Lake) (e.g., Koglin et al., 2010 and references therein). Although metamorphic/hydrothermal model has also been proposed for the Witwatersrand deposits (e.g., Phillips et al., 1987), it seems likely that the deposits indeed contain primary detrital pyrites and golds (e.g., Koglin et al., 2010; Kirk et al., 2001).

The Witwatersrand placer deposits had deposited around 2.9 Ga ago, whereas the Elliot Lake deposits did around 2.3 Ga ago. In addition, sedimentary rocks containing detrital pyrite and uraninite are very rare later than ca. 2.0 Ga (e.g., Rasmussen and Buick, 1999). According to estimation made by Grandstaff (1980), survival of detrital uraninite requires the atmospheric oxygen level below 10^{-2} to 10^{-6}. Rasmussen and Bucik (1999) also examined ca. 3.25–2.75 Ga-old fluvial sandstones from the Pilbara Craton, Australia and identified the presence of uraninite, pyrite, and siderite. Their detrital origin was argued based on multiple criteria (e.g., relative size ranges of grains related to difference in density and internal compositional zoning truncated by rounded surfaces), providing evidence for reducing environment during their weathering, transportation, and diagenesis. More quantitative estimation of the pre-GOE atmospheric O_2 is provided by Johnson et al. (2014), who studied occurrence of detrital pyrite and uraninite throughout the core of mixed siliciclastic and IF-containing sedimentary unit from the 2.4 Ga-old Koegas Subgroup, South Africa and constructed a grain erosion model involving effects of chemical weathering and physical abrasion. Based on this, the partial pressure of the Paleoproterozoic atmospheric oxygen (molecular O_2) was estimated to be $<3.2 \times 10^{-5}$ atm.

6.3.1.2 Red Beds

Red beds, in the strict sense, refer to terrigenous clastic sedimentary rocks such as sandstone and shales enriched in reddish mineral, hematite (Fe_2O_3). Red beds may occur in various depositional environments such as e.g., alluvial, lacustrine, aeolian, and passive continental margins (Turner, 1980). While various origins of red beds have been considered, three major processes are enough to be addressed to here, including (1) in-situ oxidative alteration of detrital ferromagnesian minerals and iron-bearing clay minerals in alluvial sediments, (2) lateritic weathering in source areas and deposition of detrital materials rich in ferric hydroxides at deltaic environments, and (3) diagenetic formation of hematite-coated grains or their detrital supply at desert environments (Eriksson and Cheney, 1992). Abiotic oxidation of Fe^{2+} to Fe^{3+} requires O_2 and red beds are absent in the Archean, whereas common in the Meso- to Neoproterozoic and in the Phanerozoic. It has therefore been widely believed that the onset of red bed deposition had reflected the entire oxygenation of the atmosphere.

The oldest red beds can be back to ~2.3 Ga ago. Eriksson and Cheney (1992) examined 2.3–2.1 Ga-old and <1.9 Ga-old red beds in South Africa and showed that the older red beds were characterized by red-stained matrix materials, whereas the younger ones by hematite coatings on grains. Such difference in hematite occurrence was interpreted in the context formation mechanisms and redox environment.

Hematite staining in the matrix was formed diagenetically under weakly oxidizing conditions, whereas hematite coatings on grains were formed under more oxidizing environment: it was suggested that by 1.9 Ga ago the atmospheric O_2 had reached $>10^{-2}$ atm. In ca. 2.15 Ga-old red beds in the Franceville Basin, Gabon, ferric oxides are dispersed mainly in matrices, whereas hematite coatings on grains are rare or absent (Bankole et al., 2016).

6.3.2 BLACK SHALES AND MOLYBDENUM

As described above, delivery of Mo from the continents to the oceans requires oxidative weathering of Mo-bearing sulfides, and thus, Mo concentrations of marine sediments have been used as proxy of the atmospheric oxygenation. Involvement of dissolved molybdenum oxyanion (MoO_4^{2-}) in seawater into sediments is closely related to redox environments and OM. The process of OM-related molybdenum oxyanion removal from seawater is as follows (Anbar et al., 2007 and references therein).

Sulfidic (H_2S-rich) environment is essential to this type of removal of MoO_4^{2-}. In reducing environment, sulfate is microbially reduced to be H_2S, with which molybdenum oxyanion reacts to form oxythiomolybdate (MoO_{4-xSx}^{2-}). This Mo-S complex has a close affinity with OM and is removed from seawater along with settling of OM. Development of reducing environment and deposition of OM are closely related. In stagnant and stratified water mass, high primary production and subsequent settling of OM would result in rapid consumption of dissolved O_2 and NO_3^- in the bottom water, being followed by reduction of SO_4^{2-} to H_2S. Respiration rate of sulfate reduction is much slower than aerobic and nitrate respiration, which is suitable for deposition of OM-rich sediments. Sediments enriched in OM would be eventually lithified to black shales. Thus formation of Mo-rich black shales are the consequence of oxidative weathering of MoS_2, accumulation of molybdenum oxyanion in aerobic seawater, and subsequent development of reducing environment. Therefore, if you look at the molybdenum concentration in the black shales of various ages, you would be able to find out when O_2 began to increase in the atmosphere and the oceans and even the temporal oxygenation before the GOE called a whiff of oxygen (Ye et al., 2021, Anbar et al., 2007).

6.3.3 PALEOSOL, CLUE TO OXYGEN IN THE ATMOSPHERE AND IMPLICATIONS?

6.3.3.1 Concept of Paleosol Geochemistry

Paleosol refers to fossilized soil, and its chemical composition has been used for estimating atmospheric concentrations O_2 during the Precambrian era (e.g., Holland, 1984). The main concept of paleosol studies and representative studies are introduced here.

Soils are products of weathering of parent rocks, mixed with plant-derived OM and microorganisms. The weathering is roughly classified into physical weathering, in which rocks are cracked and gradually refined due to e.g., temperature changes and the power of plant roots, and chemical weathering, in which rainwater containing O_2 and CO_2 reacts with minerals.

As described earlier, chemical weathering of Fe^{2+}-bearing minerals such as pyrite and siderite is closely related to the atmospheric level of O_2 (Equations 4.8 and 6.3). Chemical weathering of other Fe^{2+}-bearing minerals such as olivine and pyroxene, on the other hand, does not directly relate to redox condition and proceeds as hydration reaction, releasing Fe^{2+} (Equation 6.4) (Griffioen, 2017). If the environment is oxidizing, Fe^{2+} is transformed into Fe^{3+}, which precipitates as ferric hydroxide (Equation 5.3). This means retention of Fe in weathering profiles. If olivine and pyroxene are subjected to chemical weathering under reducing condition, liberated Fe^{2+} would be removed from weathering profiles.

$$(Mg, Fe)_2SiO_4 + 4CO_2 + 4H_2O \rightarrow 2(Mg^{2+}+/Fe^{2+}) + 4HCO_3^- + H_4SiO_4 \quad (6.4)$$

In modern environments, chemical compositions and their vertical profiles of soils are controlled by various factors other than chemical and mineral compositions of parent materials, atmospheric compositions and precipitations. Soil animals represented by earthworms could physically disturb soils. Organic acids such as {carboxylic acid (R-COOH)} released by heterotrophic respiration of OM would lower soil pH, occasionally resulting in redistribution of Fe and Al. On the other hand, during the Precambrian, particularly the Archean, chemical and mineral compositions of paleosols could be basically interpreted in the context of direct reaction of the atmosphere (and water) with minerals. Thus the paleosol chemical profiles could be proxy of paleoenvironment, although it cannot be excluded the possibility that microbes inhabited within surface weathered portion of terrestrial rocks (e.g., Rye and Holland, 2000; Lalonde and Konhauser, 2015).

6.3.3.2 Classical Controversy on Paleosol Records

Rye and Holland (1998) reviewed more than 50 successions described as paleosols and extracted 15 "genuine" paleosols, among which pre-2.44 Ga paleosols exhibited significant weathering-related loss of Fe, whereas paleosols formed between the 2.2 and 2.0 Ga ago and later did not show such the loss. From the data of paleosols formed between the 2.2 and 2.44 Ga, ambiguous results were obtained. Based on these observations, Rye and Holland (1998) suggested that the atmospheric pO_2 had risen to be 0.03 atm or higher during the interval from 2.2 to 2.0 Ga. Before this period, the atmospheric pO_2 was assumed to have been less than $\sim 5 \times 10^{-4}$ atm. Contrastive view has been proposed by Ohmoto (1996) from depth profiles of iron concentrations of pre- and post-2.2 Ga-old paleosols. The author examined Ti-normalized values of Fe^{2+}, Fe^{3+}, and ΣFe and showed that regardless of age, Fe^{3+}/Ti ratios of paleosols were higher than parental rocks, suggestive of oxidation of Fe^{2+} to Fe^{3+} and Fe-retention. Common loss of ΣFe in paleosol sections identified for all ages was interpreted in the context of later alteration by hydrothermal fluids or remobilization related to organic

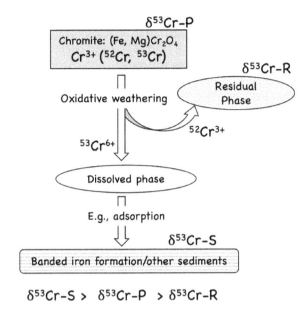

$$\delta^{53}Cr\text{-}S > \delta^{53}Cr\text{-}P > \delta^{53}Cr\text{-}R$$

FIGURE 6.4   Schematic model of oxidative weathering of Cr^{3+}-bearing minerals and related isotopic fractionation between ^{53}Cr and ^{52}C. $\delta^{53}Cr$-P: parent rock, $\delta^{53}Cr$-R: residuals (soils), $\delta^{53}Cr$-S: soluble phase. $\delta^{53}Cr = [(^{53}Cr/^{52}Cr)_{sample}/(^{53}Cr/^{52}Cr)_{standard} - 1] \times 1000$ (‰).

acids derived from terrestrial biomass. The calculated $_pO_2$ of the 3.0–2.2 Ga atmosphere was at least 1.5% PAL (Ohmoto, 1996).

6.3.3.3   Recent Controversies on Archean Oxygenic Atmosphere and Isotopic Approach to Paleosol

As shown from the brief review of the two representative paleosol studies (Rye and Holland, 1998; Ohmoto, 1996), reconstruction of the ancient atmospheric O_2 concentrations is not straightforward. This can be attributed to difficulties in identification of genuine paleosol and its exact horizon as soil sequence, later modification of chemical compositions, chemical heterogeneity of parental rocks, and use of limited redox proxies. More recently, redox-sensitive trace elements and their isotopic fractionation have been employed to gain new insights into the evolution of the atmospheric $_pO_2$. Interestingly, some results suggested the atmospheric oxygenation prior to 2.5 Ga (Philippot et al., 2013; Crowe et al., 2013; Mukhopadhyay et al., 2014). Philippot et al. (2013) examined the 2.76 Ga-old paleosol at Mount Roe Basalt in the Pilbara Craton, Western Australia. The paleosol facies were shown to contain Fe^{3+}-montmorillonite and to be enriched in Cr and Mo compared with the parental rock, suggestive of oxidative weathering of parent rock. Furthermore, sulfur isotopic signatures of sulfate inclusions in montmotillonite and calcite were shown to be distinct from those observed for Archean sulfate and sulfide characterized by S-MIF (mass independent fractionation of sulfur: see below), but similar to post-Archean ones. It was thus suggested that the atmosphere under which the paleosol were formed was oxygenated (Philippot et al., 2013). This paleosol has been previously examined by Rye and Holland (1998) and was recognized to have lost Fe, though these two examined sections were not identical to each other. These two conflicted results emphasize the difficulty in geochemical studies of paleosols.

The other studies on older paleosols suggest that oxidative weathering could be back to 2.96 and even 3.29 Ga ago (Crowe et al., 2013; Mukhopadhyay et al., 2014). Redox-dependent behaviors of rare-earth elements (REEs) were mainly examined in Mukhopadhyay et al. (2014), whereas isotopic systematics of Cr in Crowe et al. (2013), who studied 2.96 Ga-old paleosols from South Africa and claimed that the concentration level of O_2 at that time was at least 3×10^{-4} PAL. This is more than an order of magnitude higher then the GOE model indicating that the atmospheric oxygen was below 10^{-5} PAL before 2.0 Ga ago. Crow et al's assertion has been later criticized by others (as effects of modern weathering) (Heard et al., 2021). Also, onset of the oxidative weathering of the continental crust after the GOE seems to be better supported by isotopic fractionation of Mo in glacial diamictites (Greaney et al., 2020). Furthermore, redox-independent Cr isotopic could occur (Saad et al., 2017). Despite these, it seems to be worth here describing the sophisticated methodology employed by Crowe et al. (2013). Crowe et al. (2013) employed isotopic fractionation of stable isotopes of Cr (^{52}Cr and ^{53}Cr) during weathering (Figure 6.4). The solubility of Cr is different depending on its valance. Oxidized form of Cr (Cr^{6+}) is soluble, whereas Cr^{3+} is stable. Thus, oxidative weathering of Cr^{3+}-bearing minerals would release Cr^{6+} to the environment. During this process, isotopic fractionation between ^{52}Cr and ^{53}Cr occurs, resulting in relative enrichment in ^{53}Cr in dissolved phase. Thus, the oxidative weathering is expected to result in formation of soil residually enriched in ^{52}Cr relative to parental material. ^{53}Cr is, on the other hand, preferentially removed from the soil and flows into oceans. Crowe et al. (2013) demonstrated that the upper part of the examined paleosol was depleted in ^{53}Cr relative to the lower part and the parental rock, whereas the concentration of Fe was enriched, consistent with oxidative weathering. The authors also examined two facies of IFs nearly contemporaneous to the paleosol. While IF facies exclusively composed of chemical precipitates (CIF) tended to be enriched in ^{53}Cr, the other IF facies with some siliciclastic components (SIF) had δ^{53}Cr values falling within the narrow range of igneous rocks. Thus, the trends of δ^{53}Cr values of the paleosol and CIF were complementary to each other.

6.3.4 IDENTIFICATION OF SULFUR MASS-INDEPENDENT ISOTOPIC FRACTIONATION (S-MIF)

It seems now widely accepted that disappearance of the mass independent isotopic fractionation of sulfur (S-MIF) in sulfides and sulfates in the geologic record gives an important constraint on the onset of the GOE (Farquhar et al., 2000; Farquhar and Wing, 2003). Here, S-MIF is introduced in some detail.

Sulfur has four natural stable isotopes [^{32}S (95.02%), ^{33}S (0.75%), ^{34}S (4.21%), and ^{36}S (0.02%)]. During physicochemical and biochemical reactions, these isotopes are fractionated. The isotope fractionation depends on the difference in mass between isotopes. When focusing on the combination of ^{33}S and ^{32}S and the combination of

^{34}S and ^{32}S, the former has a mass difference of 1, whereas the latter has a mass difference of 2. Therefore, when looking at fractionations between these three isotopes, the ^{33}S-^{32}S always shows a smaller fractionation, which is nearly a half (precisely, 0.515) of that observed for ^{34}S-^{32}S (Figure 6.5). This is called mass-dependent isotopic fractionation. This regular isotope fractionation of S has been persistently seen for sulfides and sulfates younger than 2.4 Ga. Before this, sulfur isotopic data do not follow this systematics, which is called S-MIF (Figure 6.6).

S-MIF is thought to be attributed to the production of elemental sulfur (S_8^0) from SO_2 in the atmosphere and associated anomalous isotopic fractionation (Chapter 4). In the substantial absence of O_2 and thus the absence of ozone screen, ultraviolet (UV) penetrated deeply into the Archean atmosphere and was involved in this reaction, including photolysis of SO_2 at wavelengths <220 nm and photoexcitation at

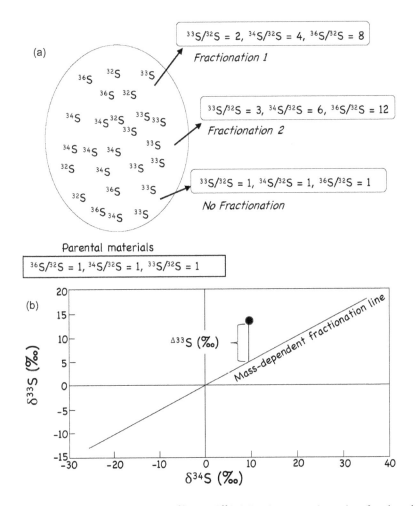

FIGURE 6.5 Relationship between δ^{34}S and δ^{33}S following mass-dependent fractionation. Δ^{33}S(‰) = δ^{33}S(‰) − δ^{34}S(‰). Also see Figure 6.6.

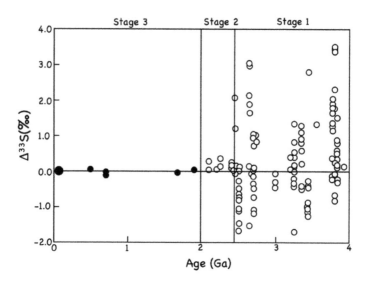

FIGURE 6.6   Secular change of $\Delta^{33}S$ (‰) of sulfates and sulfides. $\Delta^{33}S$ (%) shows the deviation from the straight line shown in Figure 6.5. (This is drawn based on Farquhar and Wing (2003).)

wavelengths 240–340 nm (Ono, 2017 and references therein). Pavlov and Kasting (2002), using a one-dimensional photochemical model, showed that sulfur of different oxidation states (S_8^0 and SO_2) with different isotopic signatures could have conservatively been removed from the atmosphere with $O_2 < 10^{-5}$ PAL (the present atmospheric level) to the oceans. By contrast, in the atmosphere with $O_2 > 10^{-5}$ PAL, any MIF signatures would have been lost due to oxidation of all sulfur compounds to SO_4^{2-}, which was homogenized and stored in the oceanic reservoir. Elemental sulfur and residual SO_2 (finally oxidized to SO_4^{2-}) produced by photochemical reactions in the anoxic atmosphere would have been washed out to the oceans, where sulfur metabolizing bacteria, represented by sulfate reducing bacteria and sulfur oxidizing bacteria, utilized these compounds. For example, H_2S produced by microbial sulfate reduction reacted with Fe^{2+} to be precipitated as pyrite (FeS_2), and SO_4^{2-} would have been precipitated as barite ($BaSO_4$). These minerals preserved in sedimentary rock records archived secular change in sulfur isotopic systematics in the atmosphere closely related to the state of oxygenation. The authors concluded that the atmospheric O_2 concentration mush have been less than 10^{-5} PAL prior to 2.3 Ga ago (Pavlov and Kasting, 2002).

6.4   SEDIMENTARY RECORDS OF OXYGENIC PHOTOSYNTHESIS: STROMATOLITE AND MISS

Sedimentary records for the evolution of oxygenic photosynthesis include stromatolite and microbially induced sedimentary structures (MISS), which are interpreted in detail in the next chapter. Thus, readers may better read Chapter 7 first and back to here. As discussed there, stromatolite (microbial mats) representing sedimentary bio-signature is not always a product of cyanobacterial activities. However, some types of microbial mats in the Archean were likely produced by cyanobacteria. Flannery and Walter (2012) reported tufted microbial mats from the 2.72 Ga-old Tumbiana Formation of the Fortescue Group in Western Australia. The millimeter-scale tufted microbial mat structures occur as coniform pseudocolumnar and columnar stro-matolites (Figure 6.7). Apical zones of the structures display filamentous palimp-sest fabrics and fenestrae. The zones extend in one to three dimensions, producing cones, ridges, and reticulations, the latter of which forms polygons ~1–1.5 cm width. Similar microbial mat structures have been reported from modern hypersaline, hydrothermal, and cryogenic environments. Despite environmental variations, the modern tufted microbial mat structures are "composed almost exclusively of verti-cally aligned bundles of filamentous cyanobacteira" (Flannery and Walter, 2012). Although the tufted microbial mat structures may not be solely attributed to micro-bial phototactic behaviors (e.g., Shepard and Sumner, 2010), it is still promising that the Tumbiana tufted microbial mats were formed by cyanobacteria. This seems to be consistent with morphological study of conical stromatolites conducted by Bosak et al. (2009). They noted that modern conical microbial mats produced by cyano-bacteria are characterized by contorted laminae with enmeshed oxygen bubbles sub-millimeter to millimeter across, which are occasionally well preserved in fossilized

FIGURE 6.7 Tufted microbial mat (stromatolite) from the ca. 2.7 Ga-old Tumbiana Formation (Flannery and Walter, 2012). (Photograph courtesy of D. Flannery.)

stromatolites. Based on criteria including contorted laminae, a distinct axial zone, and bubble-like structures, Archean conical stromatolites were examined. Some conical stromatolites 2.7 Ga-old and younger met the criteria and thus were likely produced by cyanobacteria. (Bosak et al., 2009)

Some fossilized microbial mats in older sedimentary successions have been interpreted as possible products by cyanobacteria. Noffke et al. (2003, 2008) reported MISS from the 2.9 Ga-old Pongola Supergroup, South Africa from which morphologically diverse stromatolites had already been described (Beukes and Lowe, 1989). The structures that occur in siliciclastic sedimentary units of tidal facies include erosional remnants and pockets, polygonal oscillation cracks and gas domes. Filamentous laminae containing OM enriched in ^{12}C and trapped oriented detrital grains were identified under the microscope. From the same Pongola Group, Siahi et al. (2016) described stromatolitic carbonates of various morphologies, microfabrics, and sizes that potentially corresponded to distinct microbial assemblages responding to different environmental conditions. Some of these newly described Pongola stromatolites (Figure 13 in Siahi et al., 2016) are similar to those by Flannery and Walter (2012). Although the authors did not mention the possibility that the Pongola stromatolites were built by cyanobacteria, its possibility could be considered. From the 3.22 Ga-old Moodies Group of the Barberton greenstone belt, South Africa, Homann et al. (2015) reported microbial mat structures with carbonaceous laminae from siliciclastic tidal deposits. The structures include tufted microbial mats associated with gas domes, gas- or fluid-escape structures, and shrinkage cracks (Figure 6.8). This association was restricted to the upper inter- to supratidal facies. Depositional settings and architectures of these mats are consistent with those of modern microbial mats produced by cyanobacteria.

COLUMN: OXYGEN IS A DOUBLE-EDGED SWORD

Molecular oxygen is itself reactive and can be further converted to more reactive oxygen species such as superoxide (O_2^-), hydroxyl radical ($\cdot OH$), hydrogen peroxide (H_2O_2), and singlet oxygen ($^1O_2$) through e.g., reaction with UV, effectively in association with catalysts represented by TiO_2. Reactive oxygen species are also produced within cells during aerobic respiration. They are very harmful and representative DNA damaging agent. Therefore, organisms have enzymes such as superoxide dismutase (SOD) that catalyzes disproportionation of superoxide to hydrogen peroxide, which is eventually converted by another enzyme to be H_2O and O_2. It is often said that when O_2 began to increase on the Earth, previously flourished prokaryotic obligate anaerobes presumably without enzymes detoxifying reactive oxygen species faced extinction, called oxygen catastrophe. However, the author is skeptic to this story, because concentrations of O_2 is quite various even on the modern Earth. Anaerobic and aerobic environments could coexist with each other within soils, sediments and water masses (e.g., stratified water). Also, many prokaryotes have the ability to form dormant endospore resistant to environmental risks such as UV radiation, desiccation, and toxic chemicals. Thus, it is more likely that many anaerobic microbes had survived and explored new niches during oxygenation of the Earth.

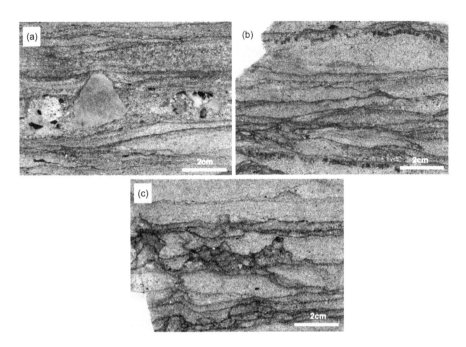

FIGURE 6.8 Photographs of cross-sectional views of polished sandstone slabs from the ca. 3.22 Ga-old Moodies Group, Barberton Greenstone Belt, South Africa, representing different paleoenvironments (Homann et al., 2015). (a) Planar mats at coastal floodplain. (b) Wavy-crinkly mats and small microbial domes at intertidal zone. (c) Tufted microbial mats with vertically stacked cones at upper inter- to supratidal zone. (Photographs courtesy of M. Homann.)

REFERENCES

Anbar, A.D., Duan, Y., Lyons, T.W., Arnold, G.L., Kendall, B., Creaser, R.A., Kaufman, A.J., Gordon, G.W., Scott, C., Garvin, J., Buick, R. 2007. A whiff of oxygen before the Great Oxidation Event? *Science* 317, 1903–1906.

Bachan, A., Kump, L.R. 2015. The rise of oxygen and siderite oxidation during the Lomagundi Event. *Proceedings of the National Academy of Sciences of the United States of America* 112, 6562–6567.

Bankole, O.M., El Albani, A., Meunier, A., Rouxel, O., Gauthier-Lafaye, F., Bekker, A. 2016. Origin of red beds in the Paleoproterozoic Franceville Basin, Gabon, and implications for sandstone-hosted uranium mineralization. *American Journal of Science* 316, 839–872.

Bekker, A., Holland, H.D. 2012. Oxygen overshoot and recovery during the early Paleoproterozoic. *Earth and Planetary Science Letters* 317–318, 295–304.

Beukes, N.J., Lowe, D.R. 1989. Environmental control on diverse stromatolite morphologies in the 3000 Myr Pongola Supergroup, South Africa. *Sedimentology* 36, 383–397.

Boer, J.L., Mulrooney, S.B., Hausinger, R.P. 2014. Nickel-dependent metalloenzymes. *Archives of Biochemistry and Biophysics* 544, 142–152.

Bosak, T., Liang, B., Sim, M.S., Petroff, A.P. 2009. Morphological record of oxygenic photosynthesis in conical stromatolites. *Proceedings of the National Academy of Sciences of the United States of America* 106, 10939–10943.

Crowe, S.A., DØssing, L.N., Beukes, N.J., Bau, M., Kruger, S.J., Frei, R., Canfield, D.E. 2013. Atmospheric oxygenation three billion years ago. *Nature* 501, 535–538.

Eriksson, P.G., Cheney, E.S. 1992. Evidence for the transition to an oxygen-rich atmosphere during the evolution of red beds in the Lower Proterozoic sequences of southern Africa. *Precambrian Research* 54, 257–269.

Farquhar, J., Bao, H., Thiemens, M. 2000. Atmospheric influence of Earth's earliest sulfur cycle. *Science* 289, 756–758.

Farquhar, J., Wing, B.A. 2003. Multiple sulufr isotopes and the evolution of the atmosphere. *Earth and Planetary Science Letters* 213, 1–13.

Fischer, W.W., Hemp, J., Johnson, J.E. 2016. Evolution of oxygenic photosynthesis. *Annual Review of Earth and Planetary Sciences* 44, 647–683.

Flannery, D.T., Walter, M.R. 2012. Archean tufted microbial mats and the Great Oxidation Event: New insights into an ancient problem. *Australian Journal of Earth Sciences* 59, 1–11.

Franscoli, F., Hudson-Edwards, K.A. 2018. Geochemistry, mineralogy and microbiology of molybdenum in mining-affected environments. *Minerals* 8, 42. doi: 10.3390/min8020042.

Garcia-Pichel, F., Lombard, J., Soule, T., Dunaj, S., Wu, S.H., Wojciechowski, M.F. 2019. Timing the evolutionary advent of cyanobacteria and the later Great Oxidation Event using gene phylogenies of a sunscreen. *mBio* 10, e00561-19. doi:10.1128/mBio.00561-19.

Grandstaff, D.E. 1980. Origin of uraniferous conglomerates at Elliot Lake, Canada and Witwatersrand, South Africa: Implications for oxygen in the Precambrian atmosphere. *Precambrian Research* 13, 1–26.

Greaney, A.T., Rudnick, R.L., Romaniello, S.J., Johnson, A.C., Gaschnig, R.M., Anbar, A.D. 2020. Molybdenum isotope fractionation in glacial diamictites tracks the onset of oxidative weathering of the continental crust. *Earth and Planetary Science Letters* 534, 116083.

Griffioen, J. 2017. Enhanced weathering of olivine in seawater: The efficiency as revealed by thermodynamic scenario analysis. *Science of the Total Environment* 575, 536–544.

Hamilton, T.L., Klatt, J.M., de Beer, D., Macalady, J.L. 2018. Cyanobacterial photosynthesis under sulfidic conditions: Insights from the isolate *Leptolyngbya* sp. *strain Hensonii*. *The ISME Journal* 12, 568–584.

Heard, A.W., Aarons, S.M., Hofmann, A., He, X., Ireland, T., Bekker, A., Qin, L., Dauphas, N. 2021. Anoxic continental surface weathering recorded by the 2.95 Ga Denny Dalton Paleosol (Pongola Supergroup, South Africa). *Geochimica et Cosmochimica Acta* 295, 1–23.

Hohmann-Marriott, M.F., Blankenship, R.E. 2011. Evolution of photosynthesis. *Annual Review of Plant Biology* 62, 515–548.

Holland, H.D. 1984. *The Chemical Evolution of the Atmosphere and Oceans*. Princeton University Press, Princeton, 582p.

Holland, H.D. 2006. The oxygenation of the atmosphere and oceans. *Philosophical Transactions of the Royal Society B* 361, 903–915.

Homann, M., Heubeck, C., Airo, A., Tice, M.M. 2015. Morphological adaptations of 3.22 Ga-old tufted microbial mats to Archean coastal habitats (Moodies Group, Barberton Greenstone Belt, South Africa). *Precambrian Research* 266, 47–64.

Hug, L.A., Baker, B.J., Anantharaman, K., Brown, C.T., Probst, A.J., Castelle, C.J., Butterfield, C.N., Hernsdorf, A.W., Amano, Y., Ise, K. and Suzuki, Y., Dudek, N., Relman, D.A., Finstad, K.M., Amundson, R., Thomas, B.C., Banfield, J.F. 2016. A new view of the tree of life. *Nature Microbiology* 1, 16048.

Johnson, J.E., Gerpheide, A., Lamb, M.P., Fischer, W.W. 2014. O_2 constraints from Paleoproterozoic detrital pyrite and uraninite. *Geological Society of America Bulletin* 126, 813–830.

Johnson, J.E., Webb, S.M., Thomas, K., Ono, S., Kirschvink, J.L., Fischer, W.W. 2013. Manganese-oxidizing photosynthesis before the rise of cyanobacteria. *Proceedings of the National Academy of Sciences of the United States of America* 110, 11238–11243.

Jones, C., Nomosatryo, S., Crowe, S.A., Bjerrum, C.J., Canfield, D.E. 2015. Iron oxides, divalent cations, silica, and the early earth phosphorus crisis. *Geology* 43, 135–138.

Kasting, J.F., Siefert, J.L. 2002. Life and the evolution of Earth's atmosphere. *Science* 296, 1066–1068.

Kirk, J., Ruiz, J., Chesley, J., Titley, S., Walshe, J. 2001. A detrital model for the origin of gold and sulfides in the Witwatersrand basin based on Re-Os isotopes. *Geochimica et Cosmochimica Acta* 65, 2149–2159.

Koglin, N., Frimmel H.E., Minter, W.E.L., Brätz, H. 2010. Trace-element characteristics of different pyrite types in Mesoarchaean to Palaeoproterozoic placer deposits. *Mineralium Deposita* 45, 259–280.

Konhauser, K.O., Pecoits, E., Lalonde, S.V., Papineau, D., Nisbet, E.G., Barley, M.E., Arndt, N.T., Zahnle, K., Kamber, B.S. 2009. Oceanic nickel depletion and a methanogen famine before the Great Oxidation Event. *Nature* 458, 750–753.

Konhauser, K.O., Robbins, L.J., Pecoits, E., Peacock, C., Kappler, A., Lalonde, S.V. 2015. The Archean nickel famine revisited. *Astrobiology* 15, 804–815. doi:10.1089/ast.2015.1301.

Lalonde, S.V., Konhauser, K.O. 2015. Benthic perspective on Earth's oldest evidence for oxygenic photosynthesis. *Proceedings of the National Academy of Sciences of the United States of America* 112, 995–1000.

Martin, W.F., Bryant, D.A., Beatty, J.T. 2018. A physiological perspective on the origin and evolution of photosynthesis. *FEMS Microbiology Reviews* 42, 205–231.

Mukhopadhyay, J., Crowley, Q.G., Ghosh, S., Ghosh, G., Chakrabarti, K., Misra1, B., Heron, K., Bose, S. 2014. Oxygenation of the Archean atmosphere: New paleosol constraints from eastern India. *Geology* 42, 923–926. doi:10.1130/G36091.1.

Mus, F., Colman, D.R., Peters, J.W., Boyd, E.S. 2019. Geobiological feedbacks, oxygen, and the evolution of nitrogenase. *Free Radical Biology and Medicine* 140, 250–259. doi:10.1016/j.freeradbiomed.2019.01.050.

Navarro-Gonzalez, R., McKay, C.P., Mvondo, D.N. 2001. A possible nitrogen crisis for Archaean life due to reduced nitrogen fixation by lightning. *Nature* 412, 61–64.

Noffke, N., Beukes, N., Bower, D., Hazen, R.M., Swift, D.J.P. 2008. An actualistic perspective into Archean worlds – (cyano-)bacterially induced sedimentary structures in the siliciclastic Nhlazatse Section, 2.9 Ga Pongola Supergroup, South Africa. *Geobiology* 6, 5–20.

Noffke, N., Hazen, R., Nhleko, N. 2003. Earth's earliest microbial mats in a siliciclastic marine environment (2.9 Ga Mozaan Group, South Africa). *Geology* 31, 673–676.

Nürnberg, D.J., Morton, J., Santabarbara, S., Telfer, A., Joliot, P., Antonaru, L.A., Ruban, A.V., Cardona, T., Krausz, E., Boussac, A., Fantuzzi, A., Rutherford, A.W. 2018. Photochemistry beyond the red limit in chlorophyll f-containing photosystems. *Science* 360, 1210–1213.

Ohmoto, H. 1996. Evidence in pre-2.2 Ga paleosols for the early evolution of atmospheric oxygen and terrestrial biota. *Geology* 24, 1135–1138.

Ono, S. 2017. Photochemistry of sulfur dioxide and the origin of mass-independent isotope fractionation in Earth's atmosphere. *Annual Review of Earth and Planetary Sciences* 45, 301–329.

Pavlov, A.A., Kasting, J.F. 2002. Mass-independent fractionation of sulfur isotopes in Archean sediments: Strong evidence for an anoxic Archean atmosphere. *Astrobiology* 2, 27–41.

Philippot, P., Teitler, Y., Gérard, M., Cartigny, P., Muller, E., Assayag, N., Le Hir, G., Fluteau, F. 2013. Isotopic and mineralogical evidence for atmospheric oxygenation in 2.76 Ga old paleosols. *Mieralogical Magazine* 77(5), 1965.

Phillips, G.N., Myers, R.E., Palmer, J.A. 1987. Problems with the placer model for Witwatersrand gold. *Geology* 15, 1027–1030.

Ragsdale, S.W. 2009. Nickel-based enzyme systems. *Journal of Biological Chemistry* 284, 18571–18575.

Rasmussen, B., Buick, R. 1999. Redox state of the Archean atmosphere: Evidence from detrital heavy minerals in ca. 3250–2750 Ma sandstones from the Pilbara Craton, Australia. *Geology* 27,115–118.

Rye, R., Holland, H.D. 1998. Paleosols and the evolution of atmospheric oxygen: A critical review. *American Journal of Science* 298, 621–672.

Rye, R., Holland, H.D. 2000. Life associated with a 2.76 Ga ephemeral pond?: Evidence from Mount Roe #2 paleosol. *Geology* 28, 483–486.

Saad, E.M., Wang, X., Planavsly, N.J., Reinhard, C.T., Tang, Y. 2017. Redox-independent chromium isotope fractionation induced by ligand-promoted dissolution. *Nature Communications* 8, 1590.

Sanchez-Baracaldo, P., Cardona, T. 2019. On the origin of oxygenic photosynthesis and cyanobacteria. *New Phytologist* 225, 1440–1446.

Schirrmeister, B.E., de Vos, J.M., Antonelli, A., Bagheri, H.C. 2013. Evolution of multicellularity coincided with increased diversification of cyanobacteria and the Great Oxidation Event. *Proceedings of the National Academy of Sciences of the United States of America* 110, 1791–1796.

Shepard, R.N., Sumner, D.Y. 2010. Undirected motility of filamentous cyanobacteria produces reticulate mats. *Geobiology* 8, 179–190.

Siahi, M., Hofmann, A., Hegner, E., Master, S. 2016. Sedimentology and facies analysis of Mesoarchean stromatolitic carbonate rocks of the Pongola Supergroup, South Africa. *Precambrian Research* 278, 244–264.

Stüeken, E.E., Buick, R., Guy, B.M., Koehler, M.C. 2015. Isotopic evidence for biological nitrogen fixation by molybdenum-nitrogenase from 3.2 Gyr. *Nature* 520, 666–669.

Turner, P. 1980. Continental red beds. *Developments in Sedimentology* 29. Elsevier, New York, 562pp.

Wang, S.-J., Rudnick, R.L., Gaschnig, R.M., Wang, H., Wasylenki, L.E. 2019. Methanogenesis sustained by sulfide weathering during the Great Oxidation Event. *Nature Geoscience* 12, 296–300.

Ward, L.M., Cardona, T., Holland-Moritz, H. 2019. Evolutionary implications of anoxygenic phototrophy in the bacterial phylum *Candidatus* Eremiobacterota (WPS-2). *Frontiers in Micorbiology* 10, 1658.

Ye, Y., Zhang, S., Wang, H., Wang, X, Tan, C., Li, M., Wu, C., Canfield, D.E. 2021. Black shale Mo isotope record reveals dynamic ocean redox during the Mesoproterozoic Era. *Geochemical Perspectives Letters* 18, 16–21.

7 Biosignatures in Ancient Rocks and Related Issues

7.1 INTRODUCTION

Biosignatures are signs of life. Although representatives are megascopic body fossils, such undoubted biosignatures could not be expected for the Earth's deep time discussed in this book. Therefore, we need to explore others, including, e.g., isotopic signatures of some elements represented by C and S, organic matter (OM), biominerals, fossilized cell, and biologically mediated sedimentary structure such as stromatolite and microbially-induced sedimentary structure (MISS). Unfortunately, these biosignatures are more or less ambiguous. They have merits and demerits, and none of them can be "smoking gun". Thus, it is important for us to recognize limitations of these biosignatures. In this chapter, various biosignatures are described, citing the latest studies. Biosignatures are here classified into two types, noncellularly preserved biosignatures and cellularly preserved one. The latter gives direct images of ancient life, while its credibility has often been subjected to criticisms. Readers are also recommended to read the recent nice review by Lepot (2020), which gives useful points of view for biosignatures different from that those presented here.

7.2 ORGANIC MATTER

7.2.1 KEROGEN AND ITS ISOTOPIC COMPOSITIONS

Organic matter (OM) is relatively ubiquitous in Archean sedimentary rocks, particularly in dark gray to black cherts. Organic matter includes kerogen and bitumen; the former refers to solvent-insoluble OM, whereas the latter refers to soluble one. Kerogenous OM occurs as various forms including dispersed fine particles, loose aggregates, clots, or grains, in addition to microfossils and putative microfossils. They also often comprise laminae. Some of clots and loose aggregates of OM are likely reworked biofilms or mats produced by benthic or adhesive microbes (e.g., Tice and Lowe, 2006) (Figures 7.1) (*Column*). Cherts enriched in OM are often called carbonaceous cherts, which are the best targets for search for cellularly preserved biosignatures. Some of them, if not all, were formed by early silicification of microbial mats (e.g., Tice and Lowe, 2004, 2006). Early silicification does not appear to replace OM but rather may play an important role for its preservation (e.g., Alleon et al., 2018). During later silicification, on the other hand, OM appears to have been remobilized, being bleached or stained with iron oxides (ferruginization) (Sugitani et al., 2007).

DOI: 10.1201/9780367855208-7

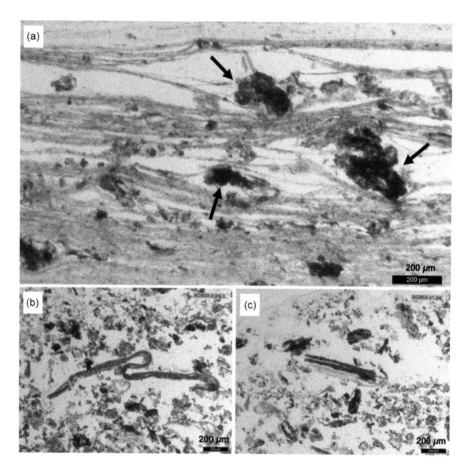

FIGURE 7.1 Occurrences of organic matter (OM) in Archean cherts. All from the ca. 3.4 Ga-old Buck Reef Chert, Barberton greenstone belt, South Africa. (a) Mat-like laminations trapping carbonaceous grains (the arrows). (b) Reworked biofilm (the center) and surrounding loose carbonaceous grains. Reworking had occurred before consolidation. (c) Probable reworked microbial mat (the center). Fragmentation and reworking occurred after consolidation.

Numerous carbon isotopic data ($\delta^{13}C_{PDB}$: Equation 4.8) of OM have been reported from various localities in the Archean cratons. Most of them are lighter than −20‰. These values, if recorded from the Neoarchean and later may readily be taken as evidence for biogenicity (Figure 7.2). However, those in Archean, particularly Paleoarchean, cannot always be interpreted as biogenic. McCollom and Seewald (2006) experimentally demonstrated that large fractionation of carbon isotopes could occur during Fischer-Tropsch-type (FTT) hydrothermal reactions. Formic acid (HCOOH) was reacted with metallic Fe at 250°C and 325 bars, producing methane and a series of C_2-C_6 hydrocarbons. All the organic products, with a few exceptions, had $\delta^{13}C_{PDB}$ values around −50.5‰. Based on

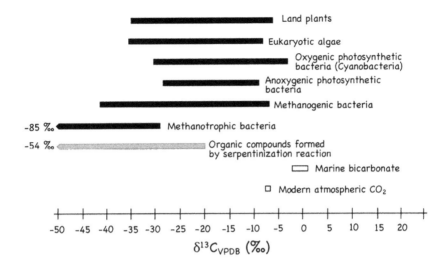

FIGURE 7.2 Carbon isotopic values of phtoautotrophs, chemoautotroph (methanogens), and chemoheterotroph (methanotrophs), with inorganic carbons. $\delta^{13}C_{sample} = \{(^{13}C/^{12}C)_{sample}/(^{13}C/^{12}C)_{standard} - 1\} \times 1000$ (‰) (Standrad = PDB: Pee Dee Belemnite as standard). Modified from Planavsky et al. (2011).

this result, the authors argued that carbon isotopic composition cannot be a diagnostic tool for identifying biogenic OM in ancient rocks. This claim is related with occurrences of OM in the Archean rocks. Black cherts rich in OM often occur as veins and dikes plausibly hydrothermal in origin. Lindsay et al. (2005), for example, showed that the black chert dikes (dike complex) underlying the ca 3.4 Ga-old Strelley Pool Formation (SPF) in the Pilbara Craton extended to "a depth of more than 2 km below the Archean paleo-seafloor". It was suggested that the dike complex was formed in relation with intrusion of granitoids. As mineralogy and geochemistry including isotopic values of OM of dikes and those of the overlying "sedimentary" SPF succession were parallel to each other, the authors casted strong doubt on biogenic origins of OM in this formation. Another negative example comes from West Greenland. Graphite (crystalline mineral composed exclusively of carbon) enriched in ^{12}C ($\delta^{13}C_{PDB} = -20$ ‰ to -50 ‰) in the ~3.8 Ga-old metamorphosed rocks from the Isua supracrustal belt (ISB) was once argued as the oldest signs of life on the Earth (Mojzsis et al., 1996). However, this discovery was later criticized by several groups. One of them led by van Zuilen identified a close association of graphite with siderite ($FeCO_3$), and magnetite (Fe_3O_4) in the ISB metacarbonate veins (van Zuilen et al., 2002). This association was interpreted to be a product of thermal disproportionation of carbonates (Equation 7.1) (van Zuilen et al., 2002). This reaction was likely induced by the metamorphic event of the ISB (the amphibolite facies: Figure 2.10). Although carbon isotopic compositions ($\delta_{13}C_{PDB}$) of graphite grains are -12‰ to -10‰, significantly heavier than previously reported values, this finding provides us with a cautious point of view on origin of graphite and precursory OM in ancient, particularly high-grade metamorphic rocks.

$$6FeCO_3 \rightarrow 2Fe_3O_4 + 5CO_2 + C \tag{7.1}$$

Tashiro et al. (2017) described graphite grains in metasedimentary rocks from the Saglek Block in northern Labrador, Canada. Syngenecity of the graphite grains was demonstrated by, e.g., their parallel occurrence to the bedding plane and Raman spectra consistent with the metamorphic temperature. FTT-synthesis via. siderite decomposition was excluded by the lack of siderite and magnetite in the host metasedimentary rocks. Premetamorphic $\delta^{13}C_{PDB}$ values of graphite were calculated to be lighter than −28.2 ‰, showing at least −25.6 ‰ fractionation from inorganic carbon source. It was suggested that the graphite grains were syngenetic and biogenic. The age of OM bearing metasedimentary rock was claimed to be over 3.95 Ga-old (the oldest evidence for life, if true) by Tashiro et al. (2017) and references therein. However, this age was later questioned from the geological point of view by Whitehouse et al. (2019).

In addition to these negative stories, positive ones are introduced. Organic matter in chert from the ~3.5 Ga-old Dresser Formation in the Pilbara Craton was analyzed by Morag et al. (2016) using secondary ion mass spectrometry (SIMS). The authors revealed that carbon isotopic compositions were specific to various microstructures, which were attributed to primary isotopic variability of OM. Such texture-specific carbon isotopic variations have been identified also in OM comprising microstructures of various morphologies in the ca. 3.4 Ga-old Strelley Pool Formation (Lepot et al., 2013), excluding the possibility of abiogenic origin of the OM. Biogenic origin of 3.7 Ga-old graphite from the ISB argued by Ohtomo et al. (2014) and Rosing (1999) is also promising. Ohtomo et al. (2014) analyzed graphite particles in the metasediments and those in hydrothermal vein, the latter of which represent abiogenic origin. In addition to isotopic analyses, high-resolution transmission electron microscope (HRTEM) analysis was performed. Carbon isotopic compositions ($\delta_{13}C_{PDB}$) of the graphite particles in the metasediments (ca. −17 ‰ on average) are significantly lighter than those in veins; the values are within the range of biogenic OM (Figure 7.2). Furthermore, the graphite particles are structurally disordered, contrastive to ordered structure of graphite in veins. The authors emphasized that such structurally disordered graphite particles were derived from complex precursory OM produced exclusively by organisms.

7.2.2 Hydrocarbons and Others – Archean Oils

Methane (CH_4), a representative hydrocarbon, is known to be produced by archaeal methanogen under a strictly anaerobic environment, although some cyanobacteria also could produce methane (Chapter 4). To my knowledge, carbon isotopic data for cyanobacterial CH_4 have not yet been described, while it is well-known that archaeal methane could be characterized by very light values of $\delta^{13}C_{PDB}$ (~ −40 ‰)=−40 to −60 ‰) (Figure 7.2). Based on this, Ueno et al. (2006a) argued that CH_4 with $\delta^{13}C_{PDB}$ of −56 to −36 ‰ contained in fluid inclusions in chert dyke of the ca. 3.5

Ga-old Dresser Formation in the Pilbara Craton, Western Australia was produced by methanogen. Namely the onset of methanogenesis could be back to ca. 3.5 Ga ago. This claim is criticized by Sherwood and McCollom (2006), saying "… an alternative, abiotic origin for the methane is equally plausible." The alternative, abiotic synthesis of isotopically light CH_4 was suggested to have been FTT reaction catalyzed by metals such as Fe-Ni alloy. In reply to this comment, Ueno et al. (2006b) emphasized that in more than 300 petrographic thin sections, they did not find any native metals but found Fe and/or Ni-bearing sulfides; also see Etiope and Sherwood (2013) for details of abiotic methane formation on the Earth.

Given that carbon isotopic compositions are not enough to argue the biogenicity of OM in Paleo- and even Mesoarchean rocks, how could we place constraints on origins of OM? Identification of molecular structures specific to compounds originated from biomolecules is one of promising approaches. Indeed, some positive results had been obtained from Paleoarchean kerogens in the Pilbara Craton. Derenne et al. (2008), for example, identified long-chain aliphatic hydrocarbons characterized by odd-over-even number predominance from kerogens extracted from carbonaceous cherts of the ca. 3.5 Ga-old Dresser Formation and concluded that they were biological in origin. De Gregorio et al. (2009) also analyzed macromolecular structure, carbon bonding, and functional group chemistry of kerogen from the ca. 3.5 Ga-old Apex chert in the Pilbara Craton, Western Australia from which highly controversial cell-like structures have been reported (Schopf, 1993) (Chapter 8), using transmission electron microscopy (TEM), synchrotron-based scanning-transmission X-ray microscopy (STXM), and SIMS. The authors suggested biogenicity of the Apex OM, based on that these signatures were similar to those identified in kerogen form the ca. 1.9 Ga-old Gunflint Chert well known to contain numerous genuine microfossils (Barghoorn and Tyler, 1965; Cloud, 1965). A similar approach was applied directly to in-situ fossil-like microstructures in the ca. 3.4 Ga.-old Strelley Pool Formation (Alleon et al., 2018): the results will be described in Chapter 11.

Lipids such as hopane and steranes were once expected to be promising molecular biosignatures (biomarker) that could place constraint on the timing of evolution of cyanobacteria and eukaryotes (Brocks et al., 1999). Steranes (cyclopentanoperhydrophenanthrene) are derived from steroids that comprise cell membrane of eukaryotic organisms. Hopane forms the central core of hopanoids that again comprise cell membranes of various phyla. One of the hopanes, 2-alpha-methylhopanes, was once considered to be derived only from cyanobacterial membrane, although it is now widely accepted that this biomarker could be derived also from other bacteria (Ricci et al., 2014 and references therein). Brocks et al. (1999) suggested the emergence of eukaryotes at least in the Neoarchean, based on detection of steranes from the 2.7 Ga-old shales in the Pilbara Craton, Western Austraila. However, this discovery was later suspected to be a result of contamination (Rasmussen et al., 2008; Brocks, 2011). French et al. (2015) analyzed cores from the Pilbara Craton that were drilled with a strict contamination control and demonstrated that the amount of detected lipid biomarkers could not be indistinguishable from the background level. This can be attributed to thermal over-maturation of OM. However, such results do not always categorically deny the future detection of indigenous biomarkers from Archean rocks, considering the presence of "oil"-bearing fluid inclusions in Archean (~3.0 Ga-old) sandstones (Dutkiewicz et al., 1998) and that thermal degradation of biomolecules could have been inhibited by early entombment within silica (Alleon, et al., 2016, 2018).

7.3   PYRITE AND SULFUR

In modern marine anoxic environments deficient in O_2, nitrate (NO_3^-), tetravalent manganese (Mn^{4+}) oxide and ferric (Fe^{3+}) oxides, sulfate (SO_4^{2-})-reducing bacteria (SRB) metabolize simple OM such as acetate ion (CH_3COO^-) (Chapter 4). In this process, SO_4^{2-} as an electron acceptor is reduced to be sulfide (H_2S), which reacts with ferrous iron (Fe^{2+}) to eventually form pyrite (FeS_2), although this process is actually more complicated. Stable isotopes of sulfur ($^{32}S, ^{33}S, ^{34}S, ^{36}S$) fractionate during this process, and fractionation between ^{34}S and ^{32}S has extensively been studied. As SRB prefers $^{32}SO_4^{2-}$ relative to $^{34}SO_4^{2-}$, the ratio of $^{34}S/^{32}S$ in pyrite tends to be smaller than that of residual SO_4^{2-} (e.g., Habicht and Canfield, 1997). It may be also noted that if this process proceeds in a closed system and primarily contained SO_4^{2-} is consumed completely, it appears that no fractionation virtually occurs. Thus, no fractionation or small fractionation of isotopes relative to parental SO_4^{2-} does not always mean no biological sulfate reduction. Additionally, we need to know sulfur isotopic compositions of SO_4^{2-} in ancient seawater or fluid in which sulfides formed. Sulfur isotopic compositions of barite ($BaSO_4$) often closely associated with pyrite could be regarded to represent those of ambient fluids including seawater.

There have been many data of sulfur isotopes on Archean pyrite. In particular, data from the 3.5 Ga-old Dresser Formation in the Pilbara Craton, Western Australia, where massive barite ($BaSO_4$) deposits containing pyrite grains occur as beds and dykes, are extensive (Figure 7.3a). In their pioneer work, Shen et al. (2001) reported a wide range of $\delta^{34}S_{VCDT}$ values with fractionations over 20‰ between microscopic pyrite grains and associated barite and suggested the early emergence of SRB (Figure 7.3b). The recent detailed studies on the Dresser materials employing analyses of quadruple sulfur isotopes ($^{32}S, ^{33}S, ^{34}S, ^{36}S$) basically support this pioneer work. Furthermore, the results suggest more complex biologically mediated sulfur cycle in the Paleoarchean, which included disproportionation of elemental sulfur (e.g., Philippot et al., 2007; Wacey et al., 2015). From the ca. 3.4 Ga-old the Strelley Pool Formation, Bontognali et al. (2012) also reported evidence for microbial disproportionation of elemental sulfur. Microbial disproportionation of elemental sulfur is a chemolithoautotrophic process that produces SO_4^{2-} and H_2S (Equation 4.10). In addition to elemental sulfur thiosulfate ($S_2O_3^{2-}$) and sulfite (SO_3^{2-}) were also utilized as both electron donor and acceptor (Finster, 2008).

Pyrite produced by SRB often occurs as a cluster called framboidal pyrite. Framboidal pyrite is spheroidal to subspheroidal cluster composed of numerous tiny euhedral pyrite grains (Ohfuji and Rickard, 2005). Nonetheless, only framboidal occurrence appears to be weak as evidence for biogenic origin of pyrite in the Archean. If framboidal occurrence is associated with negative and variable $\delta^{34}S$

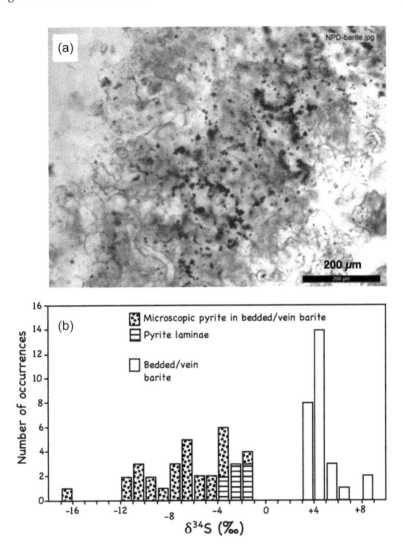

FIGURE 7.3   (a) Occurrence of pyrite micrograins in barite, the ca. 3.5 Ga-old Dresser Formation in the Pilbara Craton, Western Australia. (b) Sulfur isotopic compositions of pyrite and associated barite from the Dresser Formation. (Modified from Figure 1 of Shen et al. (2001).) Note that pyrite and barite in (a) are not identical to those analyzed by Shen et al. (2001). $\delta^{34}S_{sample} = \{(^{34}S/^{32}S)_{sample}/(^{34}S/^{32}S)_{VCDT} - 1\} \times 1000$ (‰) (VCDT = Vienna Cañon Diablo Troilite as standard).

values, their biological origin could be suggested. Though not common, plausible biogenic framboidal pyrite has been identified from the Paleoarchean sedimentary successions, exemplified by the Strelley Pool Formation (Williford et al., 2015; Duda et al., 2016). In addition, Wacey et al. (2011) suggested the early evolution of microbially mediated pyrite oxidation based on the detailed observation of detrital pyrite from the Strelley Pool Formation, which had revealed the presence of laminated carbonaceous coatings with local enrichment of N and of various surface alteration features such as spherical, chained, or channeled pits.

7.4 SEDIMENTARY STRUCTURES AND DEPOSITS

Various types of sedimentary structures have been recognized and explored for searching ancient life, including microbially induced sedimentary structure (MISS), microbialites including stromatolites, and sinters. In this section, MISS and stromatolites are focused and described in detail. Microbialites and sinters are described briefly here. Microbialites are defined as "organosedimentary deposits formed through interaction between benthic microbial communities and detrital or chemical sediments (Burne and Moore, 1987)". Stromatolite comprises microbialites together with thombolite amd leiolite, the latter two of which are characterized by peloidal fabric and no fabric, respectively (Flügel, 2010). Despite the definition, microbialites have been applied to carbonate sediments. Sinters refer to deposits precipitated from hot springs and geysers (terrestrial hydrothermal systems). They are either calcareous or siliceous. They were once considered to be purely abiogenic precipitates due to cooling and evaporation of hot springs and resultant oversaturation with respect to carbonate and silica (Walter, 1976). Recently, however, it becomes widely realized that microbial activities are involved in their formations. Particularly, geyserite (a type of siliceous sinter), has recently been paid special attention to in the context of ancient record of life on land and even Mars (Djokic et al., 2017; Ruff and Farmer, 2016) (also see *Column* in Chapter 10).

7.4.1 Microbially-Induced Sedimentary Structures

Microbially induced sedimentary structures (MISS) are sedimentary structures indirectly created by microbial mat (biofilm). There have been some confusions about the difference between MISS and stromatolite. Noffke and Awramik (2013) tried to give an answer to this confusion, saying "Binding, biostabilization, baffling, and trapping of sediment particles by microorganisms result in the formation of MISS: however, if carbonate precipitation occurs in EPS (Extracellular Polymeric Substances by K.S), and these processes happen in a repetitive manner, a multilayered build-up can form—stromatolites". This interpretation may still lacks precision in definition of stromatolite, because some well-described ancient and modern stromatolites are not associated with carbonate precipitation. In order to avoid confusion, it may be emphasized that MISS represent structures formed on the surface of siliciclastic sediments and therefore could be observed on the bedding surface (Noffke and Awramik, 2013). Originally loose and unconsolidated siliciclastic sediments could be stabilized by microbial mats, producing various MISS. Noffke (2010) identified 17 main types of MISS from a wide variety of modern environments including shelves, tidal flats, lagoons, sabkhas, and so on. Types of MISS are classified into four end members based on controlling factors such as growth, biostabilization, binding, and

FIGURE 7.4 Modern microbially induced sedimentary structures (MISS) from Tunisian coast of Mediterranean Sea (Lakhdar et al., 2020). (a) Roll-ups resulting from desiccation. (b) Dome produced by respiratory gas production below the microbial mat. (Photographs courtesy by R. Lakhdar.)

baffling/trapping. Roll-ups, mat chips, shrinkage cracks, mat fabrics, gas domes, polygonal cracks, and erosional remnants and pockets could be representative for MISS (Figure 7.4).

Archean MISS were first described from the 2.9 Ga-old Mozaan Group, South Africa (Noffke et al., 2003), and the oldest one is from the ca. 3.5 Ga-old Dresser Formation in the Pilbara Craton, Western Australia (Noffke et al., 2013) (Figure 7.5a). Noffke et al. (2013) identified polygonal cracks and gas domes, erosional remnants and pockets, and mat chips as macroscopic MISS and tufts, and laminae fabrics characterized by primary OM and trapped and bound grains as microscopic MISS. These MISS fabrics are interpreted to have been formed at the subtidal to supratidal zones and lagoon of coastal sabkha. Although the formation of modern coastal MISS is closely related to cyanobacterial communities, it is equivocal whether microbes involved in the formation of the Dresser MISS were cyanobacteria. Anoxygenic

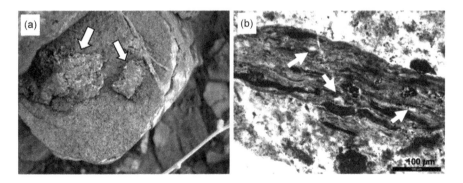

FIGURE 7.5 (a) Possible mat chips (arrowed, extensively silicified), from the ca. 3.5 Ga-old Dresser Formation in the Pilbara Craton. (b) Fragment of laminated carbonaceous cherts with detrital grains (arrows) from the ca. 3.5 Ga-old Apex chert in the Pilbara Craton, equivalent to those interpreted as MISS by Hickman-Lewis et al. (2016). (Photograph (a) courtesy of Nora Noffke.)

phototrophs and chemoautotrophs such as *Chloroflexus* and *Begiatoa* (Bailey et al., 2009) cannot be ruled out from MISS builders (Noffke et al., 2013). In contrast to the Kaapvaal Craton, South Africa (e.g., Homann et al., 2015, Noffke, 2010 and reference therein), distinct MISS occurrence is rare in the Meso- and Paleoarchean successions in the Pilbara Craton. Hickman-Lewis et al. (2016) reported another example in the stratiform, sedimentary cherts conformably inter-bedded with the Apex Basalt (informally known as the 'Apex chert') at the Chinaman Creek Locality (Figure 7.5b). Also, Noffke (2015) described MISS-like sedimentary structures in the <3.7 Ga-old Gillespie Lake Member on Mars. However, as cautiously noted by Davies et al. (2016), interpretation of some sedimentary structures, particularly in the Paleoarchean and in other planets, as MISS may require great circumspection.

7.4.2 Stromatolites

7.4.2.1 What are Stromatolites?

Stromatolite is one of the most popular scientific terms in geobiology and astrobiology. We would find lots of sites introducing this term on the Internet. Unfortunately, it is difficult to find sites that take exact and updated definition of stromatolite. In Wikipedia, for example, stromatolites are interpreted as "layered mounds, columns, and sheet-like sedimentary rocks that were originally formed by the growth of layer upon layer of cyanobacteria, a single-celled photosynthesizing microbe." However, 16S rDNA-based phylogenetic studies and fluorescene in situ hybridization (FISH) on living stromatolites have revealed their microbial diversity. For example, Goh et al. (2009) showed that at the sequence level, representative extant stromatolites at Shark Bay, Western Australia (Figure 7.6) were dominated by α- and γ-proteobacteria (58%), which were five times of cyanobacteria (11%). While cyanobacteria dominated the surface of the stromatolite, archaea and sulfate-reducing bacteria were found to do subsurface. More than 30 years ago, Awramik and Riding (1988) showed that subtidal columnar stromatolites at Shark Bay contained significant amounts of motile diatoms and argued that these eukaryotic algae played an important role in the formation and maintenance of these stromatolites. There appears to be general image that filamentous cyanobacteria are main constructors of stromatolites, and indeed recently discovered conical stromatolites in freshwater lake in Antarctica are constructed by filamentous cyanobacteria (Andersen et al., 2011). On the other hand, Suosaari et al. (2016) demonstrated that coccoid cyanobacteria predominated in microbial community comprising lithified discrete (columnar) stromatolites at Shark Bay, whereas filamentous cyanobacteria predominated in sheet-like mats. Quite diverse in microbial communities are indeed involved in the stromatolite formation.

Malcolm R. Walter wrote nearly 40 years ago "stromatolites are structures formed by petrified, biogenic sediments. It is a structure in which lamina is formed and grown by precipitation of carbonate minerals and entrapment of particles under direct influence or active secretory activity." "(Walter, 1976). It may be emphasized that M.R. Walter did not mention any morphological features. So diverse morphological types including flat laminated ones, in addition to popular columnar and conical ones,

FIGURE 7.6 (a) Photograph of modern stromatolites in Hamelin Pool, Shark Bay, Western Australia. Heights of the stromatolites are approximately 40 cm. (b) A closer view.

have been described as stromatolites. Thus, it is difficult or does not make sense to discriminate microbial mats and stromatolites. A more recent definition provided by McLoughlin (2011) says

> Stromatolites are morphologically circumscribed accretionary growth structures with a primary lamination that is, or may be, biogenic. They form centimeter- to decimeter-scale domes, cones, columns, and planiform surfaces made of carbonate layers. Stromatolites accrete through a combination of microbially mediated sediment trapping-and-binding and by the precipitation of carbonate crusts that may be due to microbial mat growth and/or be purely abiotic in origin.

As described later, the widely accepted oldest stromatolite is described from the ca. 3.4 Ga-old Strelley Pool Formation, Western Australia. Since then, stromatolites have diversified and prevailed, but dropped sharply ca. 0.6 Ga ago, possibly due to evolution of grazing organisms (Walter and Heys, 1985). Stromatolites had been once proposed as index fossils for stratigraphic correlation (Bertrand-Sarfati and Walter, 1981). Grotzinger and Knoll (1999) also noted that Archean and Paleoproterozoic stromatolites were formed largely by in-situ precipitation of laminae, whereas younger Proterozoic stromatolites by microbial accretion (trapping and binding) of carbonate sediments, reflecting long-term change of ocean chemistry. Modern and ancient conical stromatolites are characterized by regular spacing between neighboring structures and are devoid of sediment particles. Petroff et al. (2010, 2013), based on field, laboratory, and theoretical studies, suggested that the spacing resulted from competition for nutrient between neighboring structures and that diffusive gradient within the microbial mats accelerated mineral precipitation in high curvature regions, resulting in the development of conical shape.

7.4.2.2 Skepticisms to Archean Stromatolites

Although the existence of fossil stromatolites has been known from more than 100 years ago, there have been considerable debates as to whether or not they were of biological origin. This issue has been particularly serious for Archean stromatolites, because Archean stromatolites, like many other younger equivalents, are composed exclusively of carbonate minerals such as dolomite [$CaMg (CO_3)_2$] and if silicified, microcrystalline quartz. In either case, most of them are devoid of or contain very trace amount of OM. Carbonate precipitation is thought to have been induced by heterotrophic degradation of OM comprising stromatolite builders, which increased alkalinity and triggered carbonate precipitation (e.g., Duparz et al., 2009). Therefore, the absence of microfossils and even OM is not surprising. Furthermore, recrystallization of primary microcrystalline carbonates to equant grains likely had destructed cellular microfossils. However, some researchers have tried to interpret the stromatolite formation in abiotic processes. Grotzinger and Rothman (1996) proposed a numerical model for abiogenic molphogenesis of stromatolites, which combined chemical precipitation, contribution of suspended sediments, and uncorrelated random noise. This was based on detailed analyses of the 1.9 Ga-old peak-shaped stromatolites from the Cowles Lake Formation, Canada. McLoughlin et al. (2008) also performed experiment of spray colloid deposition, and demonstrated that structures mimicking stromatolites could form abiotically. These facts should not be overlooked but do not always mean that all the described stromatolites are products of abiogenic processes. Also see Awramik and Grey (2005) for the definition of stromatolites and the textural criteria for their biogenicity. In the following, I would like to review two examples of Paleoarchean stromatolites, including those from the Strelley Pool and the Dresser formations in the Pilbara Craton, Western Australia, although the older putatuve stromatolite was reported by Numan et al. (2016).

Nutman et al. (2016) described conical and domal structures from a newly exposed outcrop of the 3.7 Ga-old rocks in the ISB, West Greenland and interpreted them as the oldest stromatolites. This interpretation has been refuted by Allwood et al. (2018), based on results from their own field trip to ISB. A. Nutman and his colloborators contended against this challenge (Nutman et al., 2019). Many of the discussions appear to be parallel, which is partly attributed to that both teams did not examine the same site: Nutman says "… Nutman et al. (2016), focusing on the lower deformation northeastern end, and Allwood et al. (2018) <4 m away on its southwestern end, in a more highly deformed and altered part of it". This is a serious problem and the ISB is too far for many researchers interested in this issue to visit.

7.4.2.3 The Oldest Stromatolites? – The 3.4 Ga-old Strelley Pool Formation

Stromatolites in the ca. 3.4 Ga-old Strelley Pool Formation, Western Australia have been extensively studied, and their biogenicity is now widely accepted (Figure 7.7). However, the settlement of arguments on their biogenicity took some time through twist and turns. The first report of the SPF stromatolites was made by Lowe (1980), who described stromatolites morphologically similar to *Conophyton* common type

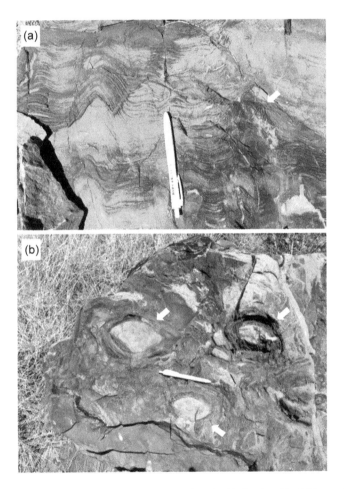

FIGURE 7.7 Conical stromatolites in the ca. 3.4 G-old Strelley Pool Formation at the Trendall Locality in the Panorama greenstone belt, North Pole region, in the Pilbara Craron. (a) Side view perpendicular to the bedding plane. (b) Cross view at the horizontal plane. Three concentric ovoids can be seen at both sides of the pen.

in the Proterozoic, from the East Strelley greenstone belt. This was followed by the discovery of new SPF stromatolite locality in the Panorama greenstone belt in 1984. The locality, well known for spectacular coniform and pseudocolumnar stromatolites, is called Trendall Locality in honor of the discoverer, A.F. Trendall. In 1994, on the other hand, D.R. Lowe (above) revised his own view and suggested abiogenic origins of some of the described Archean stromatolites older than 3.2 Ga (Lowe, 1994). Suggested potential inorganic processes constructing "stromatolites" include evaporitic precipitation and soft-sediment deformation, which caused controversy on biogenicity of Archean stromatolites (e.g., Lowe, 1995; Buick et al., 1995). Such situation has probably been changed by detailed sedimentological study of the SPF stromatolites by Hofmann et al. (1999). The examined SPF stromatolites were composed of conical and branched pseudocolumns of centimeter width and decimeter height, respectively. The pseudocolumns were composed of first-order uniform conical laminae, modified with second-order low-amplitude lamination that was corrugate and obliquely stacked. Pseudocolumns were laterally connected with spaces. Key points of their discussions related to biogenicity of the stromatolites are as follows (Hofmann et al., 1999):

1. First-order laminae comprising conical pseudocolumns and interspaces are distinct with each other. The former is much more uniform than the latter, indicating that they were formed by different processes.
2. Second-order corrugate laminae are stacked obliquely, and the convexity of these laminae is attenuated upward, suggesting their formation by upward accretion.
3. Laminae are continuous across adjacent structures of different architectures, inconsistent with being originated by strict chemical precipitation.
4. Some structures identified at the outcrop are similar to well-known Proterozoic stromatolites of plausible biogenic origin.
5. Coniform structures with the common habits and occurrences can be identified over tens of square kilometers in the Pilbara Craton and at the same time occur exclusively within specific stratigraphic unit.
6. Degree of the cone slopes are steep, higher than 40°, up to 75°. This angle does not allow accumulation of loose particles, suggesting that the cone growth is attributed to mineral precipitation.

In summary, it is unlikely that the conical pseudocolumnar structures were formed by deformation or by purely physical sedimentation processes. This work has been followed by detailed study conducted by A. Allwood and her collaborators (Allwood et al., 2006, 2009). They carefully studied the SPF stromatolites over several kilometers in the Panorama greenstone belt and described seven morphotypes, many of which were not previously described, and correlative to stromatolite taxa in younger ages. Stromatolite occurrence was restricted to the sedimentary facies representing peritidal marine carbonate platform or reef. It was poor in the facies of hydrothermal precipitates or clastic deposition and even absent in the facies of deeper marine settings. The authors also emphasized that within the platformal facies, different morphotypes of stromatolites occurred together in a small (centimeter scale) area. As physical and chemical conditions were likely uniform within such a small area, co-occurrence of morphologically diverse stromatolites requires microbial meditation. The biological origin of the SPF stromatolites is well supported by later inorganic and organic geochemical studies (Allwood et al., 2010, Wacey, 2010, Bontognali et al., 2012, Flannery et al., 2018).

Finally, I would like to briefly describe new occurrence of genuine stromatolite from the Strelley Pool Formation in the Goldsworthy greenstone belt, from which morphologically diverse microfossils have been discovered as described in detail later (Chapter 10). The new SPF stromatolites was identified from a black chert layer several cm thick interbedded with light-toned bedded to laminated cherts. The structures were not identified at the outcrop, whereas columnar structures composed of very fine black lamina appeared (Figure 7.8) when the hand specimen was cut perpendicular to the bedding plane. The columnar structure is 2 cm wide at the base and 4 cm at the apex and appears to have grown upwards and branching. A small conical structure was also found. The depressions between these structures are filled with sand grains. The sand grains were largely sedimentary rock fragments. A close observation of organic laminae shows that some parts have a fine granular structure or fibrous structure of less than 1 μm. Analysis of the carbon isotopes of OM showed light, biogenic values ($\delta^{13}C_{PDB}$ = ~ −27‰) (Sugitani et al., 2015).

The siliceous stromatolites described from the Goldsworthy greenstone belt are characterized by significant trace element characteristics that are distinct from the other SPF carbonate stromatolites and associated cherts in the Panorama greenstone belt. The latter is characterized by enrichment in heavy rare-earth elements (HREEs) relative to post-Archean Australian Shale Composite (PAAS), suggesting formation in marine environment (e.g., Van Kranendonk et al., 2003; Allwood et al., 2010). The siliceous stromatolites, on the other hand, exhibit the enrichment in light REEs (LREEs). This discrepancy cannot be attributed to lithological difference (chert vs. carbonate) (Allwood et al., 2010). Silicification may have altered primary REE signatures to some degrees (e.g., Bonnand et al., 2020), but cannot explain the observed differences. It is more likely that the siliceous stromatolites were not formed in the open marine setting. As they were probably silicified contemporaneously or soon after their formation, the water mass in which the stromatolites had grown should have been enriched in silica, probably sourced from hydrothermal activities (Sugitani et al., 2015).

7.4.2.4 The Oldest Stromatolites? – The 3.5 Ga Dresser Formation

In early 1980s, stromatolitic structures were already identified in North Pole Chert Member of the ca. 3.5 Ga-old Dresser Formation in the Pilbara Craton, Western Australia (Walter et al., 1980; Buick et al., 1981), representing Paleoarchean shallow marine to subaerial sedimentary succession with strong influence of hydrothermal activity (e.g., Buick and Dunlop, 1990; Van Kranendonk et al., 2008) (Figure 7.9).

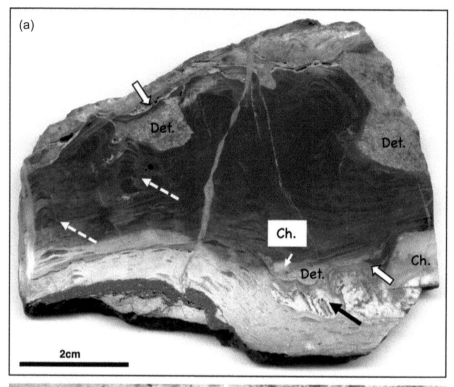

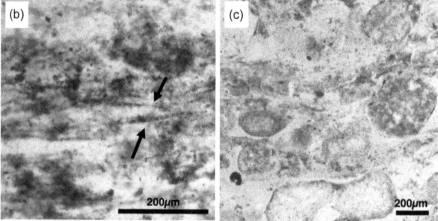

FIGURE 7.8 Cherty stromatolite preserving carbonaceous laminae from the ca. 3.4 Ga-old Strelley Pool Formation in the Goldsworthy greenstone belt, Pilbara Craton. This lithology was previously described as finely laminated carbonaceous chert. (a) Polished slab. Det.: layer or portion composed of detrital materials, Ch.: chert clast, the thick white arrow: carbonaceous laminae bridging two columns (above) and draping chert clast (below), the black arrow: molds of acicular (or platy) crystals, the dashed white arrow: internal columnar structures. (b) Blurred possible filamentous structures (the arrows). (c) Trapped detrital grains.

FIGURE 7.9   Domal stromatolite in the ca. 3.5 Ga-old Dresser Formation at the North Pole Dome in the Panorama greenstone belt, Pilbara Craton. Extensively silicified. The height is approximately 5 cm.

Walter et al. (1980) described a single domal structure, and Bucik et al. (1981) did convex structures and wrinkly lamination that were interpreted to be probable pseudocolumnar and stratiform stromatolites, respectively. Here we focus on the former example. The structure was a dome 20 cm high and 25 cm wide that protruded from a wavy laminated chert bed. Smaller wavy laminated domical structures were subsequently identified in slab made from the collected block specimen. These structures, now composed of dolomitic cherts, were thought to have been primary and sedimentary, because their relief is independent from the overlying and the adjacent beds. For example, these undeformed beds terminated at the sidewall of the structures or draped over them. The structures displayed internal wavy and wrinkle laminae of a few hundreds to dozens of micrometers in thickness. Based on these features, Walter et al. (1980) argued their biogenicity.

The Pilbara Drilling Project led by M.J. Van Kranendonk and P. Phillipot has brought breakthrough in the North Pole stromatolite study. In this project performed in 2004, drill cores unaffected by surface weathering have been obtained. Small (<1 cm) columnar structures were identified in chert-dolomite $[CaMg(CO_3)_2]$ horizon. The structures were shaped upward-broadening and/or-branching and were composed of fine wavy laminations. At margins of the columns, some laminae were steeply inclined concordant with sidewalls of the columns, whereas at flexure crests, some laminae were thickened. Individual neighboring columns were often bridged by laminae. Interestingly, the structures are composed largely by pyrite, with lesser amounts of sphalerite (ZnS), galena (PbS), dolomite[$(Ca, Mg)_2CO_3$], and chert. However, among this mineral assemblage, only dolomite was considered to have been primary phase. These sedimentological and mineralogical features are consistent with the interpretation of these structures as stromatolites

(Van Kranendonk, 2011, Van Kranendonk et al., 2008). This has been strengthened by the discovery of nano-porous pyrite that contains nitrogen-bearing OM with a $\delta^{13}C_{PDB}$ value of $-29.6 \pm 0.3\text{‰}$ and complexly twisted carbonaceous filaments and strands less than 1 μm width. (Baumgartner et al., 2019). Formation of nano-porous pyrite requires abundant OM (Wacey et al., 2014), suggesting that the stromatolitic structures were once composed largely of OM. Filamentous structures thus could be remnants of exracellular polymeric substances (EPS), major components of biofilms.

7.5 ICHNOFOSSILS IN VOLCANIC ROCKS

Endoliths represent organisms living inside various substrates such as rocks, minerals, corals, and sediments. They are not limited to prokaryotes but involve eukaryotes such as fungi, lichen, alga, and protozoa. Many of the endoliths are extremophiles that can tolerate or rather favor extreme environments generally unsuitable for life such as repeated desiccation, intense radiation of ultraviolet rays, shortage of nutrients, and boiling and freezing temperatures of water. Due to this feature, it is often suggested that this type of organisms could have first colonized on the early Earth and other planets (Ivarsson et al., 2019 and references therein). Endoliths are further classified into three types, including chasmoendolith, cryptoendolith, and euendolith. Chasmoendolith colonizes fissures and cracks of substrates, whereas cryptoendolith is defined as to colonize "structural" cavities, which are for example pore spaces of sandstones and vesicles in volcanic glasses. Contrastive to these two passive types, euendolith is active. It digs a tunnel into solid substrates by excreting chemicals to dissolve substrates such as carbonates and volcanic glasses. Here we focus on microtubular structures formed by euendolith in volcanic glasses. Microbial corroded structures including microtubes have been identified from many Cenozoic basaltic glasses in the oceanic crust. Their biological origin has been claimed based on colonial occurrence, an association with C and even nucleic acids, and light values of $\delta^{13}C_{PDB}$ values (Banerjee and Muehlenbachs, 2003 and references therein). The light $\delta^{13}C_{PDB}$ values were recorded also for carbonate minerals within basaltic glasses subjected to be microbial alteration, which was interpreted as involvement of HCO_3^- produced by decomposition of endoliths into carbonate precipitation. Such microtubular structures could be retained long after the decomposition of their builders and would be identified as "trace fossil" (Figure 7.10a, c).

Furnes et al. (2004) had first reported possible trace fossils produced by euendolith in Archean volcanic rocks (Figure 7.10b). The authors described micrometer-scale mineralized tubes in glassy rims of pillow lavas of the ~3.5 Ga-old Barberton greenstone belt, South Africa. The occasionally segmented tubes from 1 to 9 μm in width consist of titanite ($CaTiSiO_5$), a common metamorphic mineral. The tubes do not occur solitarily but comprise clusters radiating from healed fractures. The biogenic origin of the titanite tubes was argued based on similar occurrence and texture to microtubes in the Cenozoic oceanic crusts as described above and in the Cretaceous Troodos Ophiolite, Cyprus. An association of C with the microtubes

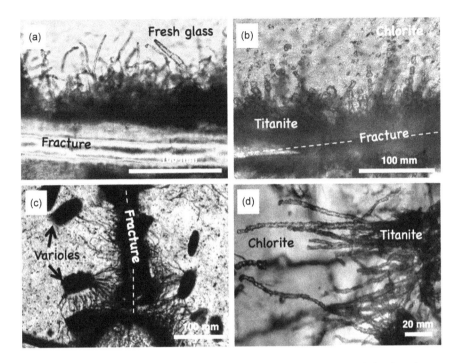

FIGURE 7.10 Microbial microtextures in fresh Cenozoic basalt glass and titanite microtextures in Archean pillow lavas metamorphosed to greenschist facies. (Modified from Staudigel et al. (2015), Figure 1). (a) Tubular biotextures in the 110-Ma-old Deep Sea Drilling Project (DSDP) sample 418A. (b) Tubular textures of the interpillow hyaloclastite from the ca. 3.5 Ga-old Hooggenoeg Formation, Barberton Greenstone Belt, South Africa. (c) Biologically generated filaments rooted in a fracture and dark brown varioles, within fresh basaltic glass (DSDP sample 418). (d) Segmented thin tubes extending from the edge of a fragmented glass now composed mainly of chlorite, from the ca. 3.4 Ga-old Euro Basalt, Pilbara Craton, Western Australia. Photograph courtesy by H. Staudigel.

was described, and disseminated carbonates in the glassy rims were characterized by isotopically light $\delta^{13}C_{PDB}$ values down to $-16.4‰$. Similar microstructures of titanite with residual organic matter have later been described from the nearly contemporaneous Euro Basalt in the Pilbara Craton (Banerjee et al., 2007) (Figure 7.10d).

The above-described Archean ichnofossils in volcanic rocks have later been criticized by N. McLoughlin and her collaborators. Grosch and McLoughlin (2014) casted a doubt on the biogenicity of Barberton microtubes in the following reasons. The age of titanite formation was determined to be ca. 2.8 Ga, which is consistent with the age of a local mafic sill but much younger than the host basalts. The occurrence of titanite microtubes was closely related to microdomains of the matrix of chlorite, alteration product of Fe- and Mg-bearing aluminosilicates. The authors suggested that the microtubes were formed related to later contact metamorphic event (retrograde metamorphism). In addition, wide morphological (size and shape) variation, lack of unique signatures such as spiraling or annulated tubes, and lack of biomarkers were emphasized.

Staudigel et al. (2015) counterargued Grosch and McLoughlin (2014). According to Staudigel et al. (2015), the retrograde metamorphic model for titanite formation had a series of weaknesses, including lack of observational and petrographic evidence for the peak temperature from which retrograde metamorphism proceeded and for the presence of precursory mineral phases such as pyroxene, quartz, and ilmenite ($FeTiO_3$). Additionally, it was noted that this alternative model could not be applied to other localities such as the Euro Basalt without experience of any contact metamorphic event. Also see Fisk et al. (2019), McCollom and Donaldson (2019), and McLoughlin et al. (2007) for further insights into the origins (both biotic and abiotic) of microgranular and microtubular structures in basaltic glasses.

7.6 SYSTEMATIC APPROACH TO BIOGENICITY ASSESSMENT OF CELL-LIKE STRUCTURES

The exploration of cellularly preserved fossils in the Earth's deep time was inspired by the discovery of microfossils from the ca. 1.9 Ga-old Gunflint Chert, Ontario, Canada (Cloud, 1965; Barghoorn and Tyler, 1965). Within 15 years later of this monumental discovery, the age of the oldest putative microfossils was dated back to 3.7 Ga, known as the east-like "microfossl" named *Isuasphaera isua,* from Isua, Greenland. As discussed in detail in the next chapter, this "microfossil" has been subjected to criticisms. Apart from the rightfulness of the criticisms, the controversy had led to the development of sophisticated approaches to microfossil analyses in the Archean as represented by Schopf and Walter (1983). The sophisticated approach involved multiple criteria, based on which biogenicty would be judged, following confirmation of Archean in age and excluding the possibility of contamination. Thus, the more the criteria are applied and satisfied, the more confidently the biogenicity can be claimed. Criteria for biogenicity should be basically tested by conventional methodologies, and it seems realistic to employ a series of prefixes such as "pseudo-", "dubio-", "possible-", "probable-", and "genuine-" in order to express the reliability of the biological origin of the object under consideration. The author does not take the "null-hypothesis" approach proposed by Brasier and coworkers (Brasier et al., 2002, 2005). In the following sections, the criteria integrated from the previous studies (Sugitani et al., 2007; Wacey, 2009 and references therein) are summarized, along with some new concepts. Also see Rouillard (2021) for the new approach for biogenicity assessment, alternative to the classical ones.

7.6.1 Geological Context

Geologic context requires that microstructures occur definitely in the Archean rocks, hopefully sedimentary rocks of low metamorphic grade, although biosignatures such as microtubes formed by euendolith could be preserved within volcanic rocks as discussed above.

It is desirable that the age of the host rock is determined directly. To my knowledge, unfortunately, there has been only a few examples of age determination of Archean sedimentary rocks (Minami et al., 1995). In general, the depositional age of Archean sedimentary rock is constrained by ages of overlying and/or underlying volcanic or volcaniclastic rocks. The age of detrital mineral grains represented by zircon contained in the sedimentary rock could place some constraints on the depositional age, although it should be cautioned that the given age is the maximum one. Even though any age data could not be provided, researchers may accept that the host rock is Archean in age, when it is unambiguously a part of well-studied Archean volcanic-sedimentary succession. In either case, it is essential that the host rock is not duricrust. Duricrusts are surface soil layers cemented by calcite (calcrete) and silica (silcrete), in the consequence of capillary action and evaporation. It is not difficult to demonstrate that the host rocks of microfossil-like structures are not duricrusts. A detailed description of lithostratigraphy of sedimentary successions containing host rocks and their petrography should be enough for this.

7.6.2 Syngenicity

In my former publications, indigenousness and syngenicity are proposed separately as criteria indicating that structures in question are neither contaminants nor later involved organisms such as endoliths (e.g., Sugitani et al., 2007). However, as syngenicty means that the structures are embedded within the primary mineral phase, but not within significantly later-formed pores, fractures, and veins, this criterion involves indigenousness in wider sense. Thus, these two criteria can be integrated into sygenicity. Oehler and Cady (2014) also describe the importance of syngenicity in detail, along with new techniques for assessing both biogenicity (as discussed below) and syngenicity.

Syngenicity is particularly important when microstructures are extracted from host rocks by acid maceration, which is an extraction method of organic-walled microfossils by HF-HCl digestion at room temperature. During this process, modern OM suspended in air such as pollens could contaminate samples. In other words, if structures under consideration are identified within petrographic thin sections, their indigenousness is basically guaranteed, although this does not always guarantee syngenicity. In order to confirm that the extracted objects are at least indigenousness, but not contaminants, Raman spectral analysis has often been used. Raman spectra of OM are sensitive to the degree of thermal maturation. With an increase in thermal maturation mainly due to metamorphism, OM becomes to be structurally ordered and eventually has a graphitic ordered structure. Potential contaminant for maceration products is modern OM, which is not yet subjected to thermal maturation. Such modern contaminated OM does not display any distinctive peaks in the Raman spectra. Organic matter in the Archean rocks, on the other hand, is thermally matured, resulting in the appearance in the spectrum specifically corresponding to graphite (G-band) in addition to that to structural disturbance and defect (D-band). If the Raman spectra of the microstructure under consideration extracted by acid maceration are identical to those of disseminated OM in the matrix of petrographic thin sections (or in macerates), the microstructure can be considered to be syngenetic to the host rock. Although the possibility that contaminants were involved in the host rock later than its formation but earlier than peak metamorphic event cannot be entirely excluded, the Raman spectral analysis provides certain levels of credibility for syngenicity.

7.6.3 BIOLOGICAL CONTEXT: SIZE AND ITS RANGE

Size criterion has two aspects including size itself and its range. As to Archean life, we may consider only about unicellular organisms. However, even unicellular organisms are so various in size. Mycoplasma (eubacteria parasitizing eukaryotes), for example, is known to have the smallest cell size down to ~200 nm, whereas *Xenophyophore* and *Caulerpa lentillifera* can be up to ~20 cm and ~1 m, respectively, although these unicellular organisms are polynuclear. In the strict sense, therefore, the size criterion merely gives the minimum value. Rather, I would like to say that very small structures in nanometer scale are not a promising paleobiological target. The smaller in size, the harder it is to distinguish between biological structures and abiological ones.

In general, narrow size distribution has been regarded as evidence for biogenicity. This is quantified as Divisional Dispersion Index (DDI), which indicates how many times of binary fissions are required to produce the smallest cell from the largest cell. Based on the measurement of modern algae, Schopf (1976) proposed that the biogenicity of microstructures requires DDI smaller than 5 and a Gaussian, normal distribution.

DDI-based discussion, however, can be applied only to putative microfossils that are assumed to have reproduced simply by binary fissions. Indeed, even prokaryotic unicellular microorganisms reproduce in various manners including multiple fissions and asymmetric division. Additionally, many prokaryotes form endospores. Therefore, a population with a wide size range does not always mean that the structures are abiogenic. Also, it should not be overlooked that fossil populations can be comprised of plural taxa (consortia) characterized by different size ranges.

Some of the filamentous microorganisms produce filamentous structure called trichome in a consequence of sequential binary division without separation; representatives are cyanobacterial taxa Nostocales, Anabaeba, Oscillatoriales, and Lyngbya. Vegetative cells tend to have similar shape and size. Thus, if the micro structure in question is segmented and segments have similar size, its biogenicity can be considered. Notably Nostocales and Anabaena occasionally have heterocyst in which nitrogen fixation is performed and akinete (dormant cell), which are different in morphology from vegetative cells. Thus segmented filamentous structures with a few units with morphologies different from the other majorities could be taken not only as biogenicity but also as cyanobacterial affinity (Chapter 13). The lack of segmented structures does not readily indicate abiogenic origin, because some of the microbes forming trichome have outer robust sheath, which has a higher potential of fossilization than vegetative cells. Such sheath does not have segmentation. It may be also noted that filamentous microbes are in most cases uniform in diameter.

7.6.4   BIOLOGICAL CONTEXT: SHAPE

Most prokaryotic cells are coccoid-, rod-, or filament-shaped, although more diverse and complex shapes have been identified. However, abiogenic processes also can produce various morphologies, as represented by silica-witherite biomorphs (e.g., Garcia-Ruiz et al. 2003). Thus, morphological similarities to extant microorganisms and post-Archean genuine microfossils should not be taken as weighted evidence for biogenicity. Cells have various elaborate morphologies and textures such as organelle, endospore, daughter cells, flagellum, and/or appendages. Elaboration is thus a key criterion for biogenicity. Namely, if you find inner elaboration in cell-like structures in rocks, it can be taken as evidence for biogenicity. Origin of elaboration is not necessarily specified, because inner elaboration could be produced by breakdown of cytoplasm. Outer and surface elaboration represented by protrusions and ornaments have been identified for many organic-walled Proterozoic microfossils and taken as evidence for eukaryotic affinity as discussed in Chapter 13 (e.g., Javaux et al. 2003). Of course, these features can be utilized also as biogenicity criteria. In collection, whatever their origins and functions are, the more elaborate the structures, the more their biogenicity would be likely. Morphological elaborations assumingly related to reproduction are strong evidence for the biogenicity of the fossil-like structure, as reproduction is one of the fundamental functions of life.

7.6.5   BIOLOGICAL CONTEXT: OCCURRENCE

Unicellular microorganisms, particularly prokaryotes, can reproduce rapidly under favorable environmental conditions, resulting in the formation of cluster called colony. If they are benthic or adhesive, the growth of colony accompanied with secretion of EPS would results in the formation of biofilm. Carbonaceous laminae observed in Archean sedimentary rocks are, if not all, interpreted as fossilized biofilm. A close spatial association of microstructures with carbonaceous laminae can be presented as strong evidence for biogenicity. Planktonic microorganisms also could comprise colonies encompassed by mucilage EPS. In either case, colonial occurrence is an essential criterion for biogenicity. Even though colonial occurrence cannot be seen, abundant occurrence of microstructures, by which statistical analyses could be done, has an advantage in assessing biogenicity. Also, taphonomic variations could be seen in colonies, as described below.

7.6.6   BIOLOGICAL CONTEXT: TAPHONOMY

Incompleteness in shape largely attributed to taphonomy could be a key feature for assessing biogenicity. Cells are enveloped by biopolymeric materials, including plasma membranes, cell walls, and cell capsules, which have plasticity. Through the process of fossilization, therefore, cells could be deformed to various degrees even within a single colony, occasionally including broken ones. Thus, the presence of some range of variability in shape could be a key part of identifying microstructures as microfossils. Biopolymers also would be degraded to smaller compounds after death of cells by hydrolysis and by heterotrophic metabolisms. Such degradation tends to progress heterogeneously, resulting in different preservation status of microfossils in the same bed. Thus, the preserved envelopes of fossilized cells could be hyaline (higher preservation) to granular (degradation) in appearance.

7.6.7   BIOLOGICAL CONTEXT: CHEMICAL AND ISOTOPIC COMPOSITIONS

In general, carbonaceous composition can be regarded as a key biogenicity criterion for the microbe-like structures. Cells are composed largely of C, N, H, S, O and P. During degradation and thermal maturation of OM, these elements other than C are eventually released. If the thermal maturation of OM related to metamorphism completely progresses to the end, graphite would be produced (e.g., Oberlin, 1984). However, below the upper greenschist facies metamorphism, this process is not completed. Thus, OM derived from microorganisms and EPS are expected to retain some essential elements, particularly N, S, and P. Indeed, an association of these elements with C in fossil-like microstructures could be taken as strong evidence for biogenicity (e.g., Oehler et al. 2010), although we need to consider the possibility that such association is a result of secondary effect such as redistribution and adsorption (e.g., Lepot et al., 2013). Another note is that OM comprising microfossils could be replaced by other materials such as hematite and sulfide (e.g., Wacey et al., 2013; Sugitani et al., 2007). Also, microbes are often encrusted by minerals: mineral encrustation of cell is a sort of biomineralization. For example, iron-oxidizing bacteria are often encrusted by iron oxides (e.g., Mori et al., 2015); this encrustation could survive long after decomposition of cells, although biogenicity of iron-oxide microtubes preserved in ancient rocks may not be easy to be demonstrated, as discussed later.

Light carbon isotopic compositions (Figure 7.2) have long been taken as a key feature in claiming the biogenicity of putative microfossils. This, however, should not be taken as "a smoking gun", because isotopic fractionation and resultant formation of OM with very light isotopic values could be produced through inorganic processes (e.g., McCollom and Seewald 2006), as described earlier. Yet, this is still an important feature for assessing biogenicity, particularly if C and other isotopic signatures associated with fossil-like microstructures are varied across micron scale and are texture-specific (e.g., Morag et al., 2016).

Advanced microscopy and spectroscopy also could provide us with new insights into chemical nature closely related to their biogenicity. As discussed in Chapter 12, Alleon et al. (2018) perfomed various microanalyses including STXM coupled with X-ray absorption near edge structure (XANES) spectroscopy on chert-embedded carbonaceous microstructures from the ca. 3.4 Ga-old Strelley Pool Formation in the Pilbara Craton. Surprisingly, despite experienced metamorphic temperature (~300°C), the microfossils retain nitrogen- and oxygen-rich organic molecules. Such preservation of organic molecules provides further evidence for the biogenicity of the microstructures.

COLUMN: BIOFILM

Biofilm refers to the structure produced by microbes that are generally attached on solid substrates. In addition, it can be formed on the water surface. Biofilm is composed of microbes and polymers secreted by themselves, which are called EPS. Polysaccharides and proteins are the main components of EPS, with others such as DNA and lipids. Though being called "film", it generally has a three-dimensional complex structure within which channels and pores are included. Construction of biofilms starts from formation of a conditioning film that is a thin layer composed of ions and dissolved OM, being followed by adhesion of microbes. The microbes increase their population and excrete EPS, by which the microbial populations would be enclosed. Biofilms could contain multiple species of microbes of different metabolic characteristics, which is enabled by the development of various microenvironments. EPS protects microbes from desiccation, UV-radiation, and chemical inhibitors and stabilize biofilm by storage and transportation of nutrients and mediating communication between microbes. Biofilms develop ubiquitously on, e.g., riverbed boulders, teeth, algae, sandy beds, and tiles in shower room. Microbial mat is a type of biofilm. Readers are recommended to Flemming et al. (2016) and references therein.

REFERENCES

Alleon, J., Bernard, S., Le Guillou, C., Beyssac, O., Sugitani, K., Robert, F. 2018. Chemical nature of the 3.4 Ga Strelley Pool microfossils. *Geochemical Perspectives Letters* 7, 37–42.

Alleon, J., Bernard, S., Le Guillou, C., Daval D., Skouri-Panet, F., Pont, S., Delbes, L., Robert, F. 2016. Early entombment within silica minimizes the molecular degradation of microorganisms during advanced diagenesis. *Chemical Geology* 437, 98–108.

Allwood, A.C., Grotzinger, J.P., Knoll, A.H., Burch, I.W., Anderson, M.S., Coleman, M.L., Kanik, I. 2009. Controls on development and diversity of Early Archean stromatolites. *Proceedings of the National Academy of Sciences of the United States of America* 106, 9548–9555.

Allwood, A.C., Walter, M.R., Kamber, B.S., Marshall, C.P., Burch, I.W. 2006. Stromatolite reef from the Early Archaean era of Australia. *Nature* 441, 714–718.

Allwood, A.C., Kamber, B.S., Walter, M.R., Burch, I.W., Kanik, I. 2010. Trace elements record depositional history of an Early Archean stromatolitic carbonate platform. *Chemical Geology* 270, 148–163.

Allwood, A.C., Rosing, M.T., Flannery, D.T., Hurowitz, J.A., Heirwegh, C.M. 2018. Reassessing evidence of life in 3,700-million-year-old rocks of Greenland. *Nature* 563, 241–244.

Andersen, D.T., Sumner, D.Y., Hawes, I., Webster-Brown, J., McKay, C.P. 2011. Discovery of large conical stromatolites in Lake Untersee, Antarctica. *Geobiology* 9, 280–293.

Awramik, S.M., Riding, R. 1988. Role of algal eukaryotes in subtidal columnar stromatolite formation. *Proceedings of National Academy of Sciences of the United States of America* 85, 1327–1329.

Awramik, S.W., Grey, K. 2005. Stromatolites: Biogenicity, biosignatures, and bioconfusion. Proceedings of SPIE–the International Society of Optics and Photonics, Astrobiology and Planetary Missions, 59060P.

Bailey, J.V., Orphan, V., Joye, S.B., Corsetti, F.A. 2009. Chemotrophic microbial mats and their potential for preservation in the rock record. *Astrobiology* 9, 843–859.

Banerjee, N.R., Muehlenbachs, K. 2003. Tuff life: Bioalteration in volcaniclastic rocks from the Ontong Java Plateau. *Geochemistry, Geophysics, Geosystems* 4. doi:10.1029/2002GC000470.

Banerjee, N.R., Simonetti, A., Furnes, H., Muehlenbachs, K., Staudigel, H., Heaman, L., Van Kranendonk, M.J. 2007. Direct dating of Archean microbial ichnofossils. *Geology* 35, 487–490.

Barghoorn, E.S., Tyler, S.A. 1965. Microorganisms from the Gunflint chert. *Science* 147, 563–577.

Baumgartner, R.J., Van Kranendonk, M.J., Wacey, D., Fiorentini, M.L., Saunders, M., Caruso, S., Pages, A., Homann, M., Guagliardo, P. 2019. Nano-porous pyrite and organic matter in 3.5-billion-year-old stromatolites record primordial life. *Geology* 47, 1039–1043.

Bertrand-Sarfati, J., Walter, M.R. 1981. Stromatolite biostratigraphy. *Precambrian Research* 15, 353–371.

Bonnand, P., Lalonde, S.V., Boyet, M., Heubeck, C., Homann, M., Nonnotte, P., Foster, I., Konhauser, K.O., Köhler, I. 2020. Post-depositional REE mobility in a Paleoarchean banded iron formation revealed by La-Ce geochronology: A cautionary tale for signals of ancient oxygenation. *Earth and Planetary Science Letters* 547, 116452.

Bontognali, T.R.R., Sessions, A.L., Allwood, A.C., Fischer, W.W., Grotzinger, J.P., Summons, R.E., Eiler, J.M. 2012. Sulfur isotopes of organic matter preserved in 3.45-billion-year-old stromatolites reveal microbial metabolism. *Proceedings of the National Academy of Sciences of the United States of America* 109, 15146–15151.

Brasier, M.D., Green, O.R., Jephcoat, A.P., Kleppe, A.K., Van Kranendonk, M.J., Lindsay, J.F., Steele, A., Grassineau, N.V. 2002. Questioning the evidence for Earth's oldest fossils. *Nature* 416, 76–81.

Brasier, M.D., Green, O.B., Lindsay, J.F., McLoughlin, N., Steele, A., Stoakes, C. 2005. Critical testing of Earth's oldest putative fossil assemblage from the ~3.5 Ga Apex chert, Chinaman Creek, Western Australia. *Precambrian Research* 140, 55–102.

Brocks, J.J., Logan, G.A., Buick, R., Summons, R.E. 1999. Archean molecular fossils and the early rise of eukaryotes. *Science* 285, 1033–1036.

Brocks, J.J. 2011. Millimeter-scale concentration gradients of hydrocarbons in Archean shales: Live-oil escape or fingerprint of contamination? *Geochimica et Cosmochimica Acta* 75, 3196–3213.

Buick, R., Dunlop, J.S.R. 1990. Evaporitic sediments of Early Archaean age from the Warrawoona Group, North Pole, Western Australia. *Sedimentology* 37, 247–277.

Buick, R., Dunlop, J.S.R., Groves, D.I. 1981. Stromatolite recognition in ancient rocks: an appraisal of irregularly laminated structures in an Early Archaean chert-barite unit from North Pole, Western Australia. *Alcheringa* 5, 161–181.

Buick, R., Groves, D.I., Dunlop, J.S.R. 1995. Abiological origin of described stromatolites older than 3.2 Ga: Comment and reply. *Geology* 23, 191–192.

Burne, R.V., Moore, L.S. 1987. Microbialites: Organosedimentary deposits of benthic microbial communities. *Palaios* 2, 241–254.

Cloud, Jr. P.E. 1965. Significance of the Gunflint (Precambrian) microflora. *Science* 148, 27–35.

Davies, N.S., Liu, A.G., Gibling, M.R., Miller, R.F. 2016. Resolving MISS conceptions and misconceptions: A geological approach to sedimentary surface textures generated by microbial and abiotic processes. *Earth-Science Reviews* 154, 210–246.

De Gregorio, B.T., Sharp, T.G., Flynn, G.J., Wirick, S., Hervig, R.L. 2009. Biogenic origin for Earth's oldest putative microfossils. *Geology* 37, 631–634.

Derenne, S., Robert, F., Skrzypczak-Bonduelle, A., Gourier, D., Binet, L., Rouzaud, J.-N., 2008. Molecular evidence for life in the 3.5 billion year old Warrawoona chert. *Earth and Planetary Science Letters* 272, 476–480.

Djokic, T., Van Kranendonk, M.J., Campbell, A., Walter, M.R., Ward, C.R. 2017. Earliest signs of life on land preserved in ca. 3.5 Ga hot spring deposits. *Nature Communications* 8, 15263. doi:10.1038/ncomms15263.

Duda, J.-P., Van Kranendonk, M.J., Thiel, V., Ionescu, D., Strauss, H., Schäfer, N., Reitner, J. 2016. A rare glimpse of Paleoarchean life: Geobiology of an exceptionally preserved microbial mat facies from the 3.4 Ga Strelley Pool Formation, Western Australia. *PLoS One* 11, e0147629. doi:10.1371/journal.pone.0147629.

Dupraz, C., Reid, R.P., Braissant, O., Decho, A.W., Norman, R.S., Visscher, P.T. 2009. Processes of carbonate precipitation in modern microbial mats. *Earth Science Reviews* 96, 141–162.

Dutkiewicz, A., Rasmussen, B., Buick, R. 1998. Oil preserved in fluid inclusions in Archaean sandstones. *Nature* 395, 885–888.

Etiope, G., Sherwood Lollar, B. 2013. Abiotic methane on Earth. *Reviews of Geophysics* 51, 276–299.

Finster, K. 2008. Microbiological disproportionation of inorganic sulfur compounds. *Journal of Sulfur Chemistry* 29, 281–292.

Fisk, M. R., Popa, R., Wacey, D. 2019. Tunnel formation in basalt glass. *Astrobiology* 19, 132–144. doi:10.1089/ast.2017.1791.

Flannery, D.T., Allwood, A.C., Summons, R.E., Williford, K.H., Abbey, W., Matys, E.D., Ferralis, N. 2018. Spatially-resolved isotopic study of carbon trapped in ~3.43 Ga Strelley Pool Formation stromatolites. *Geochimica et Cosmochimica Acta* 223, 21–35. doi:10.1016/j.gca.2017.11.028.

Flemming, H.-C., Wingender, J., Szewzyk, U., Steinberg, P., Rice, S.A., Kjelleberg, S. 2016. Biofilm: an emergent form of bacterial life. *Nature Reviews Microbiology* 14, 563–575.

Flügel, E. 2010. *Microfacies of Carbonate Rocks: Analysis, Interpretation and Application.* 2nd edition. Springer, Heidelberg. ISBN 978-3-642-03795-5.

French, K., Hallmann, C., Hope, J.M., Schoon, P.L., Zumberge, J.A., Hoshino, Y., Peters, C.A., George, S.C., Love, G.D., Brocks, J.J., Buick, R., Summons, R.E. 2015. Reappraisal of hydrocarbon biomarkers in Archean rocks. *Proceedings of National Academy of Sciences of the United States of America* 112, 5915–5920.

Furnes, H., Banerjee, N.R., Muehlenbachs, K., Staudigel, H., de Wit, M. 2004. Early life recorded in Archean pillow lavas. *Science* 304, 578–581.

Garcia-Ruiz, J.M., Hyde, S.T., Carnerup, A.M., Christy, A.G., Van Kranendonk, M.J., Welham, N.J. 2003. Self-assembled silica-carbonate structures and detection of ancient microfossils. *Science* 302, 1194–1197.

Goh, F., Allen, M.A., Leuko, S., Kawaguchi, T., Decho, A.W., Burns, B.P., Neilan, B.A. 2009. Determining the specific microbial populations and their spatial distribution within the stromatolite ecosystem of Shark Bay. *The ISME Journal* 3, 383–396.

Grosch, E.G., McLoughlin, N. 2014. Reassessing the biogenicity of Earth's oldest trace fossil with implications for biosignatures in the search for early life. *Proceedings of the National Academy of Sciences of the United States of America* 111, 8380–8385.

Grotzinger, J.P., Rothman, D.H. 1996. An abiotic model for stromatolite morphogenesis. *Nature* 383, 423–425.

Grotzinger, J.P., Knoll, A.H. 1999. Stromatolites in Precambrian carbonates: Evolutionary mileposts or environmental dipsticks? *Annual Review of Earth and Planetary Sciences* 27, 313–358.

Habicht, K.S., Canfield, D.E. 1997. Sulfur isotope fractionation during bacterial sulfate reduction in organic-rich sediments. *Geochimica et Cosmochimica Acta* 61, 5351–5361.

Hickman-Lewis, K., Garwood, R.J., Brasier, M.D., Goral, T., Jiang, H., McLoughlin, N., Wacey, D. 2016. Carbonaceous microstructures from sedimentary laminated chert within the 3.46 Ga Apex Basalt, Chinaman Creek locality, Pilbara, Western Australia. *Precambrian Research* 278, 161–178.

Hofmann, H.J., Grey, K. Hickman, A.H., Thorpe, R.I. 1999. Origin of 3.45 Ga coniform stromatolites in Warrawoona Group, Western Australia. *Geological Society of America Bulletin* 111, 1256–1262.

Homann, M., Heubeck, C., Airo, A., Tice, M.M. 2015. Morphological adaptations of 3.22 Ga-old tufted microbial mats to Archean coastal habitats (Moodies Group, Barberton Greenstone Belt, South Africa). *Precambrian Research* 266, 47–64.

Ivarsson, M., Sallstedt, T., Carlsson, D.-T. 2019. Morphological biosignatures in volcanic rocks– Applications for life detection on Mars. *Frontiers in Earth Science* 7, article 91. doi:10.3389/feart.2019.00091.

Javaux, E.J., Knoll, A.H., Walter, M. 2003. Recognizing and interpreting the fossils of early eukaryotes. *Origins of Life and Evolution of the Biosphere* 33, 75–94.

Lakhdar, R., Soussi, M., Talbi, R. 2020. Modern and Holocene microbial mats and associated microbially induced sedimentary structures (MISS) on the southeastern coast of Tunisia (Mediterranean Sea). *Quaternary Research* 100, 77–97.

Lepot, K., Williford, K.H., Ushikubo, T., Sugitani, K., Mimura, K., Spicuzza, M.J., Valley, J.W. 2013. Texture-specific isotopic compositions in 3.4 Gyr old organic matter support selective preservation in cell-like structures. *Geochimica et Cosmochimica Acta* 112, 66–86.

Lepot, K. 2020. Signatures of early microbial life from the Archean (4 to 2.5 Ga) eon. Earth-*Science Reviews* 209, 103296.

Lindsay, J.F., Brasier, M.D., McLoughlin, N., Green O.R., Fogel, M., Steele, A., Mertzman, S.A. 2005. The problem of deep carbon – An Archean paradox. *Precambrian Research* 143, 1–22.

Lowe, D.R. 1980. Stromatolites 3,400-Myr old from the Archean of Western Australia. *Nature* 284, 441–443.

Lowe, D.R. 1994. Abiological origin of described stromatolites older than 3.2 Ga. *Geology* 22, 387–390.

Lowe, D.R. 1995. Abiological origin of described stromatolites older than 3.2 Ga: Reply. *Geology* 23, 191–192.

McCollom, T. M., Donaldson, C. 2019. Experimental constraints on abiotic formation of tubules and other proposed biological structures in subsurface volcanic glass. *Astrobiology* 19, 53–63.

McCollom, T.M., Seewald, J.S. 2006. Carbon isotope composition of organic compounds produced by abiotic synthesis under hydrothermal conditions. *Earth and Planetary Science Letters* 243, 74–84.

McLoughlin, N., Brasier, M.D., Wacey, D., Green, O.R., Perry, R.S. 2007. On biogenicity criteria for endolithic microborings on early Earth and beyond. *Astrobiology* 7, 10–26.

McLoughlin, N., Wilson, L.A., Brasier, M.D. 2008. Growth of synthetic stromatolites and wrinkle structures in the absence of microbes – implications for the early fossil record. *Geobiology* 6, 95–105.

McLoughlin N. 2011. Stromatolites. In M. Gargaud et al. eds. *Encyclopedia of Astrobiology.* pp. 1603–1613. Springer, Berlin, Heidelberg. doi:10.1007/978-3-642-11274-4_1528.

Minami, M., Shimizu, H., Masuda, A., Adachi, M. 1995. Two Archean Sm-Nd ages of 3.2 and 2.5 Ga for the Marble Bar Chert, Warrawoona Group, Pilbara Block, Western Australia. *Geochemical Journal* 29, 347–362.

Morag, N., Williford, K.H., Kitajima, K., Philippot, P., Van Kranendonk, M.J., Lepot, K., Thomazo, C., Valley, J.W. 2016. Microstructure-specific carbon isotopic signatures of organic matter from~3.5 Ga cherts of the Pilbara Craton support a biologic origin. *Precambrian Research* 275, 429–449.

Mori, J.F., Neu, T.R., Lu, S., Händel, M., Totsche, K.U., Küsel, K. 2015. Iron encrustations on filamentous algae colonized by *Gallionella*-related bacteria in a metal-polluted freshwater stream. *Biogeosciences* 12, 5277–5289.

Mojzsis, S.J., Arrhenius, G., McKeegan, K.D., Harrison, T.M., Nutman, A.P., Friend, C.R. 1996. Evidence for life on Earth before 3,800 million yeras ago. *Nature* 384, 55–59.

Noffke, N., Hazen, R., Nhleko, N. 2003. Earth's earliest microbial mats in a siliciclastic marine environment (2.9 Ga Mozaan Group, South Africa). *Geology* 31, 673–676.

Noffke, N. 2010. *Geobiology: Microbial Mats in Sandy Deposits from the Archean Era to Today.* p. 208. Springer, Heiderberg.

Noffke, N., Awramik, S.M. 2013. Stromatolites and MISS—Differences between relatives. *Geological Society of America Today* 23, 4–9.

Noffke, N., Christian, D., Wacey, D., Hazen, R.M. 2013. Microbially induced sedimentary structures recording an ancient ecosystem in the ca. 3.48 billion-year-old Dresser Formation, Pilbara, Western Australia. *Astrobiology* 13, 1103–1124.

Noffke, N. 2015. Ancient sedimentary structures in the <3.7 Ga Gillespie Lake Member, Mars, that resemble macroscopic morphology, spatial associations, and temporal succession in terrestrial microbialites. *Astrobiology* 15, 169–192.

Nutman, A.P., Bennett, V.C., Friend, C.R.L., Van Kranendonk, M.J., Chivas, A.R. 2016. Rapid emergence of life shown by discovery of 3,700-million-year-old microbial structures. *Nature* 537, 535–538.

Nutman, A.P., Bennett, V.C., Friend, C.R.L., Van Kranendonk, M.J., Rothacker, L., Chivas, A.R. 2019. Cross-examining Earth's oldest stromatolites: Seeing through the effects of heterogeneous deformation, metamorphism and metasomatism affecting Isua (Greenlamd) ~3700 Ma sedimentary rocks. *Precambrian Research* 331, 105347.

Oberlin, A. 1984. Carbonization and graphitization. *Carbon* 22, 521–541.

Oehler, D.Z., Robert, F., Walter, M.R., Sugitani, K., Meibom, A., Mostefaoui, S., Gibson, E.K. 2010. Diversity in the Archean biosphere: New insights from nanoSIMS. *Astrobiology* 10, 413–424.

Oehler, D.Z., Cady, S.L. 2014. Biogenicity and syngeneity of organic matter in ancient sedimentary rocks: Recent advances in the search for evidence of past life. *Challenges* 5, 260–283.

Ohfuji, H., Rickard, D. 2005. Experimental syntheses of framboids – A review. *Earth-Science Reviews* 71, 147–170.

Ohtomo, Y., Kakegawa, T., Ishida, A., Nagase, T., Rosing, M.T. 2014. Evidence for biogenic graphite in early Archaean Isua metasedimentary rocks. *Nature Geoscience* 7, 25–28.

Petroff, A.P., Sim, M.S., Maslov, A., Krupenin, M., Rothman, D.H., Bosak, T. 2010. Biophysical basis for the geometry of conical stromatolites. *Proceedings of the National Academy of Sciences of the United States of America* 107, 9956–9961.

Petroff, A.P., Beukes, N.J., Rothman, D.H., Bosak, T. 2013. Biofilm growth and fossil form. *Physical Review X* 3, 041012.

Philippot, P., Van Zuilen, M., Lepot, K., Thomazo, C., Farquhar, J., Van Kranendonk, M.J. 2007. Early Archean microorganisms preferred elemental sulfur, not sulfate. *Science* 317, 1534–1537.

Planavsky, N., Partin, C., Bekker, A. 2011. Carbon Isotopes as a Geochemical Tracer. In: Gargaud M. et al., eds. *Encyclopedia of Astrobiology.* Springer, Berlin, Heidelberg. doi:10.1007/978-3-642-11274-4_228

Rasmussen, B., Fletcher, I.R., Brocks, J.J., Kilburn, M.R. 2008. Reassessing the first appearance of eukaryotes and cyanobacteria. *Nature* 455, 1101–1104.

Ricci, J.N., Coleman, M.L., Welander, P.V., Sessions, A.L., Summons, R.E., Spear, J.R., Newman, D.K. 2014. Diverse capacity for 2-methylhopanoid production correlates with a specific ecological niche. *The ISME Journal* 8, 675–684.

Rosing, M.T. 1999. ^{13}C-depleted carbon microparticles in >3700-Ma sea-floor sedimentary rocks from West Greenland. *Science* 283, 674–676.

Rouillard, J., van Zuilen, M., Pisapia, C., Garcia-Ruiz, J.-M. 2021. An alternative approach for assessing biogenicity. *Astrobiology* 21, 151–164. doi:10.1089/ast.2020.2282.

Ruff, S.W., Farmer, J.D. 2016. Silica deposits on Mars with features resembling hot spring biosignatures at EL Tatio in Chile. *Nature Communications* 7, 13554. doi:10.1038/ncomms13554.

Schopf, J.W. 1976. Are the oldest 'fossils', fossils? *Origins of Life* 7, 19–36.

Schopf, J.W. Walter, M.R. 1983. Archean microfossils: new evidence of ancient microbes. In J.W. Schopf Ed. Earth's Earliest Biosphere, Its Origin and Evolution. pp, 214–239. Princeton University Press, New Jersey.

Schopf, J.W. 1993. Microfossils of the early Archean Apex Chert: New evidence of the antiquity of life. *Science* 260, 640–646.

Shen, Y., Buick, R., Canfield, D.E. 2001. Isotopic evidence for microbial sulphate reduction in the early Archaean era. *Nature* 410, 77–81.

Sherwood L.B., McCollom, T.M. 2006. Biosignatures and abiotic constraints on early life. *Nature* 444, E18.

Staudigel, H., Furnes, H., DeWit, M. 2015. Paleoarchen trace fossils in altered volcanic glass. *Proceedings of National Academy of Sciences of the United States of America* 112, 6892–6897.

Sugitani, K., Grey, K., Allwood, A., Nagaoka, T., Mimura, K., Minami, M., Marshall, C.P., Van Kranendonk, M.J., Walter, M.R. 2007. Diverse microstructures from Archaean chert from the Mount Goldsworthy––Mount Grant area, Pilbara Craton, Western Australia: Microfossils, dubiofossils, or pseudofossils? *Precambrian Research* 158, 228–262.

Sugitani, K., Mimura, K., Takeuchi, M., Yamaguchi, T., Suzuki, K., Senda, R., Asahara, Y., Wallis, S., Van Kranendonk, M.J. 2015. A Paleoarchean coastal hydrothermal field inhabited by diverse microbial communities: The Strelley Pool Formation, Pilbara Craton, Western Australia. *Geobiology* 13, 522–545.

Suosaari, E.P., Reid, R.P., Playford, P.E., Foster, J.S., Stolz, J.F., Casaburi, G., Hagan, P.D., Chirayath, V., Macintyre, I.G., Planavsky, N.J., Eberli, G.P. 2016. New multi-scale perspectives on the stromatolites of Shark Bay, Western Australia. *Scientific Reports* 6, 20557.

Tashiro, T., Ishida, A., Hori, M., Igisu, M., Koike, M., Méjean P., Takahata, N., Sano, Y., Komiya, T. 2017. Early trace of life from 3.95 Ga sedimentary rocks in Labrador, Canada. *Nature* 549, 516–518.

Tice, M.M., Lowe, D.R. 2004. Photosynthetic microbial mats in the 3,416-Myr-old ocean. *Nature* 431, 549–552.

Tice, M.M., Lowe, D.R. 2006. The origin of carbonaceous matter in pre-3.0 Ga greenstone terrains: A review and new evidence from the 3.42 Ga Buck Reef Chert. *Earth-Science Reviews* 76, 259–300.

Ueno, Y., Yamada, K., Yoshida, N., Maruyama, S., Isozaki, Y. 2006a. Evidence from fluid inclusions for microbial methanogenesis in the early Archaean era. *Nature* 440, 516–519.

Ueno, Y., Yamada, K., Yoshida, N., Maruyama, S., Isozaki, Y. 2006b. Replying to: B. Sherwood Lollar & T.M. McCollom. *Nature* 444, E18–19.

Van Kranendonk, M.J., Philippot, P., Lepot, K., Bodorkos, S., Pirajno, F. 2008. Geological setting of Earth's oldest fossils in the ca. 3.5 Ga Dresser Formation, Pilbara Craton, Western Australia. *Precambrian Research* 167, 93–124.

Van Kranendonk, M.J., Webb, G.E., Kamber, B.S. 2003. Geological and trace element evidence for a marine sedimentary environment of deposition and biogenicity of 3.45 Ga stromatolitic carbonates in the Pilbara Craton, and support for a reducing Archaean ocean. *Geobiology*, 1, 91–108.

Van Kranendonk, M.J. 2011. Morphology as an indictor of biogenicity for 3.5–3.2 Ga fossil stromatolites from the Pilbara Craton, Western Australia. In J. Reitner et al. eds *Lecture Notes in Earth Sciences* 131, pp. 537–554, Springer, Hyderberg.

van Zuilen, M.A., Lepland, A., Arrhenius, G. 2002. Reassessing the evidence for the earliest traces of life. *Nature* 418, 627–630.

Wacey, D. 2009. *Early Life on Earth: A Practical Guide.* p. 274. Springer, Heidelberg.

Wacey, D. 2010. Stromatolites in the ~3400 Ma Strelley Pool Formation, Western Australia: Examining biogenicity from the Macro- to the Nano-scale. *Astrobiology* 10. doi:10.1089/ast.2009.0423.

Wacey, D., McLoughlin, N., Kilburn, M.R., Saunders, M., Cliff, J.B., Kong, C., Barley, M.E., Brasier, M.D. 2013. Nanoscale analysis of pyritized microfossils reveals differential heterotrophic consumption in the ~1.9-Ga Gunflint chert. *Proceedings of the National Academy of Sciences of the United States of America* 110, 8020–8024.

Wacey, D., Noffke, N., Cliff, J., Barley, M.E., Farquhar, J. 2015. Micro-scale quadruple sulfur isotope analysis of pyrite from the ~3480 Ma Dresser Formation: New insights into sulfur cycling on the early Earth. *Precambrian Research* 258, 24–35.

Wacey, D., Saunders, M., Brasier, M.D., Kilburn, M.R. 2011. Earliest microbially mediated pyrite oxidation in ~3.4 billion-year-old sediments. *Earth and Planetary Science Letters* 301, 393–402.

Wacey, D., Saunders, M., Cliff, J., Kilburn, M.R., Kong, C., Barley, M.E., Brasier, M.D. 2014. Geochemistry and nano-structure of a putative ~3240 million-year-old black smoker biota, Sulphur Springs Group, Western Australia. *Preambrian Research* 249, 1–12.

Walter, M.R. 1976. Geyserites of Yellowstone National Park: An example of abiogenic "stromatolites." In M.R. Walter Ed. *Stromatolites, Developments in Sedimentology* 20, pp. 87–112. Elsevier, Amsterdam.

Walter, M.R., Buick, R., Dunlop, J.S.R. 1980. Stromatolites 3,400–3,500 Myr old from the North Pole area, Western Australia. *Nature* 284, 443–445.

Walter, M.R., Heys, G.R. 1985. Links between the rise of metazoa and the decline of stromatolites. *Precambrian Research* 29, 149–174.

Whitehouse, M.J., Dunkley, D.J., Kusiak, M.A., Wilde, S.A. 2019. On the true antiquity of Eoarchean chemofossils – Assessing the claim for Earth's oldest biogenic graphite in the Saglek Block of Labrador. *Precambrian Research* 323, 70–81.

Williford, K.W., Ushikubo, T., Sugitani, K., Lepot, K., Kitajima, K., Mimura, K., Valley, J.W. 2015. A sulfur four-isotope signature of Paleoarchean metabolism. *Astrobiology Science Conference* 7275pdf.

8 Early (Paleo- to Meso-) Archean Cellularly Preserved Biosignatures

8.1 INTRODUCTION

Cellularly preserved biosignatures, if genuine, could provide us with an invaluable image of ancient life. However, it is hard to prove their biogenicity. The deeper in time, the harder it is to identify microfossils. In this chapter, we review the early Archean cellularly-preserved biosignatures, focusing on microstructures from the Isua supracrustal belt (ISB), Greenland, the Nuvvuagittuq greenstone belt (NGB), Canada, the Kaapvaal Craton, South Africa, and the Pilbara Craton, Western Australia. My findings from the Pilbara Craton would be separately presented in the later sections.

8.2 ISUA SUPRACRUSTAL BELT, GREENLAND (DENMARK)

8.2.1 GEOLOGICAL BACKGROUND

The Isua supracrustal belt (ISB) occurs in the Godhåbsfjord region of West Greenland. This belt is composed of supracrustal rocks that refer to sedimentary and volcanic rocks (their metamorphosed products) deposited on the crustal basement. By S. Moorbath and his collaborators' pioneer works in 1970s, the ages of tonalitic gneiss and iron formation (IF) were determined to be 3.7 Ga-and 3.76 Ga-old, respectively, although these ages were later recalculated to be 3.64 Ga-and 3.71 Ga-old, respectively (Moorbath and Whitehouse, 1996). Following detailed geologic and geochronological works (e.g., Nutman and Friend, 2009) have revealed the presence of ~3.8 Ga-old rocks in the ISB. The ISB is embedded in Mesoarchean gneiss/granite terrane. According to Nutman et al. (2019), two units of different ages have been identified, including the ~3.8 Ga-old unit in the southern and the western portions and the ~3.7 Ga-old unit in the northern and median portion. Both are dominated by amphibolite-facies mafic volcanic rocks, metamorphosed felsic volcanic and volcaniclastic rocks, and metamorphosed clastic and chemical sediments.

8.2.2 CELLULARLY PRESERVED BIOSIGNATURES

H.D. Pflug described "vacuolated" ellipsoidal structures from the ca. 3.8 Ga-old unit of the ISB (Pflug, 1978) (Figure 8.1). The structures range from 10 to 40 μm in diameter. About a hundred of specimens were identified in petrographic thin sections.

DOI: 10.1201/9780367855208-8

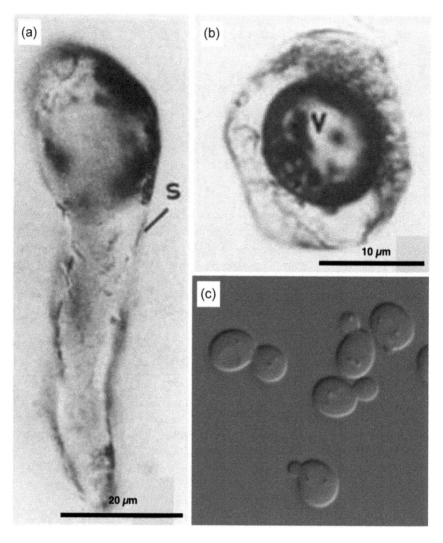

FIGURE 8.1 (a–b) *Isuasphaera isua* from the ~3.7 Ga-old Isua supracrustal belt, West Greenland (Pflug, 1978). Thin, elongated wall indicated by S in (a) is interpreted as sheath. V in (b) is interpreted as gas vacuole. (c) Yeast (eukaryotic unicellular organism) reproduced by budding. With permission from Springer Science + Business Media: *Naturwissenschaften* (Pflug, 1978); Budding yeast, *Saccharomyces cerevisiae*. Cells are around 6–7 μm across. (Photo courtesy by Erfei Bi.)

Their morphological variations including dumbbell and pair were interpreted to be in the context of cell growth and reproduction. The author claimed that the structures were biogenic. The structures were named *Isuasphaera isua*. Furthermore, the author suggested that *Isuasphaera isua* resembled recent asporogenous yeasts, based on, e.g., shape, wall structure, and structures similar to multilateral budding. Asporogenous yeasts are kinds of yeasts, which are heterotrophic eukaryotes, although the author also said, "Though our fossil finds look like yeasts they need not necessarily be yeasts in a taxonomic sense" and "Thus it seems conceivable that the Archean finds, in spite of their yeast-like features, belong to an evolutionary level which is far below that of yeasts of today". This publication is followed by Pflug and Jaeschke-Boyer (1979), who presented results of Raman spectral analyses. The material of the "cell vacuole" appeared to have preserved organic functional groups. Based on this, the authors claimed more strongly the biogenicity of *Isuasphaera isua*, retaining precaution on their biological affinity.

Isuasphaera isua, unusually complex and large microstructures as Paleoarchean life, has since been exposed to criticism. In Bridgewater et al. (1981), its biogenicity was denied in the context of geological background. The authors emphasized that the ISB has experienced repeated deformation, geochemical alteration, and metamorphic (up to amphibolite facies) events and argued that the supposed host rock for *Isuasphaera isua* was unlikely to have preserved microfossils. It was also suggested that Pflug (1978) had misinterpreted limonite (FeOOH)-stained fluid inclusions as microfossils. Roedder (1981) on the other hand suggested that *Isuasphaera* structures did not originate from fluid inclusions. The structures were reinterpreted as limonite-stained cavities (Roedder, 1981). During reviewing these papers, I had got anxious whether these authors examined the same materials (holotypes) and felt that the discussions do not fit together. Indeed, Appel et al. (2003) pointed out as follows: "One of the problems which both Pflug and his critics faced was that they did not know from which part of the IGB (=ISB by KS) the sample with the putative microfossils was collected". The authors also emphasized that the metamorphosed chert layers hosting *Isuasphaera isua* had experienced stretching deformation of an extreme degree, suggesting that spherical objects were unlikely preserved. Interestingly, the authors do not entirely deny the biogenicity of *Isuasphaera* and instead implied the possibility that the objects contained silcrete, cherty duricrust.

8.3   THE NUVVUAGITTUQ GREENSTONE BELT, CANADA

8.3.1   Geological Background

The Nuvvuagittuq greenstone belt (NGB) is located at the northwestern coast of the Ungava Peninsula, Québec, Canada, facing Hudson Bay. This greenstone belt with an area of ca. 20 km^2 is composed dominantly of metamorphic rocks of mafic (basaltic to andesitic) compositions ranging from cummingtonite amphibolite to garnet-biotite schist facies, with mafic to ultramafic sills. Iron formation and chert occur as a minor sedimentary component. The age of this greenstone belt is controversial. Cates and Mojzsis (2007) first gave an age constraint on this greenstone belt, reporting U-Pb age of 3.77 Ga for zircon from granitic intrusion. This gives a minimum age for this belt. Much older ages (>4.0 Ga) were later reported by O'Neil et al. (2012). If this is the case, NGB represents a remnant of the Hadean crust. However, this was questioned by Cates et al. (2013), who proposed the maximum age of 3.78 Ga for NGB, based on U-Pb dating of detrital zircon from quartz-biotite schists.

8.3.2 Cellularly Preserved Biosignature

Dodd and coworkers reported filaments and tubes from jaspilite (ferruginous chert) from NGB (Dodd et al. 2017) (Figure 8.2). Jaspilite comprises the NGB chemical sedimentary successions, together with banded iron formation (BIF) and banded silica formation (BSF); the occurrence of jaspilite is localized (Mloszewska et al.,

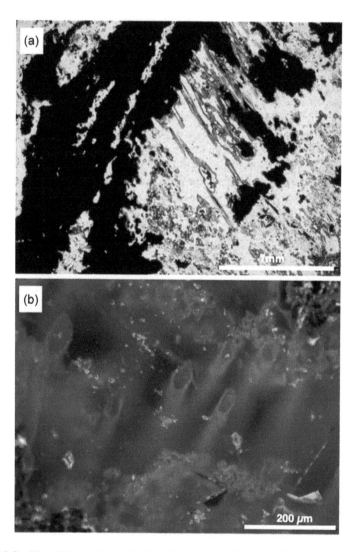

FIGURE 8.2 Hematitic putative microfossils from the ~3.8 Ga-old Nuvvuagittuq geenstone belt, Canada (Dodd et al., 2017). (a) Loose cluster of hematitic filaments and tubes in the Paleoarchean (>3.8 Ga) ferruginous sedimentary rocks from the Nuvvuagittuq greenstone belt, Canada. (b) Enlarged view of hematite tubes. Transmitted light. (Photomicrographs courtesy of M. Dodd).

2013). The structures occurring in quartz (SiO_2) layers are composed of hematite (Fe_2O_3) and range 2 to 14 μm in diameter and can be up to 500 μm in length, occasionally associated with carbonate and graphite. Some are coiled or branched. Others occur as twisted filaments radiating from terminal knob. The authors emphasized that the structures were similar in morphology and associated materials to modern and younger microfossils (e.g., Li et al., 2012). It was claimed that the structures were oxidized biomass that inhabited ancient submarine hydrothermal environment, referring to (1) the presence of other structures such as carbonate rosettes and jasper rosettes that have been reported from younger jasper deposits (Grenne and Slack, 2003), (2) the probable biological source for graphitic carbon in carbonate rosettes associated with apatite [$Ca_{10}(PO_4)_6(OH,F,Cl)$], and (3) the presence of magnetite (Fe_3O_4) granules 100–500 μm across containing carbonate, organic matter (OM), and apatite, which are potentially equivalent to those found in younger IFs with some geochemical biosignatures (Smith et al., 2012). Although other possibilities were mentioned to, the authors repeatedly emphasized that the microstructures were similar to iron-oxidizing bacteria.

While detailed descriptions and discussions presented by Dodd et al. (2017) are remarkable, the following issues may be pointed out. First, iron-oxidizing bacteria at hydrothermal vent precipitates require molecular oxygen (O_2) or nitrate (NO_3^-), if they do not utilize light. Were these electron acceptors available in Eoarchean seafloors? At least, possible metabolic pathways should have been discussed. Second, the jasper hosting the microstructures is interpreted to be hydrothermal vent precipitates, based on its occurrence in volcano-sedimentary succession and rare-earth elements (REEs) signatures (*Column*). However, the REEs signatures, characterized by depletion in light rare-earth elements (LREEs) and weak to moderate positive Eu-anomalies, differ from those found for high-temperature (>350°C) hydrothermal fluids. Rather, these features are consistent with Archean anoxic seawaters influenced by hydrothermal components, although the expected super chondritic Y/Ho ratio is equivocal.

8.4 KAAPVAAL CRATON, SOUTH AFRICA

The Kaapvaal Craton is located at the southern part of the African Shield and is known as one of the representative major Archean cratons. Microfossils and other biosignatures such as stromatolites and microbially induced sedimentary structures (MISS) have been reported from the Paleo-to Mesoarchean Barberton greenstone belt and the Neoarchean sedimentary units. The Barberton greenstone belt consists of the Barberton Supergroup (formerly called Swaziland Supergroup) (Byerly, 2015), which is comprised of the 3.55–3.27 Ga-old Onverwacht Group, the ca. 3.3 Ga-old Fig Tree Group, and the ca. 3.2 Ga-old Moodies Groups, from all of which microfossils and possible microfossils have been reported. Recently, nice reviews on biosignatures of the Barberton Supergroup have been published (Hickman-Lewis et al., 2019; Homann, 2019). Readers are recommended to read these reviews for details of the South African biosignatures. In the following, after summarizing the occurrence of cellularly preserved microfossils and possible microfossils from these three groups, selected works are reviewed.

8.4.1 THE ONVERWACHT GROUP

The dominant components of the ca. 3.3–3.5 Ga-old Onverwacht Group are komatiitic and basaltic volcanic and volcaniclastic rocks. Dacitic volcaniclastic rocks and cherts occur as minor components. This group is composed of the Sandspruit, the Theespruit, the Komati, the Hoogenoeg, the Kromberg, the Mendon, and the Weltevreden formations, in an ascending stratigraphic order (Byerly, 2015). Cellularly preserved microfossils and possible microfossils have been described from the Hoogenoeg and Kromberg formations.

The 3,472–3,435 Ma Hoogenoeg Formation is up to 3,900 m thick and composed dominantly of basalt and basaltic komatiite, capped by a thick volcanic pile of dacitic composition. This volcanic succession contains five major chert units, corresponding to the resting periods of volcanic activity. The ~3,416 to 3,334 Ma Kromberg Formation is over 2,000 m thick volcanic-sedimentary succession, again being dominated by basalts, komatiites, and mafic volcaniclastic rocks. Sedimentary rocks such as black chert, banded chert, and evaporite (silicified) occur as minor components: the lowest 150–350-m-thick unit composed of silicified carbonaceous and ferruginous sedimentary rocks is called the Buck Reef Chert (Homann, 2019 and references therein).

Cellularly preserved microfossils and other biosignatures such as microbial mats including stromatolites have been reported mostly from carbonaceous cherts. Microfossil-like carbonaceous structures of various morphologies including spheroids and filaments had been reported from this group as early as the 1960s and 1970s (e.g., Brooks et al., 1973; Engel et al., 1968; Muir and Hall, 1974; Nagy and Nagy, 1969). These findings, however, were later questioned by, e.g., Schopf (1976) and Schopf and Walter (1983). Filamentous, spheroidal, and spindle to lenticular microstructures reported by Knoll and Barghoorn (1977), Walsh and Lowe (1985), Walsh (1992), Glikson et al. (2008), and Westall et al. (2001), on the other hand, have been regarded as putative to genuine microfossils, although some raised questions to their biogenicity (e.g., Wacey, 2009). Kremer and Kaźmierczak (2017) also described spheroidal possible microfossils.

8.4.1.1 The Hoogenoeg Formation

Glikson et al. (2008) performed transmission electron microscopic (TEM) analyses on OM from chert of the Hooggenoeg Formation, which was compared to the hyperthermophilic archaeal *Methanocaldococcus jannaschii* as potential modern analogue for Archean archaeal microbes. *M. jannaschii* was cultured at optimum conditions and then subjected to degradation experiment under hydrothermal conditions (100°C, 1 atm and 132°C, 2 atm).

The authors identified three major types of OM in maceration products of the Hooggenoeg cherts, including aggregates of carbonaceous bodies with small cavities, irregularly shaped particles, and spherical to elliptic thin-walled vesicular bodies. The aggregates were interpreted to be possible equivalent to thermally degraded cells of *M. jannaschii,* from which cell walls were detached due to thermal alteration. Surprisingly, OM comprising spherical to elliptic vesicular bodies were not thermally degraded, based on measured vitrinite reflectance. This was attributed to isolation of precursory microbial cells within fluid inclusion (see Section 5.6.1) in quartz. As occurrence of fluid inclusions in host chert was not described, it is difficult to evaluate the plausibility of this interpretation.

8.4.1.2   The Kromberg Formation

Westall et al. (2001) examined laminated carbonaceous chert from the Hoogenoeg Formation using a scanning electron microscope (SEM) in addition to light optical microscope. Observation using SEM was performed for the samples etched by hydrofluoric acid (HF) vapor. Hydrofluoric acid effectively decomposes silicates, and silicon (Si) was removed by vaporized silicon tetrafluoride (SiF_4), whereas OM comprising microfossils is not affected. Chert is composed dominantly of microcrystalline quartz, and thus, HF-etching and subsequent SEM analyses could reveal microbial textures that cannot easily be identified under an optical microscope.

The authors first identified tiny spherules in petrographic thin sections. They are about 1 μm in diameter that commonly occur red as a pair, "zigzag" association, or linear chain. Rarely, four spherules comprised rectangular association. These spherules were identified also in the HF-etched samples, again showing various manners of association. Linearly connected three spherules appeared to be coated by a wrinkled thin envelope. Furthermore, populations partly embedded in pseudomorphic calcite were identified. In addition to spherules, rice-grain-like microstructures and smooth, ropy-textured films resting on bedding plane were identified in the HF-etched samples. The authors claimed that the spherules and the films represented silicified bacteria and biofilms, respectively. Major lines of presented evidence for the biogenicity of the tiny spherules include (1) similarity in shape and size to modern coccoid bacteria, (2) presence of their associations similar to cell divisions, and (3) colony-like three-dimensional occurrence. The biogenicity of the Kromberg spherules, however, was questioned by Wacey (2009), saying "Reliant morphology which is too simple to be attributed uniquely to microbes".

Walsh (1992) reported carbonaceous microstructures of various morphologies from three remote (>5 km) localities of this formation (Figure 8.3a–c). These microstructures were found within laminated black chert or black and white banded chert. The reported microstructures were spheroid, ellipsoid, spindle, and filament. Furthermore, they can be classified into subtypes, based on, e.g., size, wall texture, and occurrence. For example, filaments are either solid or hollow, are either unbranched or branched, and occur either solitarily or as a cluster. Walsh (1992) assessed the biogenicity of the Kromberg microstructures based on the criteria proposed by Buick (1990) and regarded only clustered hollow filaments as microfossils (Figure 8.3a). The supposedly biogenic hollow filaments range from approximately 1.5 to 2.5 μm in width and from 10 to 150 μm in length, and are commonly arranged subparallel to bedding and occasionally show downward extension or radiation. Solid filaments and solitary filament were judged to have been possible microfossils and dubiofossil, respectively. The other morphotypes, including spheroids, ellipsoids, spindles, and lenticular structures (Figure 8.3b and c) were judged to have been probable microfossils. The described "spindles" were later rediscovered by the author (Figure 8.3d): their biogenicity was confirmed by individual carbon isotopic analyses (Oehler et al., 2017). Walsh's study is important for several reasons. The first, diverse morphotypes

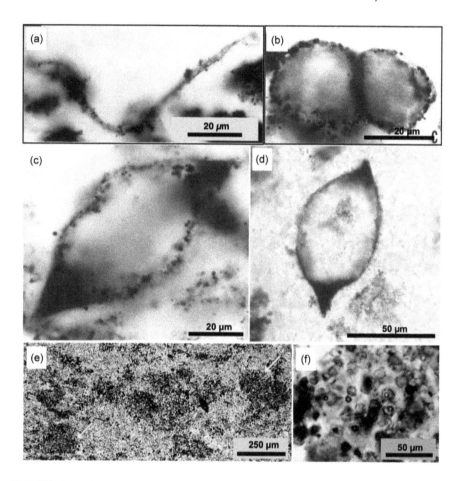

FIGURE 8.3 Archean microfossils and possible microfossils from the ca. 3.4 Ga-old Kromberg Formation of the Barberton greenstone belt. (a) A solitary hollow filament, slightly twisted, (b) paired spheroids with granular wall, and (c) a hollow lens. {(a–c) are reproduced from Walsh (1992), with permission from Elsevier.} (d) A lens independently discovered by the author (K.S.). Transmitted light. (e) Small spheroids comprising colonial clusters. (f) Enlarged image of (e) showing remains of variously degraded colonies of cyanobacterial-like microbes. {(e) and (f) are reproduced from Kremer and Kaźmierczak (2017), with permission from Elsevier.}

were discovered from multiple localities in the same formation. The second, some spheroids, ellipsoids, and spindles are unusually large as prokaryotic spheroid and rod-shaped cells, which are generally less than 5 μm across. The major dimensions of the spheroids and ellipsoids are around 20 μm and can be up to 70 μm. The spindles can be up to 140 μm across. The third, spindles, which may represent the equatorial view of lenticular structures, are morphologically equivalent to those reported from the 3.4 Ga-old Strelley Pool Formation and the 3.0 Ga-old Farrel Quartzite in the Pilbara Craton, Western Australia, as described later (e.g., Sugitani et al., 2007, 2010).

The fourth, many of these unusually large microfossils and possible microfossils have thick walls. Although biological affinities of these microfossils and possible microfossils were not specifically suggested, Walsh (1992) proposed the possibility that these large microstructures represented prokaryotic envelopes containing endospores tolerant against peculiar harsh environments of the early Earth, represented by assumed frequent large asteroid impact events (e.g., Lowe et al., 2014).

Kremer and Kaźmierczak (2017) also described small organic-walled bodies ranging from 3 to 12 μm across, from massive to weakly laminated black cherts of the Kromberg Formation (Figure 8.3e, f). The two chert samples containing fossil-like microstructures had $\delta^{13}C_{PDB}$ values of −24.27 ‰ and −26.46 ‰ respectively. The microstructures display morphological variations from fully rounded to flattened or angular-shaped, and occasionally contain dark granules of various sizes inside. Whereas they are embedded within the matrix composed of almost pure chert, the cell-like bodies are closely associated with Al-K-Mg-Fe silicates. The cell-like bodies comprise clusters up to 150 μm across.

The authors suggested that the microstructures might represent variously degraded microbial cells. Morphological variations and occasional occurrence of inner bodies were interpreted in the context of taphonomy, postmortem degradation, and selective preservation of cell components (Golubic and Hofmann, 1976). Hollow cell-like bodies were interpreted to represent the resistant polysaccharidal envelope (sheaths), whereas dark granules inside represent coalesced cell components. The authors also suggested that the association of aluminosilicates with cell-like microstructures was equivalent to that found for modern and fossil microbial mats (e.g., Douglas, 2005). Kaźmierczak and Kremer (2019) further gave in-depth biological interpretation of their benthic-planktonic life cycle and pattern of cell divisions, suggesting the possibility of cyanobacterial affinity. This interpretation cannot readily be accepted, but the possibility should be retained.

8.4.2 THE FIG TREE GROUP

The ca. 3.3 Ga-old Fig Tree Group consists mainly of sedimentary rocks with subordinate dacitic to rhyodacitic volcaniclastic and volcanic rocks. The Fig Tree sedimentary succession is composed mainly of graywackes, conglomerate, shales, chert, jasper, and ironstone. Microbe-like structures in chert have been reported by Barghoorn and Schopf (1966), Schopf and Barghoorn (1967), and Pflug (1967). In Barghoorn and Schopf (1966), surface replicas of HF-etched thin sections were examined using SEM. Identified rod-shaped and thin-walled circular microstructures were interpreted to be "comparable in size, shape, complexity of structure, and isolated habit to many modern bacillar bacteria". Much larger carbonaceous structures were also identified in black chert from the same locality (Schopf and Barghoorn, 1967). They were basically spheroidal, although distorted and partially flattened specimens were also identified. Their size range was narrow, from 15 to 23 μm in diameter (n = 28).

In addition to these relatively simple morphologies, complex and in some cases much larger structures were reported from cherts and shales of this group by Pflug (1967), who identified various morphological types of carbonaceous structures, including, e.g., ellipsoidal bodies, disc- or lens-shaped bodies, globular bodies, and

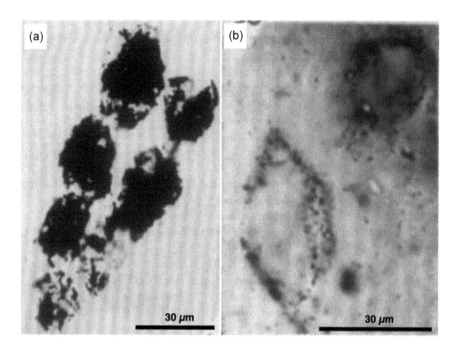

FIGURE 8.4 Possible microfossils from the ca. 3.3 Ga-old Fig Tree Group, South Africa. (a) Structures interpreted as colonies of globular type, in thin section. (b) Structure interpreted as globular type, in thin section. They can be interpreted as lenticular structures (Chapters 9 and 10). {Reproduced from Pflug (1967), with permission from Elsevier.}

filamentous structures in thi sections and mceration products. Globular type objects, for example, were generally larger than 20 μm and up to 70 μm across. Though being highly degraded and thus details are not known, the specimens shown in Figure 8.4 are similar to lenticular microfossils reported from the Kromberg Formation, South Africa (Figure 8.3c and d) (Walsh, 1992; Oehler et al., 2017) and the 3.0 Ga-old Farrel Quartzite and the 3.4 Ga-old Strelley Pool Formation, Western Australia as described in Chapters 9 and 10.

8.4.3 THE MOODIES GROUP

The ca. 3.2 Ga-old Moodies Group consists dominantly of sandstone with subordinate conglomerate, shale, siltstone, IF, and volcanic rock, representing alluvial to shallow-marine depositional environment (Heubeck et al., 2013). One of the oldest possible records of a tide has been reported (Eriksson and Simpson, 2000) (Figure 2.7). Occurrences of microfossils in shale, siltstone, and chert lenses in sandstone are known to date (Homann et al., 2016; Javaux et al., 2010).

In lens-shaped chert in siliciclastic rock of this group, Homann et al. (2016) identified abundant cylindrical filamentous molds. The regularly segmented "filaments" range from 0.3 μm to 0.5 μm in diameter and are bent in various directions (Figure 8.5a and b). The host lens-shaped chert has carbonaceous laminae: their carbon isotopic compositions were significantly enriched in ^{12}C ($\delta^{13}C_{PDB} = -26.5‰$, on average). The lens-shaped chert was interpreted as early silicified cavities, which was formed beneath microbial mats. The filamentous molds were interpreted as traces of cavity-dwelling microbes.

From gray shales and siltstones of the lowermost Clutha Formation of this group, Javaux et al. (2010) reported spheroidal microfossils (Figure 8.5c and d). The carbonaceous microstructures occur as compressed masses and can be extracted by acid maceration. Extracted microstructures range from 31 µm to 298 µm with a mean of 122 µm (n = 98), in an apparent diameter. Although the spherical shape is simple and biological occurrence such as colony-like cluster cannot be seen, the biogenicity of the carbonaceous spheroids was clamed based on (1) preserved organic wall, (2) presence of wrinkled and rolled up textures attributed to taphonomy, (3) relatively narrow size range, and (4) presence of paired specimen potentially corresponding to the stage of cell division. Large cell with acid-resistant wall appears to imply their eukaryotic affinity. However, the authors did not take this interpretation because the cell wall lacks micron-scale polygonal ornament and any appendage. Analysis using

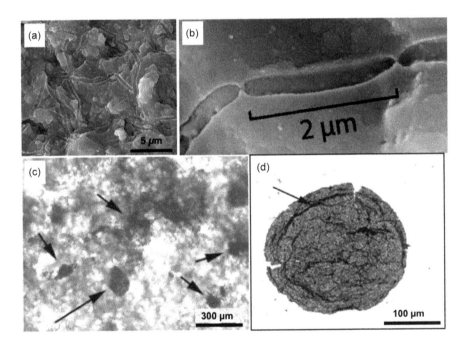

FIGURE 8.5 Microfossils and possible microfossils from the ca. 3.2 Ga-old Moodies Group, South Africa. (a) Meshwork of filamentous molds embedded in chert of the Moodies Group. Image obtained by SEM (Homann et al., 2016). (b) Enlarged view of (a). Photomicrographs courtesy of M. Homann. (c) and (d) Organic-walled spheroids from the Clutha Formation (Javaux et al., 2010). Transmitted light. (c) shows compressed spheroids (the arrows) in a petrographic section made perpendicular to the bedding. (d) Spheroid extracted from the host rock by acid maceration. The arrow shows concentric fold. Reproduced from Javaux et al. (2010), with permission from Springer.

TEM revealed that the preserved wall was single-layered, whereas well-defined eukaryotic spheroid microfossils are characterized by multilayered cell walls (see Chapter 13 for criteria of eukaryotic microfossils). So, what biological affinity is likely for these large spheroids? The authors implied the possibility of cyanobacterial affinity, based on the probable shallow water depositional environment (namely photic zone) and no signs for hydrothermal inputs and thus no sulfidic environment.

8.5 PILBARA CRATON, WESTERN AUSTRALIA

The Archean Pilbara Craton is composed of the East Pilbara, Regal, Karratha, Sholl and Kurrana terranes and the five tectonic basins including the Gorge Creek, the Mallina, the Mosiquite Creek and the Hamersley basins and the De Drey Superbasin (e.g., Van Kranendonk et al., 2006; Hickman, 2012). Most of the microfossils and possible microfossils have been reported from the Pilbara Supergroup in the East Pilbara Terrane. The supergroup is composed of the Warrawoona Group, the Sulfur Springs Group, the Strelley Pool Formation, the Kelly Group, and other independent units. The overlying De Grey Supergroup is composed of the Whim Creek, Nullagine, Gorge Creek, Croydon, and Bookingarra groups, according to the Australian Stratigraphic Unit Database (ASUD). Reviews on the microfossil assemblage described from the Pilbara Craton have been given in Wacey (2011) and Sugitani (2019a, b). In this section, some selected specimens are described, excluding those discovered by the author, which will be described in Chapters 9 and 10.

8.5.1 The Warrawoona Group

The 3.52–3.42 Ga-old Warrawoona Group is dominated by mafic to ultramafic volcanic rocks. Subordinate components include felsic volcanic rocks and sedimentary rocks such as chert, evaporite, carbonates, siliciclastic rocks and volcaniclastic rocks (e.g., Van Kranendonk et al. 2006). Microfossils and possible microfossils have been reported from four different rock units, including the Dresser Formation (~3,490 Ma), the Mount Ada Basalt (~3,470 Ma), the Apex Basalt (~3,460 Ma), and the Panorama Formation (3,449–3427 Ma) (Hickman, 2012 and references therein).

8.5.1.1 The Dresser Formation

The Dresser Formation is composed dominantly of varicolored bedded cherts, with minor carbonates and barite. Komatiitic basalt and chert veins also occur (Van Kranendonk et al., 2008). The depositional environment of this formation has been somewhat controversial. Buick and Dunlop (1990) suggested its deposition in a closed to semi-closed shallow coastal basin, whereas Van Kranendonk (2006) claimed an active volcanic caldera setting. N. Noffke and coworkers suggested a shallow to subaerial environment including sabkha (Noffke et al., 2013). More recently, terrestrial hydrothermal (hot spring) setting has also been proposed (Djokic et al., 2017).

From carbonaceous cherts of this formation, Dunlop et al. (1978) extracted small (1.2–12 μm in diameter) carbonaceous spheroids by acid maceration. The spheroids are solitary or comprise pairs and rarely chains. Some specimens appear to be splitting and rupturing (Figure 8.6a). The biogenicity of these tiny spheroids was claimed

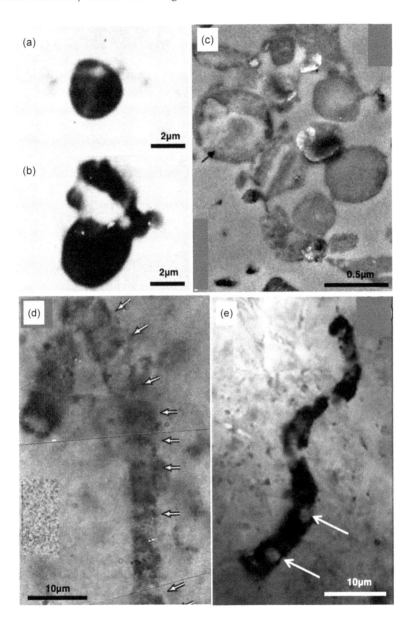

FIGURE 8.6 Possible microfossils from the Pilbara Craton. (a) and (b) Tiny carbonaceous spheroids extracted from chert of the ca. 3.5 Ga Dresser Formation. {Reproduced from Dunlop et al. (1978), with permission from Springer.} (b) shows ruptured-like morphology. (c) TEM image of clustered submicron spheres of various sizes, maceration product of chert from the Dresser Formation in the Pilbara Craton. {Reproduced from Glikson et al. (2008), with permission from Elsevier.} (d) Acute folded relatively broad septate filament from the chert of ca. 3.5 Ga Mount Ada Basalt. {Reproduced from Awramik et al. (1983), with permission from Elsevier.} (e) Sinuous relatively broad filament from chert of the ca. 3.5 Ga Apex Basalt in the Pilbara Craton. {Reproduced from Schopf et al. (2007), with permission from Elsevier.}

based on carbonaceous composition, morphological variations comparable to taphonomy and reproduction, and size distribution. These structures are now generally
considered to be not genuine microfossils, but, e.g., viscous bitumen droplets (e.g.,
Buick, 1990).

Bacterial-like carbonaceous filaments have also been reported (Ueno et al., 2001).
The structures, including solitary spiral filaments, radiating clusters of threads, and
tubular filaments, were described from hydrothermal chert veins and bedded chert.
Carbon isotopic compositions of individual filaments were analyzed using Secondary
Ion Mass Spectrometry (SIMS), demonstrating that the structures were significantly
enriched in ^{12}C ($\delta^{13}C_{PDB} < -30‰$). As described earlier, Ueno et al. (2006) also
reported various ^{13}C-depleted ($\delta^{13}C_{PDB} = -7‰ \sim -56‰$) methane ($CH_4$) from fluid
inclusions in chert veins of this formation, which was interpreted as probable metabolic products of archaeal methanogens. Glikson et al. (2008) also performed acid
maceration of bedded black cherts. They identified carbonaceous cell-like microstructures of several morphological types, including tiny (<3 μm) cell-like bodies (Figure 8.6c). It is hard to fully demonstrate that these tiny and simple-shaped
objects are genuine microfossils, although their biogenicity and Ueno et al.'s ones
may be indirectly supported by the presence of stromatolite (Van Kranendonk et al.,
2008) in the same formation (see Section 7.4.2.4) and the extraction of long-chain
aliphatic hydrocarbons with biogenic traits such as odd-over-even carbon number
dominance (Derenne et al., 2008) (see Section 7.2.2).

8.5.1.2 The Mount Ada Basalt

Awramik et al. (1983) described cell-like carbonaceous spheroids and filaments in
bedded chert collected from the Mount Ada Basalt in the North Pole area. Some of
them were claimed to be biogenic. One of the described Mount Ada filaments is of
particular interest, because the relatively thick (~5 μm) specimen is regularly septate
and acutely folded (Figure 8.6d). The former represents cellular elaboration, and the
latter represents taphonomy. The sample locality and syngenicity of the structures
were questioned by Buick (1984), being followed by some debates (Buick, 1988;
Awramik et al., 1988). Grey et al. (2010) later resolved the uncertainty in the sampling locality.

8.5.1.3 The Apex Basalt

The Apex Basalt is composed dominantly of basalt, komatiitic basalt, and serpen-
tinized peridotite, and chert occurs as minor components. Carbonaceous filamentous microstructures were reported from chert at the Chinaman Creek locality
near Marble Bar Pool (Schopf and Packer, 1987; Schopf, 1993) (Figure 8.6e). The
structures appear to be septate and were morphometrically classified into 11 taxa
(Schopf, 1993), implying the possibility of their cyanobacterial (oscillatorian)
affinity. This implication was also based on association with stromatolitic structure
and assumed shallow-water depositional environment. However, the subsequent
studies had revealed that the host rock was collected from chert dyke cutting the
Apex Basalt (Brasier et al., 2002). This chert dyke was interpreted to have been
formed by high-temperature hydrothermal circulation, unsuitable for microbial habitats. Additionally, morphologies of some of the structures appeared to have been

controlled by secondary crystal growth and associated reorganization of OM. In short, the structures had become deemed to be more likely abiogenic products (e.g., Brasier et al., 2006). This negative argument has been followed by many researchers. Pinti et al. (2009) demonstrated repeated hydrothermal alterations of the Apex chert, unsuitable for the preservation of delicate microfossil structures. Marshall et al. (2011) examined segmented (septate) structures in a different Apex chert sample, morphologically equivalent to those described by Schopf (1993), and revealed that they were quartz- and hematite-filled fractures. Marshall et al. (2012) also revealed that OM in the Apex chert had multiple generations. Brasier et al. (2015) and Bower et al. (2016) claimed that some of described specimens such as *Archaeoscillatoriopsis*, *Primaevifilum*, and *Eoleptonema apex* were pseudofossils, being composed of vermiform phyllosilicates or intragranular crack, with adsorbed or infilled OM.

My reviews on the Apex microstructures may lead many of the readers to suspect that all the described Apex structures are pseudofossils. However, it should not be overlooked that the specimens analyzed by these critics, except for that by Bower et al. (2016), were not the holotype specimens. Using Confocal Laser Scanning Microscope (CLSM), Schopf and Kudryavtsev (2009) demonstrated that one of the termed specimens, *Primaevifilum amoenum*, was a hollow cylindrical carbonaceous structure composed of a sequence of compartments. Later, Schopf and his colleagues performed individual measurements of carbon isotopic compositions on the five taxa including *Primaevifilum amoenum* (Schopf et al., 2018) and demonstrated that they had $\delta^{13}C_{PDB}$ values from −31‰ to −39 ‰, significantly lighter than the bulk value (−27‰) and associated dispersed particulate OM in the matrix (−25 ± 10 ‰), excluding the possibility that the structures were formed by remobilization and reorganization of OM. Thus, some of the Apex microstructures seem to have now regained their honor as genuine microfossils. This is supported by organo-chemical features of OM from the Apex chert that are similar to those comprising bona fide microfossils, such as structural amorphousness and the presence of aromatic domains (De Gregorio et al., 2009).

8.5.1.4　The Panorama Formation

The Panorama Formation of the Sulgash Subgroup is 1.3 to 1.5 km thick and is composed of rhyolite, felsic volcaniclastic rocks, quartz-rich sandstone and chert, according to ASUD. Westall and coworkers identified microstructures on the vapor HF-etched surface of cherts collected from the so-called Kitty's Gap chert of this formation (Westall et al., 2006, 2011). The cherts are silicified volcaniclastic sediments, cut by numerous chert veins formed by hydrothermal circulations. Identified carbonaceous microstructures are thin (<0.3 µm in width) filaments and tiny (0.4–0.8 µm in diameter) coccoids (Figure 8.7a). Rods ~1 µm long were also described. These microstructures comprise clusters associated with film-like OM that was interpreted as extracellular polymeric substances (EPS) (Figure 8.7b). The shape, size, composition, and occurrence of these structures appear to point to their biogenicity. However, the absence of corresponding structures in petrographic thin sections may not be overlooked, as noted by Wacey (2009). Although the structures are indeed tiny, some of them could be detectable under the conventional petrographic microscope (× 1,000 magnification).

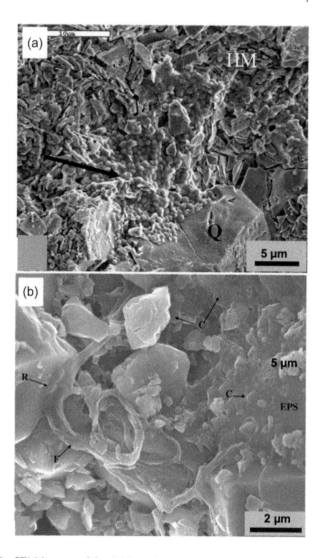

FIGURE 8.7 SEM-images of fossil-like microstructures that appeared on the HF-etched surfaces of volcaniclastic rock from the Panorama Formation. (a) Possible silicified colony of coccoidal microorganisms at the edge of the clast (HM – hydromuscovite) (the arrow). Q: quartz. (b) Silicified film composed of multicomponents on the indurated sediment surface. EPS: a thin layer of alveolar-textured extracellular polymeric substances, F: degraded filaments, C: dividing coccoids, R: rod-shaped structures. (Reproduced from Westall et al. (2011), with permission from Elsevier.)

8.5.2 THE STRELLEY POOL FORMATION

The 3,426–3350 Ma Strelley Pool Formation (SPF) is composed of siliciclastic sedimentary rocks, carbonates (mainly bedded and stromatolitic dolomite), cherts, and volcaniclastic rocks. It has been identified widely in the East Pilbara Terrane. The first description of microbe-like structures in this formation was made by Schopf and Packer (1987). The structures were carbonaceous and were composed of sheathed several spheroids. The overall diameter is around 50 μm. Such morphological elaboration appears to be consistent with the biogenicity; relocation and recollection of the same structures are coveted. In 2010, the author (KS) and colleagues discovered morphologically diverse microstructures from several localities (e.g., Sugitani et al., 2010), which will be described in Chapter 10 and discussed for their biogenicity in Chapter 11. Another assemblage of microfossils was reported from the basal sandstone of this formation at the East Strelley greenstone belt (Wacey et al., 2011). The three identified morphotypes were coccoids (~10 μm across), sheath-like tubes (~10 μm in diameter), and envelope-like spheroids (~80 μm across). Microanalyses including a Focused Ion Beam (FIB)-TEM and SIMS were performed to reveal ultrastructures and isotopic compositions of C and S. Based on the results, it was claimed that the microstructures were fossilized sulfur metabolizing bacteria. This claim was partially declined by Wacey et al. (2018), who reinterpreted the large spheroids as probable volcanic vesicles.

8.5.3 THE SULFUR SPRINGS GROUP

The ca. 3.24 Ga-old Sulfur Springs Group is comprised dominantly of volcanic and volcaniclastic rocks from felsic to ultramafic compositions, with minor sedimentary rocks. Microbe-like structures occur in the Kangaroo Caves Formation of this group, which consists of chemical sediments such as BIF and chert, breccia, siliciclastic rocks, and volcaniclastic rocks of intermediate to felsic composition, in association with hydrothermal black chert veins and massive sulfide deposits. Filamentous and spheroid structures were described from the massive sulfide deposits and the sedimentary rocks (Rasmussen, 2000; Duck et al., 2007).

Rasmussen (2000) described pyritic filaments in the massive sulfide deposits that were interpreted to have been precipitated from high (~300°C) temperature hydrothermal fluids at depths (>1,000 m) (Vearncombe et al., 1995). The identified sinuous filaments, up to 300 μm in length and uniformly 0.7–0.9 μm in diameter. They were intertwined with each other and oriented variously, which was interpreted in the context of microbial behavioral variations. Rasmussen (2000) suggested that the filaments represented pyritized microfossils of possible hyperthermophilic and chemoautotrophic affinity. This interpretation has been supported by the presence of carbon and nitrogen patches and nanopores (Wacey et al., 2014).

Duck et al. (2007) performed acid maceration of the sedimentary rock overlying the massive sulfide deposit and obtained bundles of carbonaceous filamentous and tubular structures less than 1μm in width (cross section) and nanospheres <50–100 nm in diameter. Bulk carbon isotopic compositions ($\delta^{13}C_{PDB}$) range from −26.8 ‰ to −34.0 ‰. It was claimed that the carbonaceous microstructures were similar in appearance to microbes and microfossils obtained from modern and ancient seafloor hydrothermal systems. However, the establishment of biogenicity of such tiny structures does not seem to be straightforward.

8.5.4 The Gorge Creek Group and Others

The ca. 3.0 Ga-old (but ~3.27 Ga-old according to ASUD, also provisional as discussed in the Chapter 11) Gorge Creek Group comprises the De Grey Supergroup, together with the Bookingarra, Croydon, Nullagine, and Whim Creek groups: the De Grey Supergroup unconformably overlies the Pilbara Supergroup. The Gorge Creek Group is further divided into the Farrel Quartzite, the Coondamar Formation, and the Cleaverville Formation. Microfossils and possible microfossils have been described from the Farrel Quartzite and the Dixon Island Formation (not official).

8.5.4.1 The Farrel Quartzite

From the Farrel Quartzite in the Goldsworthy greenstone belt in the northeastern Pilbara Craton, morphologically diverse microstructures have been reported by the author (KS) and colleagues (e.g., Sugitani et al., 2007). They will be described in detail in Chapter 9 and their biogenicity in Chapter 11.

8.5.4.2 The Dixon Island Formation (Not Official)

From the drill core of sedimentary succession at Dixon Island in the western Pilbara Craton, microbe-like carbonaceous structures have been described by Kiyokawa et al. (2006). Black chert and other variously colored cherts, rhyolitic tuff, and black shale comprise sedimentary succession ~350m thick that overlies basalt exhibiting a pillow structure. The succession was thought to had deposited at a paleodepth of ~500 to 2,000m in close association with hydrothermal activities (Kiyokawa et al., 2006, 2014). The succession dated ca. 3.2 Ga was termed as the Dixon Island Formation of the Cleaverville Group (Kiyokawa et al., 2006), though not official (see ASUD). Spheroidal and filamentous microbe-like structures have been described. Iron-rich spheroids occurred in stromatolite-like ferruginous beds, whereas carbonaceous spheroids did in black cherts. Filamentous structures described from black cherts were diverse in morphology and occurrence, including spiral- or rod-shaped filaments, and dendritically stalked filaments. Bulk carbon isotopic compositions ($\delta^{13}C_{PDB}$) ranged from $-33‰$ to $-27‰$. Some of them, particularly hollow filaments 50–100 μm long and 10 μm wide, were likely biogenic, although biogenicity assessment on these objects seems to be not sufficient.

COLUMN: RARE-EARTH ELEMENTS AND SIGNIFICANCE OF SHALE (PAAS)-NORMALIZATION

Rare-earth elements (REEs) are a series of elements including lanthanides, Sc, and Y. Lanthanides are composed of 15 elements with continuous atomic numbers from $_{57}La$ to $_{71}Lu$. Except Eu and Ce, their valance is commonly 0 or 3+, suggesting similar chemical feature; Eu and Ce has a valance of 2+ and 4+, respectively, in addition to 3+. Atomic radius gradually decreases from La to Lu, which causes fractionation between lanthanides during, for example, partial melting and fractional crystallization of magma. Since the abundances of REEs are highly variable in natural samples, normalized values are generally used to describe their features.

Normalization is generally made against chondrite, primitive mantle, and shale. Normalization against shale (NASC; North American Shale Composite, PAAS; Post-Archean Australian Shale Composite; MUQ; Mud from Queensland) is generally applied to fine-grained sediments and sedimentary rocks and chemical precipitates (chert and carbonate). Here, shale compositions are regarded to represent the upper continental crustal compositions as a source of REEs supplied to the oceans.

Signatures of REEs in Archean chemical sediments have often been used to judge their marine or nonmarine origin (e.g., Allwood et al., 2010); based on that, modern seawater and its derivatives are characterized by (1) enrichment of HREE (Gd-Lu) relative to LREE (La-Eu), due to selective retention of HREE in seawater, and preferential coupling of HREE with carbonate ions compared to LREE (Lee and Byrne, 1992), (2) positive La anomalies on the basis of shale-normalization, explained by the W-type tetrad effect in seawater and chemical sediments of seawater origin (Masuda et al., 1987), and (3) Y/Ho values significantly larger than the value (30) of chondrite (super-chondritic value, up to 70), explained by preferential scavenging of lanthanides by particulate matter in water column and/or near hydrothermal vent settings (Nozaki et al., 1997). It has been widely accepted that Archean seawater was characteristically enriched in Eu. Positive Eu-anomalies often observed in Archean chemical sediments have generally been regarded as a signature of enrichment of the hydrothermal component in seawater of the same age (e.g., Sugitani, 1992; Bolhar et al., 2005).

REFERENCES

Allwood, A.C., Kamber, B.S., Walter, M.R., Burch, I.W., Kanik, I. 2010. Trace elements record depositional history of an Early Archean stromatolitic carbonate platform. *Chemical Geology* 270, 148–163.

Appel, P.W.U., Moorbath, S., Touret, J.L.R. 2003, Early Archaean processes and the Isua greenstone belt, West Greenland. *Precambrian Research* 126, 173–179,

Awramik, S.M., Schopf, J.W., Walter, M.R. 1983. Filamentous fossil bacteria from the Archean of Western Australia. *Precambrian Research* 20, 357–374.

Awramik, S.M., Schopf, J.W., Walter, M.R. 1988. Carbonaceous filaments from North Pole, Western Australia: Are they fossil bacteria in Archean stromatolites? A discussion. *Precambrian Research* 39, 303–309.

Barghoorn, E.S., Schopf, J.W. 1966. Microorganisms three billion years old from Precambrian of South Africa. *Science* 152, 758–763.

Bolhar, R., Van Kranendonk, M.J., Kamber, B.S. 2005. A trace element study of siderite–jasper banded iron formation in the 3.45 Ga Warrawoona Group, Pilbara Craton—Formation from hydrothermal fluids and shallow seawater. *Precambrian Research* 137, 93–114.

Bower, D.M., Steele, A., Fries, M.D., Green, O.R., Lindsay, J.F. 2016. Raman imaging spectroscopy of a putative microfossil from the ~3.46 Ga Apex chert: insights from quartz grain orientation. *Astrobiology* 16, 169–180.

Brasier, M.D., Antcliffe, J., Saunders, M., Wacey, D. 2015. Changing the picture of Earth's earliest fossils (3.5-1.9 Ga) with new approaches and new discoveries. *Proceedings of the National Academy of Sciences United States of America* 112, 4859–4864.

Brasier, M.D., Green, O.R., Jephcoat, A.P., Kleppe, A.K., Van Kranendonk, M.J., Lindsay, J.F., Steele, A., Grassineau, N.V. 2002. Questioning the evidence for Earth's oldest fossils. *Nature* 416, 76–81.

Brasier, M., McLoughlin, N., Green, O., Wacey, D. 2006. A fresh look at the fossil evidence for early Archaean cellular life. *Philosiphical Transactions of the Royal Soceity B361*, 887–902.

Bridgewater, D., Allaart, J.H., Schopf, J.W., Klein, C., Walter, M.R., Barghoorn, E.S., Strother, P., Knoll, A.H., Gorman, B.E. 1981. Microfossil-like objects from the Archaean of Greenland: a cautionary note. *Nature* 289, 51–53.

Brooks, J., Muir, M.D., Shaw, G., 1973. Chemistry and morphology of Precambrian microorganisms. *Nature* 244, 215–217.

Buick, R., 1984. Carbonaceous filaments from North Pole, Western Australia: Are they fossil bacteria in Archaean stromatolites? *Preacmbrian Research* 24, 157–172.

Buick, R., 1988. Carbonaceous filaments from North Pole, Western Australia: Are they fossil bacteria in archaean stromatolites? *A reply Precambrian Research* 39, 311–317.

Buick, R., 1990. Microfossil recognition in Archean rocks: an appraisal of spheroids and filaments from a 3500 M.Y. old chert-barite unit at North Pole, Western Australia. *Palaios* 5, 441–459.

Buick, R., Dunlop, J.S.R., 1990. Evaporitic sediments of Early Archaean age from the Warrawoona Group, North Pole, Western Australia. *Sedimentology* 37, 247–277.

Byerly, G.R., 2015. Onverwacht Group. *Encyclopedia of Astrobiology*. doi:10.1007/978-3-662-44185-5_5128.

Cates, N.L., Mojzsis, S.J. 2007. Pre-3750 Ma supracrustal rocks from the Nuvvuagittuq supracrustal belt, northern Quebec. *Earth and Planetary Science Letters* 255, 9–21.

Cates, N. L., Ziegler, K., Schmitt, A. K., Mojzsis, S. J. 2013. Reduced, reused and recycled: Detrital zircons define a maximum age for the Eoarchean (ca. 3750–3780 Ma) Nuvvuagittuq Supracrustal Belt, Québec (Canada). *Earth and Planetary Science Letters* 362, 283–293.

De Gregorio, B.T., Sharp, T.G., Flynn, G.J. Wirick, S., Hervig, R.L. 2009. Biogenic origin for Earth's oldest putative microfossils. *Geology* 37, 631–634.

Derenne, S., Robert, F., Skrzypczak-Bonduelle, A., Gourier, D., Binet, L., Rouzaud, J.-N. 2008. Molecular evidence for life in the 3.5 billion year old Warrawoona chert. *Earth and Planetary Science Letters* 272, 476–480.

Djokic, T., Van Kranendonk, M.J., Campbell, K.A., Walter, M.R., Ward, C.R. 2017. Earliest signs of life on land preserved in ca. 3.5 Ga hot spring deposits. *Nature Communications* 8, 1–9.

Dodd, M.S., Papineau, D., Grenne, T., Slack, J.F., Rittner, M., Pirajno, F., O'Neil, J., Little, C.T.S. 2017. Evidence for early life in Earth'sb oldest hydrothermal vent precipitates. *Nature* 543, 60–64.

Douglas, S. 2005. Mineralogical footprints of microbial life. *American Journal of Science* 305, 503–525.

Duck, L.J., Glikson, M., Golding, S.D., Webb, R.E. 2007. Microbial remains and other carbonaceous forms from the 3.24 Ga Sulphur Springs black smoker deposit, Western Australia. *Precambrian Research* 154, 205–220.

Dunlop, J.S.R., Milne, V.A., Groves, D.I., Muir, M.D. 1978. A new microfossil assemblage from the Archaean of Western Australia. *Nature* 274, 676–678.

Engel, A.E.J., Nagy, B., Nagy, L.A., Engel, C.G., Kremp, G.O.W., Drew, C.M. 1968. Alga-like forms in Onverwacht series, South Africa: Oldest recognized lifelike forms on Earth. *Science* 161, 1005–1008.

Eriksson, K.A., Simpson, E.L. 2000. Quantifying the oldest tidal record: The 3.2 Ga Moodies Group, Barberton Greenstone Belt, South Africa. *Geology* 28, 831–834.

Glikson, M., Duck, L.J., Golding, S.D., Hofmann, A., Bolhar, R., Webb, R., Baiano, J.C.F., Sly, L.I. 2008. Microbial remains in some earliest Earth rocks: Comparison with a potential modern analogue. *Precambrian Research* 164, 187–200.

Golubic, S., Hofmann, H.J. 1976. Comparison of Holocene and Mid-Precambrian entophysalidaceae (Cyanophyta) in stromatolitic algal mats: cell division and degradation. *Journal of Paleontology* 50, 1074–1082.

Grenne, T., Slack, J.F. 2003. Bedded jaspeers of the Ordovician Løkken ophiolite, Norway: seafloor deposition and diagenetic maturation of hydrothermal plume-derived silica-iron gels. *Mineralium Deposita* 38, 625–639.

Grey, K., Roberts, F.I., Freeman, M.J., Hickman, A.H., Van Kranendonk, M.J., Bevan, A.W.R. 2010. Management plan for state geoheritage reserves. Geological Survey of Western Australia, Record 2010/13.

Heubeck, C., Engelhardt, J., Byerly, G.R., Zeh, A., Sell, B., Luber, T., Lowe, D.R. 2013. Timing of deposition and deformation of the Moodies Group (Barberton Greenstone Belt, South Africa): Very-high-resolution of Archaean surface processes. *Precambrian Research* 231, 236–262.

Hickman, A.H. 2012. Review of the Pilbara Craton and Fortescue Basin, Western Australia: Crustal evolution providing environments for early life. *Island Arc.* doi:10.1111/j.1440-1738.2011.00783.x.

Hickman-Lewis, K., Westall, F., Cavalazzi, B. 2019. Traces of early life from the Barberton greenstone belt, South Africa. In M.J. Van Kranendonk, V. Bennett, E. Hoffmann eds. *Earth's Oldest Rocks.* pp.1029–1058. Elsevier, Amsterdam.

Homann M. 2019. Earliest life on Earth: Evidence from the Barberton greenstone belt, South Africa. *Earth-Science Reviews* 196, 102888.

Homann, M., Heubeck, C., Bontognali, T.R.R., Bouvier, A.-S., Baumgartner, L.S., Airo, A. 2016. Evidence for cavity-dwelling microbial life in 3.22 Ga tidal deposits. *Geology* 44, 51–54.

Javaux, E.J., Marshall, C.P., Bekker, A. 2010. Organic-walled microfossils in 3.2-billion-year-old shallow-marine siliciclastic deposits. *Nature* 463, 934–938.

Kaźmierczak J., Kremer, B. 2019. Pattern of cell division in ~3.4 Ga-old microbes from South Africa. *Precambrian Research* 331, 105357

Kiyokawa, S., Ito, T., Ikehara, M., Kitajima, F. 2006. Middle Archean volcano-hydrothermal sequence: Bacterial microfossil-bearing 3.2 Ga Dixon Island Formation, coastal Pilbara terrane, Australia. *Geological Society of America Bulletin* 118, 3–22.

Kiyokawa, S., Koge, S., Ito, T., Ikehara, M. 2014. An ocean-floor carbonaceous sedimentary sequence in the 3.2-Ga Dixon Island Formation, coastal Pilbara terrane, Western Australia. *Precambrian Research* 255, 124–143.

Knoll, A.H., Barghoorn, E.S. 1977. Archean microfossils showing cell division from the Swaziland system of South Africa. *Science* 198, 396–398.

Kremer, B., Kaźmierczak, J. 2017. Cellulary preserved microbial fossils from ~3.4 Ga deposits of South Africa: A testimony of early appearance of oxygenic life? *Precambrian Research* 295, 117–129.

Lee J.H., Byrne R.H. 1992. Examination of comparative rare earth element complexation behavior using linear free-energy relationships. *Geochimica et Cosmochimica Acta* 56, 1127–1137.

Li, J., Zhou, H., Peng, X., Wu, Z., Chen, S., Fang, J. 2012. Microbial diversity and biomineralization in low temperature hydrothermal iron-silicate-rich precipitates of the Lau Basin hydrothermal field. *FEMS Microbiology Ecology* 81, 206–216.

Lowe, D.R., Byerly, G.R., Kyte, F.T. 2014. Recently discovered 3.42–3.23 Ga impact layers, Barberton Belt, South Africa: 3.8 Ga detrital zircons, Archean impact history, and tectonic implications. *Geology* 42, 747–750.

Marshall, C.P., Emry, J.R., Marshall, A.O. 2011. Haematite pseudomicrofossils present in the 3.5-billion-year-old Apex Chert. *Nature Geoscience* 4, 240–243.

Marshall, A.O., Emry, J.R., Marshall, C.P. 2012. Multiple generations of carbon in the Apex Chert and implications for preservation of microfossils. *Astrobiology* 12, 160–166.

Masuda, A., Kawakami, O., Dohmoto, Y., Takenaka, T. 1987. Lanthanide tetrad effects in nature: two mutually opposite types, W and M. *Geochemical Journal* 21, 119–124.

Mloszewska, A.M., Mojzsis, S.J., Pecoits, E., Papineau, D., Dauphas, N., Konhauser, K.O. 2013. Chemical sedimentary protoliths in the > 3.75 Ga Nuvvuagittuq supracrustal belt (Quebec, Canada). *Gondwana Research* 23, 574–594.

Moorbath, S., Whitehouse, M.J. 1996. Sm-Nd isotopic data and Earth's evolution. *Science* 273, 1878.

Muir, M.D., Hall, D.O. 1974. Diverse microfossils in Precambrian Onverwacht group rocks of South Africa. *Nature* 252, 376–378.

Nagy, B., Nagy, L.A., 1969. Early Pre-Cambrian Onverwacht microstructures: Possibly the oldest fossils on Earth? *Nature* 223, 1226–1229.

Noffke, N., Christian, D., Wacey, D., Hazen, R.M. 2013. Microbially induced sedimentary structures recording an ancient ecosystem in the ca. 3.48 billion-year-old Dresser Formation, Pilbara, Western Australia. *Astrobiology* 13, 1103–1124. doi:10.1089/ast.2013.1030.

Nozaki, Y., Zhang, J., Amakawa, H. 1997. The fractionation between Y and Ho in the marine environment. *Earth and Planetary Science Letters* 148, 329–340.

Nutman, A.P., Bennett, V.C., Friend, C.R.L., Van Kranendonk, M.J., Rothacker, L., Chivas, A.R. 2019. Cross-examining Earth's oldest stromatolites: Seeing through the effects of heterogeneous deformation, metamorphism and metasomatism affecting Isua (Greenland) ~3700 Ma sedimentary rocks. *Precambrian Research* 331, 105347.

Nutman, A.P., Friend, C.R.L. 2009. New 1:20,000 scale geological maps, synthesis and history of investigation of the Isua supracrustal belt and adjacent orthogneisses, southern West Greenland: A glimpse of Eoarchaean crust formation and orogeny. *Precambrian Research* 172, 189–211.

O'Neil, J., Carlson, R. W., Paquette, J.-L., Francis, D. 2012. Formation age and metamorphic history of the Nuvvuagittuq greenstone belt. *Precambrian Research* 220–221, 23–44.

Oehler, D.Z., Walsh, M.W., Sugitani, K., Liu, M.-C., House, C.H. 2017. Large and robust lenticular microorganisms on the young Earth. *Precambrian Research* 296, 112–119.

Pflug, H.D. 1967. Structured organic remains from the Fig Tree Series (Precambrian) of the Barberton mountain land (South Africa). *Review of Palaeobotany and Palynology* 5, 9–29.

Pflug, H.D. 1978. Yeast-like microfossils detected in oldest sediments of the earth. *Naturwissenschaften* 65, 611–615.

Pflug, H.D., Jaeschke-Boyer, H. 1979. Combined structural and chemical analysis of 3,800-Myr-old microfossils. *Nature* 280, 483–486.

Pinti, D.L., Mineau, R., Clement, V. 2009. Hydrothermal alteration and microfossil artefacts of the 3,465-million-year-old Apex chert. *Nature Geoscience* 2, 640–643.

Rasmussen, B. 2000. Filamentous microfossils in a 3,235-million-year-old volcanogenic massive sulphide deposit. *Nature* 405, 676–679.

Roedder, E. 1981. Are the 3,800-Myr-old Isua objects microfossils, limonite-stained fluid inclusions, or neither? *Nature* 293, 459–462.

Schopf, J.W. 1976. Are the oldest 'fossils', fossils? *Origins of Life* 7, 19–36.

Schopf, J.W. 1993. Microfossils of the early Archean Apex chert: New evidence of the antiquity of life. *Science* 260, 640–646.

Schopf, J.W., Barghoorn, E.S. 1967. Algal-like fossils from the early Precambrian of South Africa. *Science* 156, 508–512

Schopf, J.W., Kudryavtsev, A.B. 2009. Confocal laser scanning microscopy and Raman imagery of ancient microscopic fossils. *Precambrian Research* 173, 39–49.

Schopf, J.W., Kudryavtsev, A.B., Czaja, A.D., Tripathi, A.B. 2007. Evidence of Archean life: Stromatolites and microfossils. *Precambrian Research* 158, 141–155.

Schopf, J.W., Kudryavstev, A.B., Osterhout, J.T., Williford, K.H., Kitajima, K., Valley, J.W., Sugitani, K. 2018. An anaerobic ~3400 Ma shallow-water microbial consortium: presumptive evidence of Earth's Paleoarchean anoxic atmosphere. *Precambrian Research* 299, 309–318.

Schopf, J.W., Packer, B.M. 1987. Early Archean (3.3-billion to 3.5-billion-year-old) microfossils from Warrawoona Group, Australia. *Science* 237, 70–73.

Schopf, J.W., Walter, M.R. 1983. *Earth's Earliest Biosphere: Its Origin and Evolution.* J.W. Schopf, ed. pp. 543. Princeton University Press, Princeton, New Jersey.

Smith, A.J.B., Beukes, N.J., Gutzmer, J., Johnson, C.M., Czaja, A.D. 2012. Iron isotope fractionation in stromatolitic oncoidal iron formation, Mesoarchean Witwatersrand-Mozaan Basin, South Africa. *Goldscmidt* 2012, Abstracts 2384.

Sugitani, K. 1992. Geochemical characteristics of Archean cherts and other sedimentary rocks in the Pilbara Block, Western Australia: evidence for Archean seawater enriched in hydrothermally-derived iron and silica. *Precambrian Research* 57, 21–47.

Sugitani, K. 2019a. Early Archean (pre-3.0 Ga) cellularly preserved microfossils and microfossil-like structures from the Pilbara Craton, Western Australia – A review. In M.J. Van Kranendonk et al. Eds. *Earth's Oldest Rocks* 2nd Edtion. Elsevier, Amsterdam, 1007–1028p.

Sugitani, K. 2019b. Fossils of ancient microorganisms. In V.M. Kolb Ed. *Handbook of Astrobiology.* CRC Press.

Sugitani, K., Grey, K., Allwood, A., Nagaoka, T., Mimura, K., Minami, M., Marshall, C.P., Van Kranendonk, M.J., Walter, M.R. 2007. Diverse microstructures from Archaean chert from the Mount Goldsworthy-Mount Grant area, pilbara craton, western australia: Microfossils, dubiofossils, or pseudofossils? *Precambrian Research* 158, 228–262.

Sugitani, K., Lepot, K., Nagaoka, T., Mimura, K., Van Kranendonk, M., Oehler, D.Z., Walter, M.R. 2010. Biogenicity of morphologically diverse carbonaceous microstructures from the ca. 3400 Ma Strelley Pool Formation, in the Pilbara Craton, Western Australia. *Astrobiology* 10, 899–920.

Ueno, Y., Isozaki, Y., Yurimoto, H., Maruyama, S. 2001. Carbon isotopic signatures of individual Archean microfossils (?) from Western Australia. *International Geology Review* 43, 196–212.

Ueno, Y., Yamada, K., Yoshida, N., Maruyama, S., Isozaki, Y. 2006. Evidence from fluid inclusions for microbial methanogenesis in the Archaean era. *Nature* 440, 516–519.

Van Kranendonk, M.J. 2006. Volcanic degassing, hydrothermal circulation and the flourishing of early life on Earth: A review of the evidence from c. 3490-3240 Ma rocks of the Pilbara Supergroup, Pilbara Craton, Western Australia. *Earth-Science Reviews* 74, 197–240.

Van Kranendonk, M.J., Philippot, P., Lepot, K., Bodorkosm S., Pirajno, F. 2008. Geological setting of Earth's oldest fossils in the ca. 3.5 Ga Dresser Formation, Pilbara Craton, Western Australia. *Precambrian Research* 167, 93–124.

Van Kranendonk, M.J., Smithies, R.H., Hickman, A.H., Chanpion, D.C. 2006. Review: secular tectonic evolution of Archean continental crust: interplay between horizontal and vertical processes in the formation of the Pilbara Craton, Australia. *Terra Nova* 19, 1–38.

Vearncombe, S.E., Barley, M.E., Groves, D.I., McNaughton, N.J., Mikucki, E.J., Vearncombe, J.R. 1995. 3.26 Ga black smoker-type mineralization in the Strelley Belt, Pilbara Craton, Western Australia. *Geological Society of London Journal* 152, 587–590.

Wacey, D. 2009. *Early Life on Earth: A Practical Guide.* pp. 274, Springer.

Wacey, D., 2011. Earliest evidence for life on Earth: An Australian perspective. *Australian Journal of Earth Sciences* 59, 153–166.

Wacey, D., Kilburn, M.R., Saunders, M., Cliff, J., Brasier, M.D. 2011. Microfossils of sulphur-metabolizing cells in 3.4-billion-year-old rocks of Western Australia. *Nature Geoscience* 4, 698–702.

Wacey, D., Saunders, M., Cliff, J., Kilburn, M.R., Kong, C., Barley, M.E., Brasier, M.D. 2014, Geochemistry and nano-structure of a putative ~3240 million-year-old black smoker biota, Sulphur Springs Group, Western Australia. *Precambrian Research* 249, 1–12. doi:10.1016/j.precamres.2014.04.016.

Wacey, D., Saunders, M., Kong, C. 2018. Remarkably preserved tephra from the 3430 Ma Strelley Pool Formation, Western Australia: Implications for the interpretation of Precambrian microfossils. *Earth and Planetary Science Letters* 487, 33–43.

Walsh, M.M., 1992. Microfossils and possible microfossils from the early Archean Onverwacht Group, Barberton Mountain Land, South Africa. *Precambrian Research* 54, 271–293.

Walsh, M.M., Lowe, D.R. 1985. Filamentous microfossils from the 3,500-Myr-old Onverwacht Group, Barberton Mountain Land, South Africa. *Nature* 314, 530–532.

Westall, F., de Wit, M.J., Dann, J., van der Gaast, S., de Ronde, C.E.J., Gerneke, D., 2001. Early Archean fossil bacteria and biofilms in hydrothermally-influenced sediments from the Barberton Greenstone belt, South Africa. *Precambrian Research* 106, 93–116.

Westall, F., Foucher, F., Cavalazzi, B., de Vries, S.T., Nijman, W., Pearson, V., Watson, J., Verchovsky, A., Wright, I., Rouzaud, J.-N., Marchesini, D., Anne, S. 2011. Volcaniclastic habitats for early life on Earth and Mars: A case study from ~3.5 Ga-old rocks from the Pilbara, Australia. *Planetary and Space Science* 59, 1093–1106.

Westall, F., de Vries, S.T., Nijman, W., Rouchon, V., Orberger, B., Pearson, V., Watson, J., Verchovsky, A., Wright, I., Rouzaud, J.-N., Marchesini, D., Severine, A. 2006. The 3.455 Ga "Kitty's Gap Chert", an early Archean microbial ecosystem. In: Reimold, W.U. and Gibson, R.L. eds. *Processes on the Early Earth*. Geological Society of America Special Paper, 405. *Geological Society of America*, pp. 105–131.

9 Overview of the Pilbara Microstructures 1
The Farrel Quartzite Assemblage

9.1 INTRODUCTION

My first encounter with Archean life was back to 2001. I found various microstructures in a petrographic thin section of black chert collected from the ca. 3.0 Ga-old Farrel Quartzite (FQ) at the Mount Grant of the Goldsworthy greenstone belt, about 100 km east of Port Hedland, as mentioned in Preface. This luck was followed by discoveries of similar assemblages from three remote localities of the ca. 3.4 Ga-old Strelley Pool Formation (SPF). The microfossil assemblages described from these two formations with assumed significant time gap (400 million years) are basically similar to each other, being composed of spheroids, films, lenses, and filaments. In a broad sense, most of them are morphologically equivalents to previously described Archean microfossils, including putative ones. However, their unexpected exellent preservation states, morphological variations, and occurrence of diverse colonial aggregates are unparalleled to any other Archean microfossil assemblages. In this chapter, I briefly describe the FQ microfossil assemblage, with detailed information about lithostratigraphy and petrography of the volcanic sedimentary successions in the southern part of the Goldsworthy greenstone belt. Interpretations of these assemblages would be given in Chapters 11–13. My intention to give accurate geological information had resulted in too detailed geological descriptions, which may bother general readers. So, readers who are not unfamiliar with the geological side of things can skip the related sections. This is the same for the next chapter.

9.2 LOCAL GEOLOGY AND LITHOSTRATIGRAPHY OF THE GOLDSWORTHY GREENSTONE BELT

9.2.1 LOCAL GEOLOGY

The Goldsworthy greenstone belt located at northeastern edge of the Pilbara Craton (Figure 9.1) is composed of Mount Grant and Mount Goldsworthy and unnamed ridges (Figure 9.2). According to the geological map issued from Geological Survey of Western Australia (GSWA) (Australia 1: 100,000 Geological Series De

DOI: 10.1201/9780367855208-9

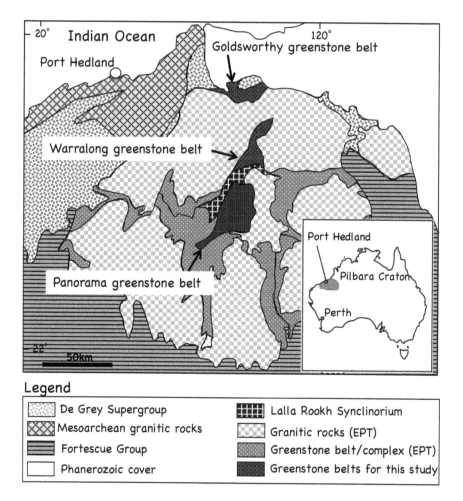

FIGURE 9.1   Geology of the East Pilbara Terrane. After Hickman (2008).

Grey, Smithies et al., 2004), Archean volcanic-sedimentary successions of wide age ranges occur in this greenstone belt, including unassigned mafic to ultramafic volcanic rocks, the ca. 3.5 Ga-old Panorama Formation, the ca. 3.4 Ga-old Strelley Pool Formation, the 3.3 Ga-old Euro Basalt, and the 3.02 to 2.93 Ga-old De Grey Supergroup (Figure 9.2). The ca. 3.0 Ga-old Gorge Creek Group, a lower part of the De Grey Supergroup, is here composed of the Farrel Quartzite and the overlying Cleaverville Formation. The Farrel Quartzite is underlain by the Euro Basalt or unassigned mafic-ultramafic volcanic rocks, with unconformity. Age data obtained from this greenstone belt is limited to two groups. One is from an unnamed ridge located in the northern portion, which is composed of felsic volcaniclastic sandstone. Two populations of detrital zircon grains gave U-Pb ages of $3,458 \pm 9$ Ma and $3,513 \pm 2$ Ma, respectively (Hickman, 2008 and references therein). The other is from sandstone of the Farrel Quartzite in the southeastern Goldsworthy greenstone belt.

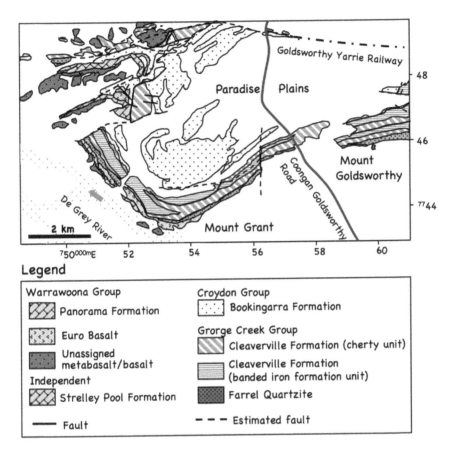

FIGURE 9.2 Local geology of the Mount Goldsworthy greenstone belt, after Australia 1:100,000 Geological Series "De Grey" (Smithies et al., 2004). In this map, the Strelley Pool Formation and the Panorama Formation are indistinguishable.

Two detrital zircon grains gave CHIME (Chemical Isochron Method: Suzuki, 2005) ages of around 3.7 Ga (Sugitani et al., 2003).

9.2.2 OVERVIEW OF LITHOSTRATIGRAPHY

Mount Goldsworthy and Mount Grant are composed basically of the common volcanic-sedimentary succession (Figure 9.3). The succession dipping mostly 90° comprise the east-west trending straight ridge at Mount Goldsworthy, whereas the neighboring Mount Grant is vent to the north. As suggested from the geological map (Figure 9.2), the sedimentary successions comprising Mount Grant exhibit a synclinal structure. The structure was half broken probably by tectonic processes, leaving a complex mass now present in the northern part of this greenstone belt.

The lower sedimentary succession of the Goldsworthy greenstone belt is composed of the Farrel Quartzite and the Cleaverville Formation in most portions. At the

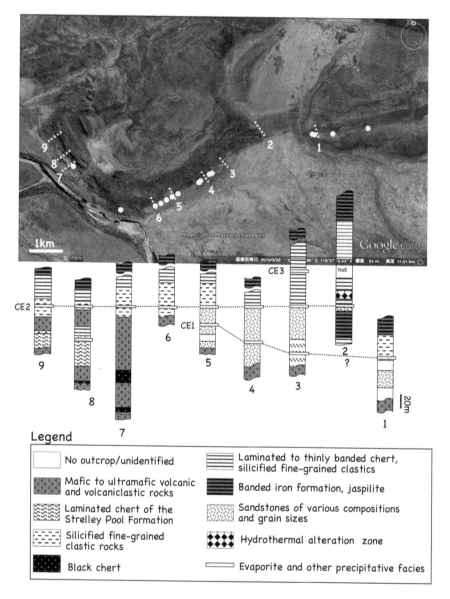

FIGURE 9.3 Stratigraphic columns measured at representative localities of the Mount Goldsworthy greenstone belt. White circles in the Google image indicate localities from which carbonaceous black cherts containing fossil-like microstructures were collected.

western portion, a thin sedimentary unit correlative to the Strelley Pool Formation occurs below the Euro Basalt and above the unassigned volcanic unit.

The sedimentary succession was previously subdivided into the lower, the middle, and the upper units based on the sections measured at Mount Goldsworthy

(Sugitani et al., 2003, 2006). The former two correspond to the siliciclastic rocks of the Farrel Quartzite, whereas the rest corresponds to the Cleaverville Formation composed mostly of chert and banded iron formation (BIF), respectively. In my previous studies, two beds of silicified evaporite (*Column*) characterized by vertically to subvertically oriented giant columnar crystals were identified in the Farrel Quartzite, including the lower bed, named CE1 (Figure 9.4a), and the upper bed, named CE2 (Figure 9.4b). The latter occurs at the higher horizon of the Farrel Quartzite close to the boundary with the Cleaverville Formation: the boundary is conformable and traditional. During the later fieldtrips, a single bed of possible evaporite origin (CE3) was identified also in the Cleaverville Formation (Figures 9.3, and 9.4c, d). In many places, this bed appears to be massive, but locally displays relics of columnar crystals. This bed can be traced at least 1.5 km along the strike.

FIGURE 9.4   Evaporitic facies of the Farrel Quartzite and the Cleaverville Formation. (a) Evaporite of CE1 at Mt. Goldsworthy. (b) Evaporite of CE2 at northwestern Mt. Grant. (c) Possible evaporite in the Cleaverville Formation at eastern Mt. Grant. The thickness of the bed is around 30 cm. (d) Close-up view of the Cleaverville possible evaporite.

9.3   SILICICLASTIC UNIT (THE FARREL QUARTZITE)

9.3.1   Assignment to the Farrel Quartzite – Its Story and Remained Problem

The lower siliciclastic unit at the Goldsworthy greenstone belt, from which diverse fossil-like microstructures have been discovered, was first interpreted as to be the Corboy Formation of the Gorge Creek Group, according to the geological map issued from Geological Survey of Western Australia (GSWA) (Sugitani et al., 2003). At that time, on the other hand, the mapping project of the Pilbara Craton that aimed revision of lithostratigraphy of the Pilbara Supergroup was conducted by GSWA. Under such situation, our research team had an idea that the rock unit containing the microstructures could be correlative to the 3.4 Ga-old Strelley Pool Formation (Sugitani et al., 2006). This was so attractive for us, because the older the microfossils were, the higher their significance was. However, the lack of definitive evidence for assignment to the Strelley Pool Formation had brought us confusion about the age of microstructures discovered from the Mount Grant and subsequently from Mount Goldsworthy. The idea of assignment to the Strelley Pool Formation was eventually discarded, and our controversies had converged to that the microstructure-bearing chert is a part of the Farrel Quartzite (Sugitani et al., 2007), newly defined formation of the Gorge Creek Group, although this assignment may still need to be provisional (Chapter 11).

9.3.2   Lithostratigraphy and Sedimentary Geology

The Farrel Quartzite is composed mainly of mudstone, siltstone, sandstone, conglomerate, volcaniclastic rocks, and chert. These rocks have been subjected to greenschist facies metamorphism and pervasive silicification. At the central portion of Mount Grant (the type locality), the succession is composed of several subunits based on lithological variations, which are here denoted as the lower clastic unit (CU1), the lower chert-evaporite unit (CE1), the middle clastic unit (CU2), and the upper clastic unit (CU3) and the upper chert-evaporite unit (CE2): this subdivision is not always conservative along the strike. Detailed lithostratigraphy of CE1 and CU2 is shown in Figure 9.5.

CU1 at the type locality is composed of very coarse to fine grained sandstone. Very coarse-grained sandstone is matrix-poor and rich in volcanic fragments and cherts (Figure 9.6a). Volcanic fragments are rounded to angular and are texturally various, although they are composed exclusively of microcrystalline quartz and fine TiO_2-oxides such as anatase, due to silicification. Some fragments are characterized by micron-sized spinifex texture, suggesting ultramafic composition, whereas others are vitric grains. Fragments composed dominantly of microcrystalline quartz and interlocked mica are also present: they have possibly been derived from felsic volcanic rocks. Additionally, there can be seen monocrystalline quartz grains, which are possibly derived from granitic or felsic igneous rocks. Medium- to fine-grained sandstone is compositionally distinct from coarser-grained sandstones. They are rich in matrix composed of microcrystalline quartz and mica. Framework grains tend to be dominated by angular-shaped quartz of possible first-cycle volcanic in

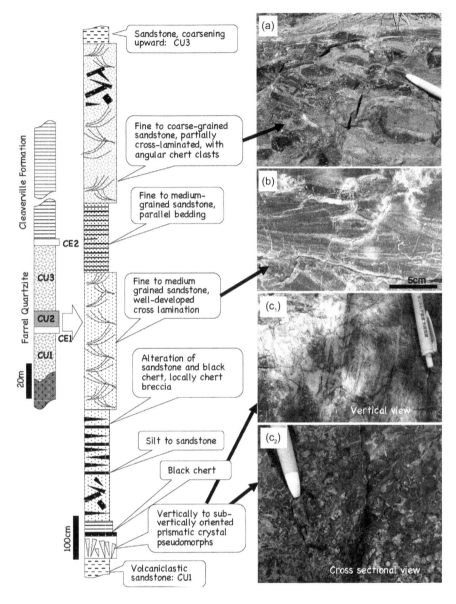

FIGURE 9.5 Detailed lithostratigraphy of the middle clastic unit (CU2) of the Farrel Quartzite and CE1.

origin, subordinately with mica and mixture of mica and microcrystalline quartz (Figure 9.6b). The lower evaporite-chert unit (CE1), ca. 2 km traceable along the strike at the central Mount Grant, is composed of massive chert with giant crystal pseudomorphs (Figure 9.5c) and overlying black-dark gray beds, including carbonaceous cherts characterized by fine lamination and detrital carbonaceous grains

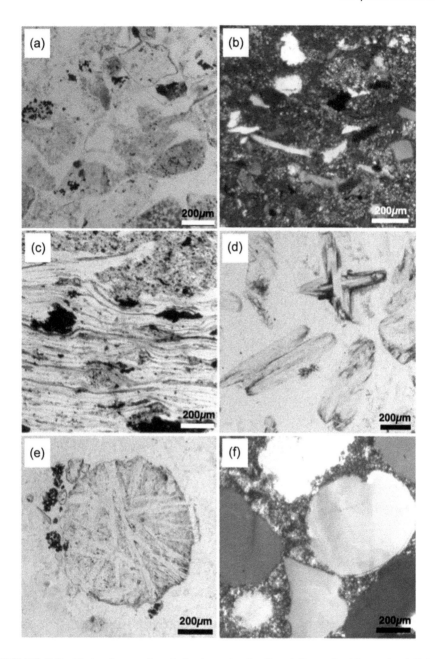

FIGURE 9.6 Photomicrographs of clastic rocks the Farrel Quartzite. Transmitted light except for (b) and (f) (crossed nicols). (a) Poorly sorted lithic sandstone enriched in volcanic fragments from CU1. (b) Medium-grained sandstone from CU1, characterized by very angular monocrystalline quartz grains. (c) Finely laminated carbonaceous chert with carbonaceous grains from CE1. (d) Aggregates of columnar crystals in a cross-laminated sandstone from CU2. (e) Spherule characterized by crystal skeletons in a cross-laminated sandstone from CU2. (f) Well-rounded monocrystalline quartz in very-coarse grained sandstone from CU3.

(Figure 9.6c) and fine-grained clastic rocks. The middle clastic unit (CU2) is composed of bedded fine- to very coarse-grained sandstone with minor chert, characterized by cross-lamination in lower horizon (Figure 9.5b) and by tabular chert clasts (rip-up clasts) in upper horizon (Figure 9.5a). Cross-laminated sandstone partially contains extensively silicified reworked columnar crystal aggregates (Figure 9.6d), which are admixed with trace amounts of spherules of possible impact origin (Figure 9.6e) (Chapter 5). Coarser-grained sandstone tends to be poorer in the matrix. Framework grains are dominated by monocrystalline quartz, with subordinate polycrystalline quartz. Many of these quartz grains are well rounded. CU3 is poorly bedded and shows a coarsening upward trend. The compositions of very coarse-grained sandstones of this unit vary from dominance of rounded quartz grains (Figure 9.6f) to that of angular lithic fragments. Locally, lenticular conglomerate body composed of mafic to ultramafic volcanic clasts is intercalated. The uppermost portion of CU3 at the type section is characterized by massive and quartz-rich matured sandstone, which is directly overlain by CE2 that is comprised of carbonaceous black chert containing fossil-like microstructures, massive chert with giant crystal pseudomorphs, terrigenous clastic rocks, and volcaniclastic rocks.

As shown by the measured sections (Figure 9.3), lateral variation in lithostratigraphy and thickness of the Farrel Quartzite is conspicuous. The Farrel Quartzite is up to 80 m thick at the central portion of Mount Grant, whereas becomes to be thinner to the east and the west. At the outcrop along the Coongan Goldsworthy Road (Figure 9.2), there can be seen no siliciclastic rocks correlative to the Farrel Quartzite; the basal volcanic unit appears to be overlain directly by the cherty unit correlative to the Cleaverville Formation. Also, at the northwestern end of the Mount Grant, the siliciclastic unit occurs only as thin silty layer. Though details are omitted here, significant lateral variation in thickness is observed also at Mount Goldsworthy (Sugitani et al., 2003).

9.3.3 Sources of Detrital Materials

Chemistry and mineralogy of siliciclastic rocks are controlled preliminarily by compositions and degree of weathering of source rocks and subsequently by sedimentation processes such as mechanical sorting, decomposition, and diagenesis (e.g., Nesbitt et al., 1996). Chemical compositions of immature siliciclastic sediments such as shale and graywacke have often been used to infer provenance and tectonic setting (e.g., Bhatia and Crook, 1986). In addition, indices of chemical weathering could provide us with implications of the ancient climate and the atmospheric composition (e.g., Fedo et al., 1996; Sugitani et al., 1996). As it has been argued by several researchers that in the Pilbara Craton, continental crust had evolved as early as >3.0 Ga ago (e.g., Buick et al., 1995; Green et al., 2000), siliciclastic rocks of the ca. 3.0 Ga-old Farrel Quartzite are expected to provide information on the early evolution of continental crusts.

In this context, petrographic and chemical analyses of siliciclastic rocks collected from the Mount Goldsworthy were performed (Sugitani et al., 2006). Key feature is that some of the coarse-grained sandstone contain abundant, very well-rounded monocrystalline quartz. They can be derived from either felsic volcanic rocks that

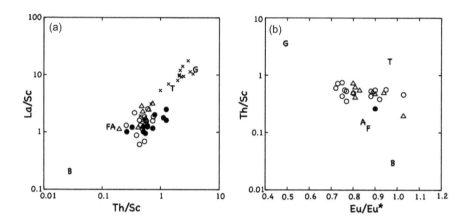

FIGURE 9.7 (a) Th/Sc plotted against La/Sc. (b) Th/Sc plotted against Eu/Eu* (Cullers and Podkovyrov, 2002). Open circles: medium-grained sandstone; black circles: coarse- to very coarse-grained sandstone; open triangles: fine-grained sandstone from the middle unit. Crosses in (a) are volcanic and volcaniclastic rocks of the Panorama Formation (Cullers et al., 1993). G: granite; T: tonalite-trondhjemite-granodiorite; F: felsic volcanic rock; A: andesite; B: basalt. Data for average values of these early Archean igneous rocks are sourced from Condie (1993). (Reproduced from Sugitani et al. (2006), with permission from Elsevier.)

could contain monocrystalline quartz as phenocrysts or granitic rocks, important components of the continental crust, although megaquartz-veins and metamorphic rocks such as quartzite could be alternative sources. In order to overcome this ambiguity, we employed geochemical approach using Zr, Ti, Th, Al, Cr, and rare earth elements (REEs). However, straightforward utilization of these elements could result in misinterpretation. Interelement relationships between Al, Ti, and Zr suggested that the source area for the siliciclastic rocks was subjected to acid hydrothermal alteration. It was proposed that this alteration had resulted in Al-dissolution and the formation of Zr- and Ti-enriched residual detrital material. Such detritus could be enriched in heavy rare earth elements (HREEs), because zircon ($ZrSiO_4$) is enriched in HREEs. Contribution of such detrital materials could modify pristine geochemical features of source rocks. Fortunately, however, Th/Sc, La/Sc, and Eu/Eu* (Eu-anomaly), which are good provenance indicators, appeared to have been conservative. Values of La/Sc and Th/Sc tend to be distinct depending on igneous rock types on magnitude orders, whereas Eu/Eu* varies from ca. 0.5 for granitic rocks to ca. 1.0 for basaltic rocks. Plots on the discrimination diagrams using these indices clearly show that the FQ siliciclastic rocks contain detrital materials originated from granitic rocks (Figure 9.7), providing additional evidence for the Archean continental growth.

9.3.4 DETAILED DESCRIPTIONS OF CE2

9.3.4.1 Lithostratigraphy

CE2 is 2–3 m thick and exhibits drastic vertical changes of lithofacies (Figure 9.8). This unit can be traced 7 km along the strike. The basal portion is composed of

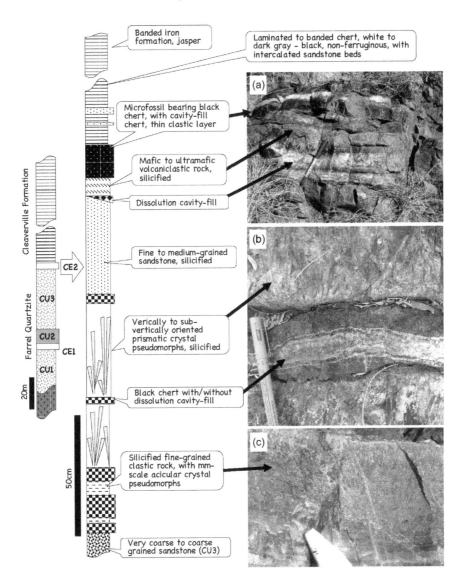

FIGURE 9.8 Detailed lithostratigraphy of the CE2 (see Figure 9.2) of the Farrel Quartzite.

~15-cm-thick black chert bed with a few thin (~1 cm) fine-grained clastic layers containing silicified acicular crystals (Figure 9.8c), which contains barite ($BaSO_4$) inclusions. This subunit is overlain by two sets of massive chert beds that contain subvertically to vertically oriented crystals, identical to CE1 in appearance. The lower and upper massive cherts are ~20 and ~40 cm thick, respectively, and are separated by a thin (~3 cm) black chert (Figure 9.8b), which is occasionally associated with a pure chert layer characterized by botryoidal texture, probably representing dissolution cavity fill. This subunit is followed upward by sandstone, light gray chert, white to translucent chert characterized by stalactite-like dissolution cavity-fill texture, greenish to reddish bed and finally black chert. This black chert is carbonaceous and hosts fossil-like microstructures (Figure 9.8a).

9.3.4.2 Petrography of Black Chert

Although CE2 contains several black chert layers, fossil-like microstructures have been identified only from the uppermost black chert layer. This black chert is mostly less than 20 cm thick (Figure 9.9a). The black chert displays a few different types of microscopic fabrics, including, massive fabric (*Mf*), fenestrae-like fabric (*Fs*), and pseudo-lamination fabric (*Pl*) (Figure 9.9b) (Sugitani et al., 2011). In *Mf*, fine carbonaceous particles are distributed mostly homogenously, with dispersed irregularly shaped masses composed of pure chert. *Fs* refers to portion where irregularly shaped pure chert masses composed of microcrystalline quartz are well developed. Fenestrae-like microcrystalline quartz masses display domain to chalcedonic extinction figure. *Pl* is used here to describe portion in which carbonaceous laminae 100–300 μm thick alternate with pure cherts. The carbonaceous laminae appear to be connected with each other or branched. Therefore, this fabric is not formed by particle settling, although the exact formation process is not known. In addition to these fabrics of carbonaceous composition, stratiform pure chert layers (Cf) of various thicknesses are identified. This layer is characterized by botryoidal structures composed of microcrystalline quartz both at the bottom and at the roof, with megaquartz mosaic infill at the center. Plausible interpretation is that this layer represents postdepositional cavity-fill silica precipitates that formed after dissolution of soluble mineral layers.

At glance, the black chert appears to be free from detritus. However, fragmented (thus detrital) film-like structures (Figure 9.9b) are abundant. In addition, thin intercalated clastic layers and carbonaceous grains with fabrics different from the above described three types are identified (Figure 9.9c). Some of the clastic layers are composed of mafic to ultramafic volcanic fragments admixed with extensively silicified rectangular crystals (Figure 9.9d).

9.3.4.3 Rare-Earth Elements and Y Geochemistry

Analyses of REEs of the black cherts containing fossil-like microstructures collected from CE2 were performed in order to give constraints on their source fluids (Sugahara et al., 2010). Here, these data are briefly described, and their interpretation would be given in Section 9.5, together with data from the Cleaverville Formation. PAAS (Post Archean Australian Shale Composite: Taylor and McLennan, 1985)-normalized REEs signatures of the black cherts from CE2 can be summarized as follows. The samples are characterized by HREE-enrichment. La- and Ce-anomalies are variable, ranging from 0.93 to 4.0 and 0.64 to 1.72, respectively. Most samples, except for two, show weak positive Eu-anomalies (<2). Y/Ho ratios are super-chondritic but less than 40 (see *Column* in the Chapter).

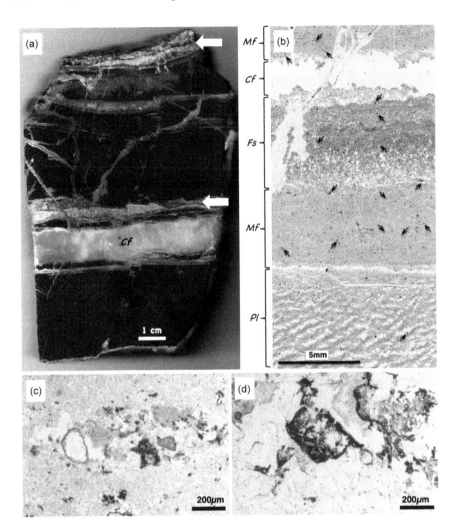

FIGURE 9.9   Lithology of microfossil-bearing carbonaceous black cherts. (a) Polished slab. The white arrows show detrital layers. (b) Vertical change of lithofacies of the black chert. Mf: massive fabric, Cf: cavity-fill facies, Fs: fenestral fabric, Pl: pseudo-lamination fabric. The arrows show film-like structures. (c) Reworked carbonaceous grains in Mf. (d) Mafic to ultramafic volcanic fragments admixed with extensively silicified rectangular crystals at the bottom and the left. Transmitted light for (b)–(d).

9.4   THE CHERT-BIF UNIT (THE CLEAVERVILLE FORMATION)

9.4.1   LITHOSTRATIGRAPHY AND PETROGRAPHY

The Farrel Quartzite in the Goldsworthy greenstone belt is conformably overlain by the Cleaverville Formation. This unit consists dominantly of parallel-laminated variegated chert, silicified mudstone or tuff, carbonaceous black chert, jaspilite, and BIF.

At the type section at the Mount Grant, relatively iron-poor cherts or fine-grained clastic rocks dominate the lower horizon of this unit, in which thin (<30 cm) medium to coarse-grained terrigenous clastic layers or lenses are intercalated. Iron-rich facies (jaspilite and BIF) comprise the upper horizon of this unit. At Mount Goldsworthy, jaspilite, and BIF occur from the base of this unit. Also, relatively thick (~1 m) coarse-grained sandstone bed with platy chert clasts (rip-up clasts) locally occurs (Figure 5.7). As described earlier, evaporitic layer composed of massive chert bed with poorly preserved crystal pseudomorphs (CE3) occurs locally in this unit (Figure 9.4c and d).

Cherts and BIFs of the Cleaverville Formation are various in petrographic features. In cherts, planar to undulating laminae dominate, although anastomosed one is locally observed. Such the laminae or bands are composed of carbonaceous matter (CM), carbonate, hematite (Fe_2O_3), and/or unknown minute minerals. Carbonaceous laminated to banded cherts occasionally have biomat-like fine lamination with ripped up texture (Figure 9.10a). Non-carbonaceous laminated cherts tend to be relatively enriched in fine carbonate particles. These components often comprise granules or peloids (Figure 9.10b), which rarely have concentric textures (Figure 9.10c). Carbonate mineralization associated with carbonaceous laminae can be seen (Figure 9.10d). Banded iron formation is characterized by parallel lamination and banding (Figure 9.10e). Goethite [FeO(OH)], products of weathering or alteration of iron bearing carbonates, occurs in places and locally dominates laminae and bands. In the least-altered portion, however, tiny hematite particles and associated carbonates can be seen (Figure 9.10f).

9.4.2 Rare-Earth Elements and Y Geochemistry

PAAS-normalized REEs patterns of cherts and BIFs from the Cleaverville Formation have been reported by Sugahara et al. (2010). The key characteristic are: (1) relative enrichment of HREE; (2) positive La anomalies; (3) a wide range of Y/Ho, from chondritic to super-chondritic values (up to 70), and (3) positive Eu-anomalies (up to 2.56).

9.5 EVOLUTION OF THE DEPOSITIONAL BASIN OF THE FARREL QUARTZITE – THE CLEAVERVILLE FORMATION

9.5.1 Lithostratigraphic Constraints

The clastic sedimentary succession in the Goldsworthy greenstone belt, now assigned to the Farrel Quartzite, is composed of poorly sorted volcaniclastic sandstone derived largely from mafic to ultramafic rocks (CU1), a thin unit composed of carbonaceous chert, massive chert with giant crystal pseudomorphs and minor siliciclastic rocks (CE1), bedded quartz-rich or silicified crystal aggregate-rich sandstone characterized by cross lamination and tabular chert clasts (rip-up clasts) (CU2), coarsening upward massive sandstone locally with lenticular conglomerate rich in mafic to ultramafic clasts (CU3), and again a thin unit composed of carbonaceous chert, massive chert with giant crystal pseudomorphs, and minor siliciclastic rocks (CE2). Key points to understand the depositional environment of the Farrell Quartzite are as follows:

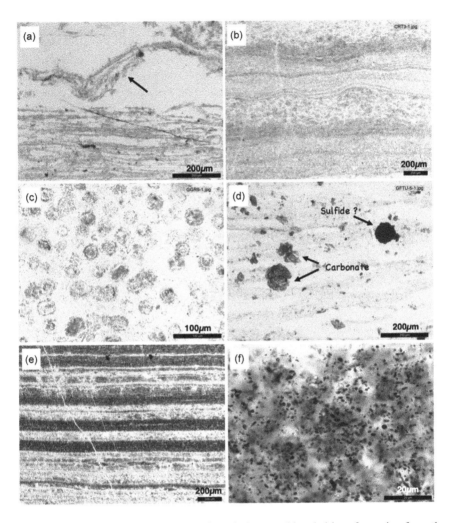

FIGURE 9.10 Photomicrographs of laminated cherts and banded iron formation from the Cleaverville Formation. Transmitted light. (a) Carbonaceous black chert characterized by fine carbonaceous laminae (probably biomat) with ripped-up texture (the arrow). (b) Finely laminated and granular-textured chert, being composed of fine carbonate particles and unknown tiny acicular crystals, and oxides (altered products). (c) Granules composed of carbonate particles showing concentric structure. (d) Fine discontinuous carbonaceous laminae and associated carbonate rhombs and clusters. Possible sulfide granules are also present. (e) Parallel lamination of banded iron formation. (f) Tiny hematite particles and associated carbonate particles with high reflective index.

1. Vertical change of lithofacies is drastic, and lateral variation of thickness and lithostratigraphy are significant.
2. The first cycle deposition of clastic materials related to erosion of basement possibly including the Euro Basalt was locally interrupted by the deposition of chemical/biogenic sediments including carbonaceous chert and massive chert (CE1).

3. The second cycle clastic deposition (CU2) includes high-energy events as indicated by rip-up clasts.
4. The third cycle clastic deposition (CU3) is characterized by increasing sedimentation of mature clastic materials represented by very-coarse grained and well-rounded quartz-rich sandstones. At the type section, the coarsening upward trend is clear.
5. This third cycle clastic deposition was once terminated by the deposition of chemical sediments represented by massive cherts with giant crystal pseudomorphs (CE2), although deposition of fine- to medium-grained clastics had continued.
6. The Cleaverville Formation represents chemical precipitation or fine-grained clastic (probably ash) deposition under subaqueous condition and conformably overlies the Farrel Quartzite. Just above CE2 at the type locality, laminated cherts are intercalated with sandstones, whose abundance appears to decrease upward.
7. Vertically- to subvertically-grown giant prismatic crystals (now pseudomorphs) in the massive cherts of CE1 and CE2 are likely evaporitic minerals. Similar deposit also occurs in the Cleaverville Formation (CE3).

The above-summarized features suggest that the Farrel Quartzite had deposited an erosion-related depression. Detrital materials first supplied to the basin were mostly immature as indicated by abundance of mafic to ultramafic volcanic fragments and angular monocrystalline quartz grains, which are the most likely first-cycle detritus. On the other hand, well-rounded quartz grains are more likely recycled detritus, probably originated from granitic rocks or felsic volcanic rocks due to deeper erosion. This is consistent with the presence of recycled zircon aged around 3.7 Ga (See Section 9.2.1) and trace element geochemistry of siliciclastic rocks collected from Mount Goldsworthy as discussed above (Sugitani et al., 2006). It is suggested that the supply of various detritus had filled the depression to form a relatively flat surface with shallow water body. This event had occurred at least twice, as represented by CE1 and CE2. This water body was evaporated, resulting in precipitation of vertically to subvertically oriented giant crystals interpreted as nahcolite ($NaHCO_3$) (Sugitani et al., 2003). This event was associated with deposition of carbonaceous cherts.

At the type section in Mount Grant, the deposition of the Farrel Quartzite was terminated by CE2 formation, being followed by deposition of nonferruginols cherts or fine-grained clastic rocks (possibly tuff) with intercalated sandstone beds and then ferruginous cherts and BIFs. The observed lithostratigraphic trend indicates that the water depth at the site of deposition had increased gradually. An increase in the water depth may have been related to an increase in influence of seawater, as discussed in detail in the next section.

9.5.2 TRACE ELEMENT CONSTRAINTS

The observed HREE-enrichment, positive La-anomalies, positive Eu-anomalies, and super-chondritic Y/Ho values suggest the contribution of seawater to laminated chert and BIF of the Cleaverville Formation. It may be noted here that the variations

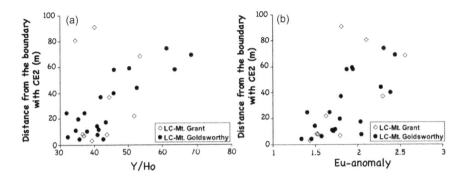

FIGURE 9.11   Stratigraphic trends of Y/Ho ratio (a) and Eu-anomaly {[Eu/(2/3Sm+1/3Tb)] PAAS} (b) in laminated cherts and banded iron formations of the Cleaverville Formation from Mount Grant and Mount Goldsworthy. Vertical axis shows distance from the boundary between the Farrel Quartzite and the Cleaverville Formation. (Reproduced from Sugahara et al. (2010), with permission from Elsevier.)

in Y/Ho ratio and degree of Eu-anomaly are not random but stratigraphically controlled (Figure 9.11). The plots of samples collected from the Mount Grant and the Mount Goldsworthy display similar trends over 100 m interval both for Y/Ho and Eu-anomaly, which basically increase upward. These trends are neither attributed to adsorption to Fe-oxides nor contamination of clastic materials, as evidenced by no positive correlations with Fe_2O_3 and TiO_2 (Figures 5 and 8 in Sugahara et al., 2010).

It is widely regarded that Archean seawaters were characterized by HREE-enrichment and Y/Ho values (~60) much higher than chondritic values (~30), like modern seawater, and were by positive Eu-anomaly, unlike modern seawaters (e.g., Allwood et al., 2010 and references therein). The positive Eu-anomalies can be attributed to higher contribution of hydrothermal fluids and/or anoxic deep marine waters at that time. Thus, the stratigraphic trends may reflect an increasing contribution of Archean seawater, which might have been related to the basin subsidence. It may be noted that the uppermost samples of the Cleaverville Formation at Mount Grant are out of the overall trends (Figure 9.11). These data should not be discarded as outliers. As described earlier, the evaporite bed (CE3) deposition also occurs within the Cleaverville Formation (Figures 9.3 and 9.4c, d). This suggests that the basin was again isolated from open-ocean and the evaporitic environment temporarily developed. More importantly, the observed REEs trends suggest that the basin water was supplied from other sources than the Archean open oceans. The other sources may included continental run-off and low-temperature hydrothermal fluids (Sugahara et al., 2010), because the samples collected from the lower horizons of the Cleaverville Formation are characterized by smaller positive Eu-anomalies and smaller Y/Ho values.

9.5.3   DEPOSITIONAL ENVIRONMENT OF THE BLACK CHERT IN CE2

The basin evolution related to the deposition of the Cleaverville Formation could place constraints on origins of the black chert containing fossil-like microstructures at CE2. Figure 9.12 shows distributions of all the samples analyzed by Sugahara et al. (2010)

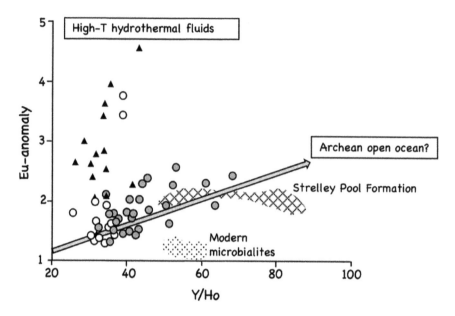

FIGURE 9.12 Binary plot of Y/Ho ratio and Eu-anomaly $\{[Eu/(2/3Sm+1/3Tb)]_{PAAS}\}$. There is a positive correlation observed among samples from Farrel Quartzite and Cleaverville Formation. White circles: black cherts from CE2, gray circles: cherts and BIFs from the Cleaverville Formation, black triangle: BIFs from the Panorama Formation. (Bolhar et al., 2005). This correlation implies a change of basin water from nonmarine water to seawater (also see text). Data for modern microbialites and the ~3.4 Ga Strelley Pool Formation are from Webb and Kamber (2000) and Van Kranendonk et al. (2003), respectively. (Modified from Sugahara et al. (2010), with permission from Elsevier.)

and other reference data on the Eu-anomaly vs. Y/Ho binary plot. The arrow in this figure shows the general stratigraphic trend of the Cleaverville samples and gives a possible endmember of Archean open ocean water. It may be noted that possible Archean high temperature hydrothermal deposits (BIF from the Panorama Formation) are plotted away from this trend and are characterized by relatively low Y/Ho ratio and variously but higher values of positive Eu-anomalies up to 4.5. On this plot, the black cherts from CE2, except for two, are plotted close to the endmember opposed to the supposed Archean open ocean endmember. Thus, these cherts were not precipitated from water mass entirely sourced by seawater, although HREE enrichment and super-chondritic Y/Ho values (but less than 40) suggest some seawater contribution. Considering the presence of the two samples having signatures consistent with high-T hydrothermal fluids, the basin water in which the black cherts had precipitated was fueled intermittently and/or locally by such fluids. In either case, the cherts are not direct precipitates from high-T hydrothermal fluids, but from modified high-T hydrothermal fluids away from their vents or low-temperature hydrothermal fluids, mixed with seawater and potentially continental runoff. An association with evaporites may give further constraint on the depositional environment. Namely, the depositional site for the black cherts was a coastal closed to semi-closed basin or expanded shallow to subaerial flat area, more specifically sabkha-like setting, coastal supratidal mudflat and sandflat.

9.6   FOSSIL-LIKE MICROSTRUCTURES

Various types of fossil-like microstructures have been identified, including films, filaments, small and large spheroids, and lenses (Sugitani et al., 2007). Their biogenicity has been established by multidisciplinary studies (Grey and Sugitani, 2009; Sugitani et al., 2011; House, 2013; Schopf et al., 2010; Sugitani et al., 2009; Oehler et al., 2009, 2010). Here, only brief descriptions of representative microstructures identified from the Farrel Quartzite are given. Detailed descriptions and biogenicity discussions would be presented in Chapter 11, together with the similar structures from the Strelley Pool Formation.

9.6.1   Spheroids

9.6.1.1   Small Spheroids

Small spheroids (small spheroid structures) approximately less than 15 μm across are abundant. They are generally hollow, being filled with pure cherts and have hyaline (transparent), solid (dense and opaque) to granular wall. Spheroids with broken habit are common. They often comprise colony-like clusters composed of several to hundreds of individuals, which are sometimes associated with film-like structures or misty carbonaceous materials. The shape of colony is various, from irregular to circular (Figure 9.13a).

9.6.1.2   Large Spheroids

Large spheroids (large spheroid structures) are defined for structures larger than 15 μm across. They are various in shape and wall architecture. Walls are either robust or flexible, either thick or thin, and either smooth or wrinkled/dimpled. Most of the large spheroids are spherical: oblate spheroids are present as minority. Based on these features, four submorphotypes have been identified, including (1) flexible-walled large spheroids up to 270 μm in diameter, characterized by compressed appearance (Figure 9.13b), (2) robust and thin-walled large spherical spheroids up to 60 μm in diameter, with various wall architectures (Figure 9.13c and d), (3) robust and thick-walled large spherical spheroids up to 30 μm in diameter, (Figure 9.13e), and (4) robust-walled large oblate spheroids (Figure 9.13f) up to 40 μm along the major dimension. Flexible-walled large spheroids occur exclusively solitarily, whereas the others occasionally comprise colonial clusters.

9.6.2   Lenses

Lenses (lenticular structures) range mostly from 20 to 60 μm across the major dimension, and rarely can be over 100 μm (Figure 9.14a). This morphological type accounts for the majority of identified fossil-like microstructures. The structures are characterized by flange, a sheet-like appendage surrounding the central body. The central body is either translucent or solid in appearance. Substructures and objects are occasionally identified inside the central bodies. The lenticular structures occur solitarily or comprise colonial clusters (Figure 9.14b), often together with spheroids. Lenticular structures are various in the style of flange attachment, flange width, and flange textures. These morphological and textural variabilities will be described and discussed in Chapter 12.

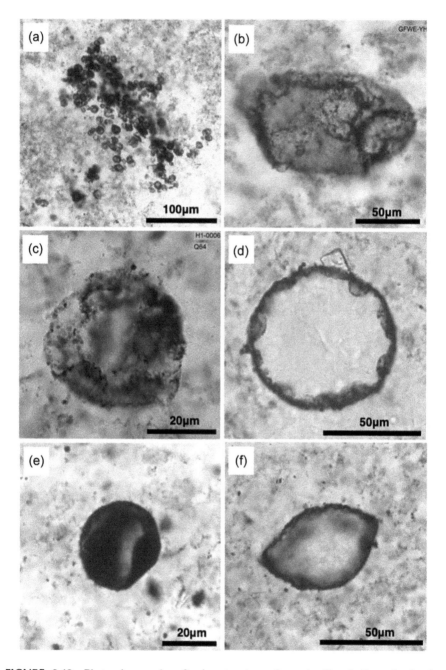

FIGURE 9.13 Photomicrographs of microstructures from the Farrel Quartzite in the Goldsworthy greenstone belt. All in the petrographic thin section. Transmitted light. (a) Clusterd small spheroids. (b) Flexible-walled large spheroid. (c) Robust and thin-walled large spherical spheroid characterized by highkly wrinkled wall. (d) Robust and thin-walled large spherical spheroid characterized by dimpled wall. (e) Robust and thick-walled large spherical spheroid. (f) Robust-walled large oblate spheroid.

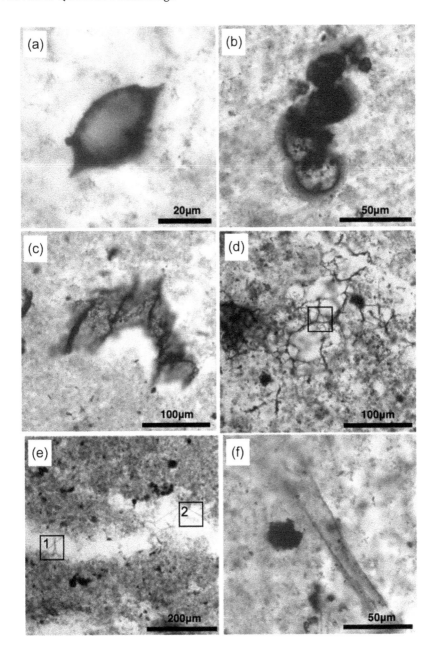

FIGURE 9.14 Photomicrographs of microstructures from the Farrel Quartzite in the Goldsworthy greenstone belt. All in the petrographic thin section. Transmitted light. (a) An equatorial view of a representative lenticular structure. (b) Colonial cluster composed of lenticular structures. Note the polar view of a specimen with hyaline flange. (c) Wrinkled single-sheeted film. (d) Complex network composed of carbonaceous threads. Enlarged view of the inset is shown in Figure 11.13a. (e) Dispersedly distributed thin (a few microns across) filaments in fractures. Enlarged views of the insets 1 and 2 are shown in Figure 11.13b and c. (f) Thick (~20 μm) and hollow filament.

9.6.3 Films

Films (film-like structures) are common (Figure 9.14c). The structures have completely random orientations. They are embedded in the cloudy chert matrix, not being controlled by crystal boundaries. Most of them are single-sheeted, often with parallel wrinkles. They are folded, crumped, and twisted, to various degrees, and generally have notched edges. Some edges are found rolled up. Under high magnification ($\times$ 1,000), the films are generally granular in appearance. Hyaline (glassy) specimens are present, though rare. Highly granular films tend to be thicker than 2–3 µm, whereas smoother and hyaline ones tend to be less than 1µm thick. Due to deformation, their total length and width cannot always be measured precisely. However, the structures can be larger than 500 µm. In addition to the dominant single-layered films, some films appear to be multilayered or branched.

9.6.4 Filaments

Filaments (filamentous structures) are not common, but their morphology and occurrence are diverse. Nontubular, thread-like filaments occur within the colloform-textured portions of cavity fills and veins (Figure 9 in Sugitani et al., 2007). Namely, the host phases of threads are not primary but secondary. The structures are less than 1 µm in width. Length is variable and can be up to several hundred microns. These threads are occasionally twisted and entangled and appear to be branched. Though bacterial-like morphology, their biogenicity has been questioned due to the close association with the secondary phases (Sugitani et al., 2007).

A peculiar occurrence of thread is found in a specific thin layer (Figures 3 and 4 in Sugitani et al., 2011). Numerous threads are clearly identified within irregularly shaped masses of microcrystalline quartz. They are entangled and appear to penetrate into the carbonaceous matrix (Figure 9.14d). Similar filaments also occur in fractured portion (Figure 9.14e), but detailed observation have revealed that they were tubular as described later. Distinctly tubular filamentous structures are rare and occur dispersedly (Figure 9.14f). They are generally more than 10 µm in diameter. One specimen ~15 µm in diameter with a small, semi-hollow sphere inside has been reported by Sugitani et al. (2007).

COLUMN: EVAPORITE

Evaporite is a deposit generally formed by evaporation of water mass and resultant super-saturation to salts, which effectively occurs in a closed to semiclosed basin. For example, as seawater is evaporated, calcium carbonate precipitates first, followed by calcium sulfate, sodium salt, and magnesium salt. In other words, mineral assemblage of evaporite deposits in geologic records, if they originated from seawater, could be a clue to compositions of ancient seawater. In this context, evaporites identified in Archean sedimentary successions are of great importance. Unfortunately, however, all minerals supposed to be of evaporitic origin found in the Archean do not possess original compositions. Without exception, they have been silicified to be composed almost exclusively of microcrystalline quartz. Therefore, in order to identify their original mineralogy, we need to take detour way (Chapter 13). For example, the forms of individual crystals, occurrence of crystal aggregates, and interfacial angles could give some constraints on the origins of silicified evaporitic minerals. Three-dimensional images reconstructed using the micro-CT method give us precise crystal habit, and chemical mapping using a scanning electron microscope with energy dispersive X-ray spectroscopy (SEM-EDX) and an electron probe micro analyzer (EPMA) could provide information of original chemistry. Mineral inclusions in silicified crystals, if present, may provide critical information of original mineralogy, although they could have been initially involved as impurities. Even though identification of original mineral species is successfully done, it must be cautioned that the evaporite formation requires somewhat special environmental conditions such as closed to semi-closed basins (not open oceans) and various evaporitic minerals could be formed successively according to the degree of evaporation as described above. In other words, combination of evaporitic minerals could reliably constrain compositions of parental brines.

REFERENCES

Allwood, A.C., Kamber, B.S., Walter, M.R., Burch, I.W., Kanik, I. 2010. Trace elements record depositional history of an early Archean stromatolitic carbonate platform. *Chemical Geology* 270, 148–163.

Bhatia, M.R., Crook, K.A.W. 1986. Trace element characteristics of graywackes and tectonic setting discrimination of sedimentary basins. *Contributions to Mineralogy and Petrology* 92, 181–193.

Bolhar, R., Van Kranendonk, M.J., Kamber, B.S., 2005. A trace element study of siderite-jasper banded iron formation in the 3.45 Ga Warrawoona Group, Pilbara Craton-Formation from hydrothermal fluids and shallow seawater. *Precambrian Res.* 137, 93–114.

Buick, R., Thornett, J.R., McNaughton, N.J., Smith, J.B., Barley, M.E., Savage, M. 1995. Record of emergent continental crust ~3.5 billion years ago in the Pilbara craton of Australia. *Nature* 375, 574–577.

Condie, K.C. 1993. Chemical composition and evolution of the upper continental crust: Contrasting results from surface samples and shales. *Chemical Geology* 104, 1–37.

Cullers, R.L., DiMarco, M.J., Lowe, D.R., Stone, J. 1993. Geochemistry of as ilicified, felsic volcaniclastic suite from the early Archaean Panorama Formation, Pilbara Block, Western Australia: An evaluation of depositional and post-depositional processes with special emphasis on the rare-earth elements. *Precambrian Research* 60, 99–116.

Cullers, R.L., Podkovyrov, V.N. 2002. The source and origin of terrigenous sedimentary rocks in the Mesoproterozoic Ui group, southeastern Russia. *Precambrian Research* 117, 157–183.

Fedo, C.M., Eriksson, K.A., Krogstad, E.J. 1996. Geochemistry of shales from the Archean (~3.0 Ga) Buhwa greenstone belt, Zimbabwe: Implications for provenance and source-area weathering. *Geochimica et Cosmochimica Acta* 60, 1751–1763.

Green, M.G., Sylvester, P.J., Buick, R. 2000. Growth and recycling of early Archean continental crust: Geochemical evidence from the Coonterunah and Warrawoona Groups, Pilbara Craton, Australia. *Tectonophysics* 322, 69–88.

Grey, K., Sugitani, K. 2009. Palynology of Archean microfossils (c. 3.0 Ga) from the Mount Grant area, Pilbara Craton, Western Australia: Further evidence of biogenicity. *Precambrian Research* 173, 60–69.

Hickman, A.H. 2008. Regional review of the 3426-3350 Ma Strelley Pool Formation, Pilbara Craton, Western Australia. *Geological Survey of Western Australia Record* 2008/15.

House, C.H., Oehler, D.Z., Sugitani, K., Mimura, K. 2013. Carbon isotopic analyses of ca. 3.0 Ga microstructures imply planktonic autotrophs inhabited Earth's early oceans. *Geology* 41, 651–654.

Nesbitt, H.W., Young, G.M., McLennan, S.M., Keays, R.R. 1996. Effects of chemical weathering and sorting on the petrogenesis of siliciclastic sediments, with implications for provenance studies. *The Journal of Geology* 104, 525–542.

Oehler, D.Z., Robert, F., Walter, M.R., Sugitani, K., Allwood, A., Meibom, A., Mostefaoui, S., Selo, M., Thomen, A., Gibson, E.K. 2009. NanoSIMS: Insights to biogenicity and syngeneity of Archaean carbonaceous structures. *Precambrian Research* 173, 70–78.

Oehler, D.Z., Robert, F., Walter, M.R., Sugitani, K., Meibom, A., Mostefaoui, S., Gibson, E.K. 2010. Diversity in the Archean biosphere: New insights from NanoSIMS. *Astrobiology* 10, 413–424.

Schopf, J.W., Kudryavtsev, A.B., Sugitani, K., Walter, M.R. 2010. Precambrian microbe-like pseudofossils: A promising solution to the problem. *Precambrian Research* 179, 191–205.

De Grey, W.A. Sheet 2757 (version 2.0): Western Australia Geological Survey, 1:100 000 Geological Series.

Sugahara, H., Sugitani, K., Mimura, K., Yamashita, F., Yamamoto, K. 2010. A systematic rare-earth elements and yttrium study of Archean cherts at the Mount Goldsworthy greenstone belt in the Pilbara Craton: Implications for the origin of microfossil-bearing black cherts. *Precambrian Research* 177, 73–87.

Sugitani, K., Yamashita, F., Nagaoka, T., Yamamoto, K., Minami, M., Mimura, K., Suzuki, K. 2006. Geochemistry and sedimentary petrology of Archean clastic sedimentary rocks at Mt. Goldsworthy, Pilbara Craton, Western Australia: Evidence for the early evolution of continental crust and hydrothermal alteration. *Precambrian Research* 147, 124–147.

Sugitani, K., Grey, K., Allwood, A.C., Nagaoka, T., Mimura, K., Mimura, M., Marshall, C.P., Van Kranendonk, M.J., Walter, M.R. 2007. Diverse microstructures from Archaean chert from the Mount Goldsworthy – Mount Grant area, Pilbara Craton, Western Australia: Microfossils, dubiomicrofossils, or pseudofossils? *Precambrian Research* 158, 228–262.

Sugitani, K., Grey, K., Nagaoka, T., Mimura, K., Walter, M.R. 2009. Taxonomy and biogenicity of Archaean spheroidal microfossils (ca. 3.0 Ga) from the Mount Goldsworthy–Mount Grant area in the northeastern Pilbara Craton, Western Australia. *Precambrian Research* 173, 50–59.

Sugitani, K., Horiuchi, Y., Adachi, M., Sugitaki, R. 1996. Anomalously low Al_2O_3/TiO_2 values for Archean cherts from the Pilbara Block, Western Australia – Possible evidence for extensive chemical weathering on the early earth. *Precambrian Research* 80, 49–76.

Sugitani, K., Mimura, K., Suzuki, K., Nagamine, K., Sugisaki, R. 2003. Stratigraphy and sedimentary petrology of an Archean volcanic-sedimentary succession at Mt. Goldsworthy in the Pilbara Block, Western Australia: Implications of evaporite (nahcolite) and barite deposition. *Precambrian Research* 120, 55–79.

Sugitani, K., Mimura, K., Walter, M.R. 2011. Farrel Quartzite microfossils in the Goldsworthy greenstone belt, Pilbara Craton, Western Australia. Additional evidence for a diverse and evolved biota on the Archean Earth. In V. Tewari and J. Seckbach eds. *Stromatolites: Interaction of Microbes with Sediments.* Springer, Heidelberg, 117–132pp.

Suzuki, K., 2005. CHIME (Chemical Th-U-total Pb isochron method) dating on the basis of electron microprobe analysis. *The Journal of the Geological Society of Japan.* 111, 509–526.

Taylor, S.R., McLennan, S.M. 1985. *The Continental Crust: Its Composition and Its Evolution.* Blackwell, Oxford, 312 pp.

Van Kranendonk, M.J., Webb, G.E., Kamber, B.S. 2003. Geological and trace element evidence for a marine sedimentary environment of deposition and biogenicity of 3.45 Ga stromatolitic carbonates in the Pilbara Craton, and support for a reducing Archaean ocean. *Geobiology* 1, 91–108.

Webb, G.E., Kamber, B.S. 2000. Rare earth elements in Holocene reefal microbialites: A new shallow seawater proxy. *Geochimica et Cosmochimica Acta* 64, 1557–1565.

10 Overview of the Pilbara Microstructures 2

The Strelley Pool Formation Assemblage

10.1 INTRODUCTION

When the first paper describing the Farrel Quartzite (FQ) "microfossil" assemblage was published in 2007 (Sugitani et al., 2007), we realized that lenticular possible microfossils had already reported from the 3.4 Ga-old chert of the Barberton greenstone belt, South Africa (Walsh, 1992). However, I had no idea to search for microfossils in the contemporaneous Strelley Pool Formation (SPF) in the Pilbara Craton and was even skeptic to the personal communication about finding of possible lenticular microfossil in the SPF chert by D. Oehler, possibly in 2008. After that, I had information about the SPF occurrence in the Goldsworthy greenstone belt from M.J. Van Kranendonk (Figure 9.1), and my colleague, K. Mimura and I performed a fieldtrip to the so-called Waterfall locality and another SPF locality in the Warralong greenstone belt. Although I could not find any fossil-like microstructures in the first-round examination of petrographic thin sections made from the collected black chert samples, further examination led me to find microstructures similar to those already described from the Farrel Quartzite (Sugitani et al., 2010). This discovery was followed by identification of similar microstructures in carbonaceous black to dark gray cherts collected from the Panorama greenstone belt and the Warralong greenstone belt (Sugitani et al., 2013). In this chapter, the SPF fossil-like microstructures from these four localities and the host rocks and sedimentary successions are described.

The Strelley Pool Formation is a sedimentary unit composed predominantly of volcaniclastics, conglomerates, sandstones, shales, carbonates, evaporites, and cherts. Its thickness ranges from 10 to 1000 m. This formation, now identified in 11 greenstone belts in the East Pilbara Terrane (EPT) of the Pilbara Craton (Hickman, 2008), was previously called the Strelley Pool Chert (SPC) (Lowe, 1980). Although once assigned to a member of the Kelly Group (Van Kranendonk et al., 2004), the Strelley Pool Formation was later interpreted to be an independent sedimentary formation, separating the Warrawoona and the Kelly groups, well-known major volcanic units of the Pilbara Supergroup (Hickman, 2008). Its depositional age is assumed to be 3,433–3,350 Ma, being constrained by the ages of the underlying Mount Ada Basalt and the overlying Euro Basalt.

DOI: 10.1201/9780367855208-10

As reported by Hickman (2008), the SPF lithostratigraphic variations are significant both regionally and locally. Depositional environments of this formation were mostly from shallow-water to subaerial, including shallow-water marine, coastal, and estuarine and nonmarine such as lacustrine, fluviatile, and sabkha. Stromatolites of this formation are one of the most well-established and widely accepted sedimentary biosignatures as described in Chapter 5.

10.2   THE PANORAMA GREENSTONE BELT

10.2.1   Local Geology and Lithostratigraphy

In the Panorama greenstone belt (Figure 9.1), the Strelley Pool Formation is up to ca. 30 m thick and generally occurs as a thin sedimentary unit within a thick volcanic pile (Figure 10.1a). The Strelley Pool Formation there was once interpreted to have conformably to unconformably overlain the Panorama Formation (Van Kranendonk, 2000), which has later been reinterpreted as silicified mafic rocks of the Mount Ada Basalt (Hickman, 2012a). Four major members characterized by distinct lithofacies have been identified by Allwood et al. (2006, 2007), including Member 1: a basal jasper/chert conglomerate, Member 2: laminated stromatolitic carbonate/chert, Member 3: bedded black chert with silicified evaporite beds, silicified pebble conglomerate, and stromatolitic laminated ironstone, and Member 4: silicified, fining-upward clastic/volcaniclastic rocks cut by hydrothermal vein chert. Member 1 is interpreted to represent deposition on a rocky shoreline, whereas Member 2 represents deposition in an isolated peritidal carbonate

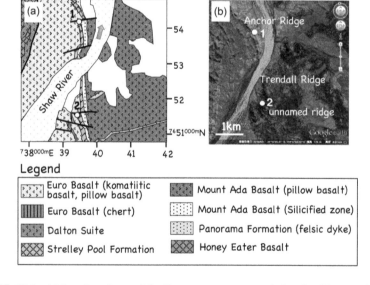

FIGURE 10.1   (a) Local geology of the Panorama greenstone belt, after Western Australia Geological Survey 1:100 000 Geological Series "North Shaw" (Hickman, 2012b). (b) The Google image for (a). 1: Panorama locality 1, 2: Panorama locality 2.

platform. Member 3 is interpreted to have deposited in a shallow restricted basin. Member 4 represents a deepening depositional environment, which was related to subsiding of fault blocks.

From two localities of the Strelley Pool Formation in the Panorama greenstone belt (Figure 10.1a and b), fossil-like microstructures have been identified (Sugitani et al., 2010, 2013). One locality is at the Anchor Ridge, where extensive mapping and description of stromatolites have been performed (e.g., Alwood et al., 2006). Black cherts were collected from the south bank of an unnamed creek (21°11′46″S, 119°18′23″S), during the field excursion organized by Australian Centre for Astrobiology in 2005. This locality was formerly designated as Anchor Ridge locality, but here is renamed Panorama locality 1. The other locality is located at unnamed ridge south of the Trendall Ridge, hereafter called Panorama locality 2 (S21°13′21.6″, E119°18′20.0″).

10.2.2 Panorama Locality 1

10.2.2.1 Lithostratigraphy and Petrography

Two black chert samples were collected from this locality (Figure 10.2a). Detailed lithostratigraphic examination of this area was made by Allwood et al. (2007). The section AR-3 measured by Allwood et al. (2007) is the closest to the Panorama locality 1. AR-3 ca. 10 m thick was interpreted as being composed only of Member 4, although the collected black chert is likely correlated to the Member 3 facies.

The black cherts are composed of a carbonaceous chert layer, cavity-fill chert layer, and detrital layer, each of which is less than 1 cm thick (Sugitani et al., 2010). The carbonaceous chert layer is further classified into two major fabric types. One type is characterized by laminae composed of disseminated minute carbonaceous matter with local irregularly shaped chert mass and by granular pyrite (FeS_2) and its aggregates (the lower half of Figure 10.2b). The other type is characterized by biomat-like fabric composed by carbonaceous bundles or filaments, with disseminated carbonates (not specified) and barite ($BaSO_4$) (the upper half of Figure 10.2b). The bundles or filaments tend to comprise three dimensional networks with spheroidal masses composed of microcrystalline quartz (the lower half of Figure 10.2c), horizontally and laterally grading into mat-like parallel lamination (the upper half of Figure 10.2c). The detrital layer is composed mainly of carbonaceous grains and chert clasts with carbonaceous rim. Both are characterized by highly irregular shape, suggesting of reworking of unconsolidated sediments (Figure 10.2d). The detrital layer also locally contains aggregates of pyrite and partially to completely silicified barite crystals.

10.2.2.2 Fossil-Like Microstructures

Filamentous microstructures have been identified from the carbonaceous layer characterized by mat-like fabrics (Figure 10.2c, details are shown in Figure 11.5). The structures are only locally present, although if present, plural structures are associated with each other. They are sinuous and can be traced at least 150 μm, while their width is constant around 3 μm. The central portion of these filaments appear to be translucent, suggesting that they are tubular. Lenticular structures, basically

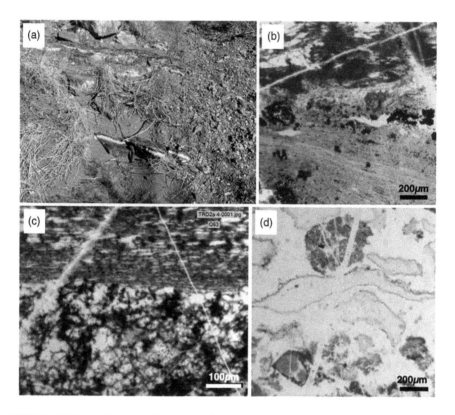

FIGURE 10.2 (a) Photograph of microfossil-bearing carbonaceous black cherts at the Panorama locality 1. (b) Two types of fabric comprising black chert. The upper half shows biomat-like two- to three-dimensional network composed of carbonaceous bundles. The lower half shows the laminated (locally massive) fabric characterized by enrichment of sulfide, with irregularly shaped microcrystalline quartz mass. (c) Details of the fabric composed of carbonaceous bundles. The upper half: two-dimensional network composed of bundles aligned parallel to the bedding plane. The lower half: three-dimensional network. (d) Facies characterized by reworked carbonaceous grains. Transmitted light for (b–d).

morphologically equivalent to those described from the Farrel Quartzite, are not found in portions displaying mat-like fabric but are present portions characterized by disseminated minute carbonaceous matter comprising laminae (the lower half of Figure 10.2b). Their abundance is low and disperse. Some are highly degraded. Relatively large spheroids (up to 60 µm across) are also present but somewhat ambiguous, whereas small spheroids have not yet been identified.

10.2.3 Panorama Locality 2

10.2.3.1 Lithostratigraphy and Petrography

Based on the fieldwork performed in 2019, the lithostratigraphy at the Panorama locality 2 preliminarily described in Sugitani et al. (2013) has been refined. The measured section ca. 4 m thick is composed of greenish gray chert, laminated to massive

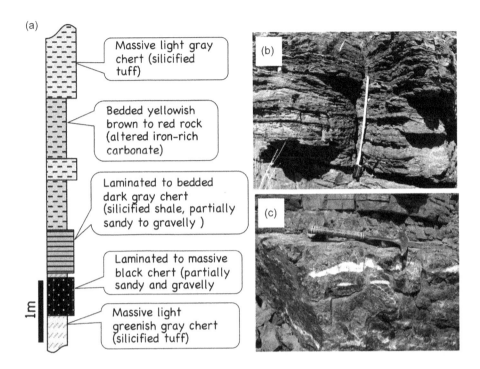

FIGURE 10.3 (a) Stratigraphic column of the Strelley Pool Formation at the Panorama locality 2. (b) Photograph of reddish to yellowish brown carbonate chert. This lithology is composed of alteration of carbonate layers (depression) and chert layers (projection) mm to cm thick. (c) Photograph of laminated to massive carbonaceous black chert, containing fossil-like microstructures and possible impact spherules.

black chert, laminated to bedded gray-black chert, reddish to yellowish brown rock (carbonate-chert), and light gray chert (Figure 10.3a). The greenish-gray chert is enriched in green Cr-rich mica (fuchsite) $[K(Al,Cr)_2(AlSi_3O10)(OH)_2]$ and minute TiO_2 such as anatase. The light gray chert is petrologically similar to the greenish-gray chert, although it contains green mica to lesser amount. No detrital quartz has been observed, whereas irregularly shaped carbonaceous grains are present. The gray chert is carbonaceous but dusty and enriched in mica. The reddish to yellowish brown rock is composed of iron oxides, carbonate, and quartz (Figure 10.3b). Carbonate occurs as rhombs of various sizes, which have been oxidized or silicified to various degrees.

Laminated to massive black chert (Figure 10.3c) is carbonaceous and a target of the paleontological study. A block sample 15 cm thick collected in 2013 is described in detail here (Figure 10.4a). This sample can be subdivided into four sublayers based on petrographic features (Figure 10.4b). The lowermost sublayer 4 is characterized by fine carbonaceous lamination and abundant sulfide particles of pyrite and sphalerite (ZnS) (Figure 10.5b). Irregularly shaped fenestrae of various sizes are also common. The sublayer 3 represents clastic facies. It contains abundant carbonaceous grains, many of which have fluffy and blurred outline. The

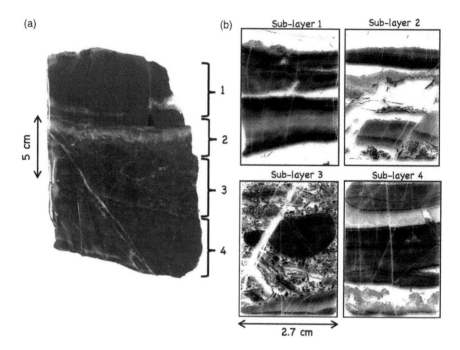

FIGURE 10.4 (a) Polished slab of the laminated to massive carbonaceous black chert from the Panorama locality 2, showing abrupt vertical facies changes. The numbers from 1 to 4 correspond to the sublayers in (b). (b) Scanned images of the thin sections of sublayers 1–4.

sublayer 2 is composed of facies similar to sublayer 4 but locally enriched in detrital materials and spherules as described below. The sublayer 1 is also similar to sublayer 4 (Figure 10.5a). Fabrics of the sublayers 1, 2, and 4 is basically equivalent to that of the carbonaceous layer containing lenticular microstructures in the black chert from the Panorama locality 1. The sublayer 3 is similar to the detrital layer in the locality 1 black chert.

Possible impact spherules, including nonspherical morphologies, have been identified from the detritus-rich portion of the sublayer 2 (Sugitani et al., in preparation) (Figure 10.5c, d). Spherules range from 20 μm to 2 mm across and are composed mainly of microcrystalline quartz, sericite (fine-grained white mica) $[KAl_2AlSi_3O_{10}(OH)_2]$, and anatase (TiO_2) (Figure 10.5d), though relative abundance is various. Many of the spherules are compositionally homogenous with outlined or rimmed by fine anatase particles, whereas zoned spherules are also present. They have a serictic core with a mantle composed of microcrystalline quartz. Impact origin of the spherules can be inferred from the presence of oxide (TiO_2)-rich rim and the occurrence of teardrop- and dumbbell-shaped spherules, although not definitive at this stage.

10.2.3.2 Fossil-Like Microstructures

Laminated to massive black cherts from the locality 2 contain fossil-like microstructures of various morphologies. Three major morphological types such as spheroid,

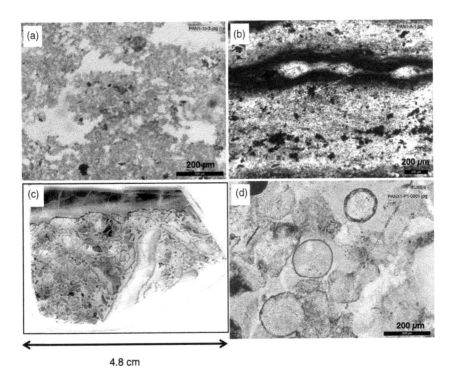

4.8 cm

FIGURE 10.5 Photomicrographs and a scanned image of laminated to massive carbonaceous black chert shown in Figure 10.4. (a) Relatively massive portion in the sublayer 1, characterized by irregularly shaped pure chert masses. (b) Fine sulfide-rich portion in sublayer 4. (c) Scanned image of a thin section of sublayer 2 cut parallel to the bedding plane, being composed of detritus of various shapes including possible impact spherules. (d) Photomicrograph of another thin section containing spherules made from the sublayer 2. Transmitted light for (a), (b) and (d).

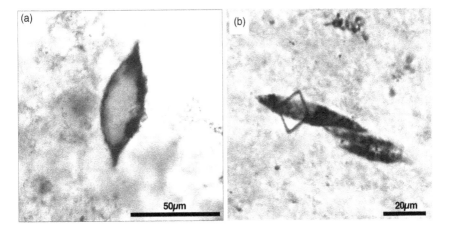

FIGURE 10.6 Fossil-like microstructures from the Panorama locality 2. (a) Lenticular structure with hollow interior. (b) Nonflanged lenticular structure. Note the associated occurrence of euhedral carbonate crystal. Tansmitted light.

lens, and filament have been identified. Additionally, film-like structures are also present.

Based on size, spheroids can be further classified into two types, including small spheroid less than 15 μm in diameter and large spheroid larger than 15μm, although the latter tends to be as large as 50 μm or more in diameter. Small spheroids often occur as clusters. Large spheroids may be further classified into three types based on wall architecture. One is characterized by wall of relatively thick appearance, whereas the other by thin translucent wall. The rest type is distinct from the others by more or less oblate in shape.

Lens (lenticular microstructure) is the most abundant morphological type and often comprises colony-like clusters. This type is basically similar to that identified from the Farrel Quartzite (Figure 10.6a), although submorphotypes without flange are common (Figure 10.6b).

Filamentous structures are rare and show isolated distribution. Very thin filaments with solid appearance had been described from colloform-textured cavity-fill portions in the previous study (Sugitani et al., 2013). As their biogenicity is dubious, they are not described here. Hollow filaments ranging from 5 to 20 μm in diameter have been identified. Rarely inner spheroids can be seen (also see Figure 6F in Sugitani et al., 2013). Film-like structures are dominated by single-sheeted type. Complex honeycomb structured film is present, although only one specimen has been identified to date (Sugitani et al., 2013).

10.2.4 Depositional Environment

Allwood et al. (2007) noted that the upper part of the member 4 lithofacies "grades upward to bedded black (carbonaceous) chert and interbedded silicified gray-green volcaniclastic (tuffaceous) mudstone and rare sandstone" (Allwood et al., 2007). This once led me to suggest that the examined succession could correspond to the Member 4. However, the presence of parallel laminated carbonate-chert possibly representing stromatolite facies suggests that the sedimentary succession might be a condensed section of Members 3 and 4, because Member 4 lacks carbonate facies.

Shallow subaqueous depositional environment of the examined unit is consistent with the lack of pseudomorphic evaporitic crystals and desiccation cracks at the locality 2. The black chert also has a layer composed largely of carbonaceous sand grains that are probably formed by reworking of unconsolidated precursors, suggesting intermittent current-induced deposition. As such the layer occasionally contains pebble-sized clast, the current was likely strong. Irregularly shaped pure chert masses in this black chert are, on the other hand, quite similar to open-space structures in carbonate rocks and bacterial mats called fenestrae-like fabric, many of which indicate supratidal, intertidal, and peritidal to inner platform environment (Flügel, 2010). As the black cherts have seawater-like REEs signatures and anomalously high heavy metal concentrations (Sugitani et al., 2018), it can be proposed that the black chert deposited in the peritidal to inner platformal marine environment, influenced by hydrothermal fluids.

10.3 THE WARRALONG GREENSTONE BELT

10.3.1 LOCAL GEOLOGY AND LITHOSTRATIGRAPHY

The Strelley Pool Formation in the Warralong greenstone belt (Figures 9.1 and 10.7a) is distinct in lithostratigraphy from the other SPF successions. It is composed dominantly of coarse- to very coarse-grained quartz-rich sandstone; its thickness can be over 200 m (Hickman, 2008) (Figure 10.7b). A thin unit containing chert and evaporite less than 10 m thick conformably overlies the sandstone (Figure 10.7c) can be traced over 500 m along strike.

At the first examined locality (the Warralong locality 1, S20°49′45.2″, E119°30′02.5″) (Sugitani et al., 2010), this unit is composed of white-light gray-black laminated chert (silicified carbonate) (Figure 10.8a), sandstone, white-light gray massive chert with vertically to subvertically oriented giant crystal pseudomorphs (silicified evaporite: Figure 10.8b) and laminated to banded white-gray-black chert (Figure 10.8c). Conglomerate containing chert pebbles of various colors overlies thus unit with erosional contact. Fossil-like microstructures have been identified from the laminated to banded white-gray-black chert. In 2019, the author explored the new locality (the Warralong locality 3, S20°49′38.2″, E119°30′04.3″) (Figure 10.7b) and measured section. Here, this section is described (Figure 10.7d). The cherty unit at this locality 3 is about 1.5 m thick overlying laminated white to light gray chert with relict stromatolitic texture (Figure 10.8d) is composed of evaporitic chert characterized by dissolution collapse structure and laminated to banded white-gray-black chert (Figure 10.8e). Sandstone with chert clasts is also present (Figure 10.8f).

10.3.2 PETROGRAPHY OF GRAY-BLACK CHERT

Fossil-like microstructures have been identified from carbonaceous gray-black chert, which is composed of thin layers of different fabrics and compositions. One type is secondary cavity fill silica. The others include detrital and laminated carbonaceous layers. Detrital layer is composed mainly of reworked carbonaceous grains, with some lithic grains and detrital quartz. Two types of laminated carbonaceous layers are identified. One is characterized by fine lamination with lenticular crystal ghosts, whereas the other by irregularly shaped masses composed of pure chert, occasionally with abundant pyrite particles.

10.3.3 DEPOSITIONAL ENVIRONMENT

The examined uppermost unit is a condensed section including stromatolite and evaporite. It is reasonable to suggest that the unit had deposited in a shallow and closed environment. It is equivocal whether the depositional site was marine or nonmarine and how hydrothermal fluids had contributed to the chert formation. Detailed geochemical study is required.

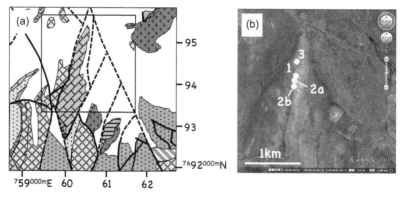

Legend

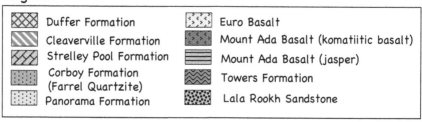

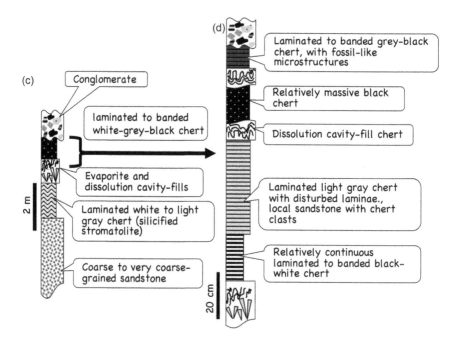

FIGURE 10.7 (a) Local geology of the Warralong greenstone belt, after Western Australia Geological Survey 1: 100 000 Geological Series "Carlindie (Van Kranendonk, 2004). (b) The Google image of the studied area {the inset in (a)}. The numbers show the localities from which microfossils have been identified. (c) Stratigraphic column at the locality 1. (d) Details of the subunit containing laminated black cherts with fossil-like microstructures, from the locality 3.

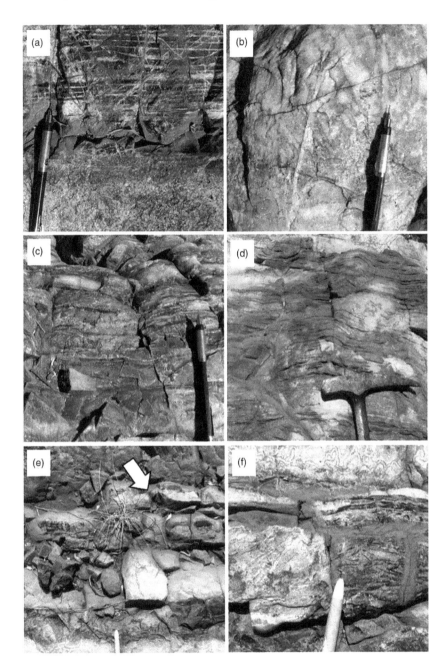

FIGURE 10.8 Representative lithofacies of the Strelley Pool Formation in the Warralong greenstone belt. (a) Very coarse-grained sandstone and overlying black-white laminated chert at locality 2. (b) Silicified evaporite at locality 2. (c) Light gray to black laminated to banded chert with fossil-like microstructures at locality 1. (d) Chert laminate locally showing concave-upward structures, probably silicified stromatolite at locality 3. (e) Laminated carbonaceous black chert with fossil-like microstructures (the arrow), at locality 3. (f) Thin tabular chert clasts in sandstone at locality 3.

10.3.4 Fossil-Like Microstructures

Fossil-like microstructures have been identified from four localities so far, although their abundance is low, and in most cases, their preservation state is poor. Identified morphological types are lens and small spheroid. Single-layered film-like structures have also been identified. The film-like structures are granular, and wrinkles are common. Carbonaceous clots or sulfide grains are locally attached on the surface.

10.4 THE GOLDSWORTHY GREENSTONE BELT

10.4.1 Local Geology and Lithostratigraphy

In the Goldsworthy greenstone belt (Figure 9.1), the Strelley Pool Formation has been mapped across the western and northern parts (Smithies et al., 2004) (Figure 9.2). At the western Mount Grant, this formation is overlain by mafic volcanic and volcaniclastic rocks of the Euro Basalt and then the Farrel Quartzite. The Strelley Pool Formation at this locality is characterized by laminated chert and associated silicified evaporite (identified by vertically to subvertically oriented crystal pseudomorphs). In the northern Goldsworthy greenstone belt, this formation is mapped within the block structurally isolated from the Mount Grant, where the volcanic-sedimentary succession comprises syncline (Figure 10.9a). The studied outcrop is located at the axis of this syncline and is informally referred to as the "Waterfall locality" (Sugitani et al., 2010). The Strelley Pool Formation there occurs as ca. 60-m-thick succession and is composed mainly of medium to coarse-grained, lithic sandstone and is overlain by the Euro Basalt (Figure 10.9c). Framework grains of lithic sandstones tend to be dominated by mafic to ultramafic volcanic fragments, with minor amounts of sedimentary fragments and quartz. Interbedded five cherty units from 1.5 to 3 m thick have been identified. The cherty units, except for the uppermost one, have two distinct lithologies. One is white to light gray, wavy laminated chert. This lithology can be interpreted as silicified stromatolitic carbonate (Figure 10.10a) (e.g., Lowe, 1983). The other is again white to light gray massive chert. This chert has subvertically radiating pseudomorphic giant crystals and can be possible silicified evaporite (Figure 10.10b) (e.g., Sugitani et al., 2003). To date, microfossils and other biosignature have been identified from the uppermost cherty unit, whose lithofacies are described in detail below.

10.4.2 Lithofacies and Petrography of the Uppermost Cherty Unit

Lithofacies of the uppermost cherty unit at the type locality are various (Figure 10.9d). From the bottom to the top, sandstone with flat cherty pebbles (Figure 10.10c), layer characterized by vertically to sub-vertically stacked platy structures (local occurrence) (Figure 10.10d), laminated chert of various colors including finely laminated black chert, black-white laminated chert, and partially reddish, dark-gray chert (Figure 10.10e), sandstone containing angular chert clasts that can be up to 50cm across (Figure 10.10e), and massive black chert occur (Figure 10.10f).

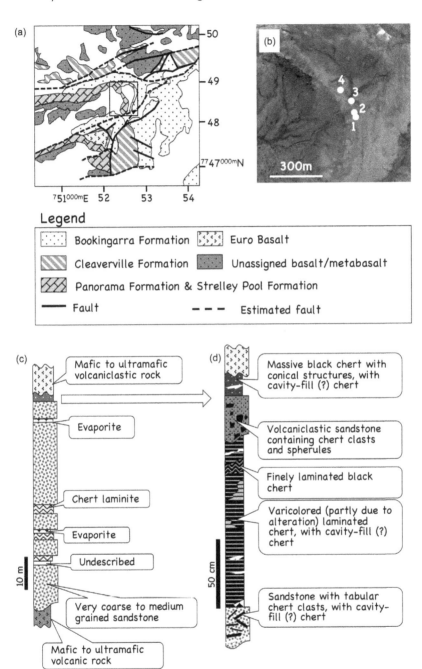

FIGURE 10.9 (a) Local geology of the northern part of the Goldsworthy greenstone belt containing the Strelley Pool Formation, after Western Australia Geological Survey 100 000 Geological Series "De Grey" (Smithies et al., 2004). (b) The Google image of the studied area {the inset in (a)}. The numbers show the localities from which fossil-like microstructures have been identified. (c) Stratigraphic column at the locality 1. (d) Details of the uppermost cherty unit containing laminated to massive carbonaceous black cherts containing fossil-like microstructures.

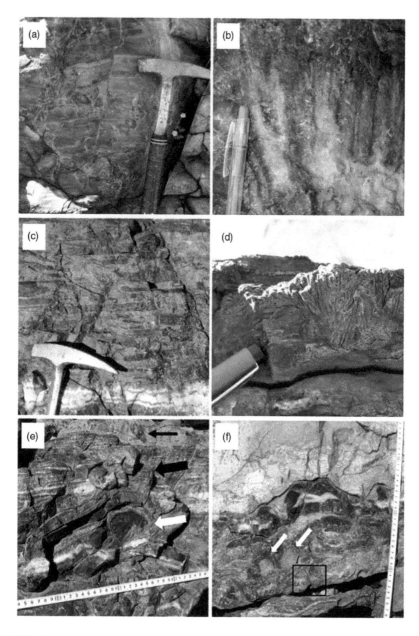

FIGURE 10.10 Photographs of representative lithologies of the Strelley Pool Formation in the Goldsworthy greenstone belt. (a) Chert laminate, probably silicified flat stromatolite. (b) Silicified evaporite characterized by vertically to sub-vertically grown large crystals. (c) Tabular chert clasts in sandstone of the uppermost cherty unit. (d) Sandstone containing possible silicified evaporite crystals of the uppermost cherty unit. (e) Varicolored laminated chert (the solid thick arrow) and finely laminated black chert (the white arrow). The thin arrow shows the overlying volcaniclastic sandstone. Of the uppermost cherty unit. (f) Massive black chert containing microfossils of the uppermost cherty unit, being characterized by conical structures (the arrows) and unconformably overlain by volcaniclastic rock.

Flamework grains of the pebbly sandstone at the base of the type section (Figure 10.10c) are mafic to ultramafic volcanic fragments, unknown dusty grains, and extensively silicified euhedral crystals (Figure 10.11a). Cherty pebbles are characteristically tabular in shape and can be up to 30 cm in apparent length and are slightly carbonaceous, being characterized by fine anastomosing laminae composed of tiny iron oxides, possible alteration products of pyrite (Figure 10.11b). This lithology can be traced along the strike and thins to the north, and the size of clasts becomes to be mm-scale. Vertically to sub-vertically stacked platy structures (Figure 10.10d) were previously interpreted as possible silicified evaporite (Sugitani et al., 2015). The platy structures, however, are variously bent (unlike crystal habit) and are composed of secondary iron oxides and voids, and their outlines are blurred (Figure 10.11c). Their matrix is composed of microcrystalline quartz, well-rounded to irregular-shaped framework grains from medium to very coarse sand size. Many of them appear to have been extensively silicified or voids. Some are oxide-rimmed and exhibit peripheral oolitic structures (Figure 10.11d). Euhedral-shaped extensively silicified crystals are also present (Figure 10.11e and f).

Finely laminated black chert is characterized by fine carbonaceous lamination (Figure 10.12a). Degraded carbonaceous filamentous structures are rarely identified (Figure 10.12b), and in many places, laminae are composed of carbonaceous blobs (Figure 10.12c). This type of chert is also laterally continuous and is interbedded within the light-colored laminated chert at the section ca. 100 m north of the type section (the locality 3 in Figure 10.7b), where this carbonaceous chert was found to comprise columnar structures, which had been interpreted as stromatolite (Sugitani et al., 2015 and see the Chapter 7). The associated black-white laminated chert is carbonaceous, with lamination-parallel secondary cavity-fill cherts (Figure 10.12d). The upper sandstone contains angular clasts of black laminated chert up to 50 cm across, though mostly less than 1 mm up to 5 cm, with an overall weak fining-upward trend. This matrix is composed exclusively of silicified volcaniclastic fragments including vesicular pyroclastic fragments, volcanic lithic fragments, rare glass shards, and euhedral crystal pseudomorphs. A new sample of this lithology collected at the uppermost horizon in 2019 contained spherules.

The massive black chert, from which abundant and diverse fossil-like microstructures have been identified, is gray to black and locally has reddish patches and lenticular pure cherts. The cherts are carbonaceous and characterized by irregularly shaped chert masses (Figure 10.13a), with local enrichment of pyrite and sphalerite (Figure 10.13b). Coniform structures up to 3 cm in height characterize this chert (Figure 10.10f). Sugitani et al. (2015) interpreted these structures as possible siliceous sinter (geyserite) (*Column*).

The above described lithostratigraphy is basically conservative through the outcrop, although the thickness of each lithology changes drastically. In addition, around the locality 3 (Figure 10.9b), the section is characterized by the abundance of light-colored (white, red, light brown to light gray) chert, which is interlayered with carbonaceous dark gray to black cherts. Some of the carbonaceous cherts are identical in fabric to the finely laminated carbonaceous chert (Figure 10.12a–c). The others are composed of poorly sorted carbonaceous grains, some of which appear to be originated from reworking of unconsolidated precursor of finely laminated carbonaceous chert.

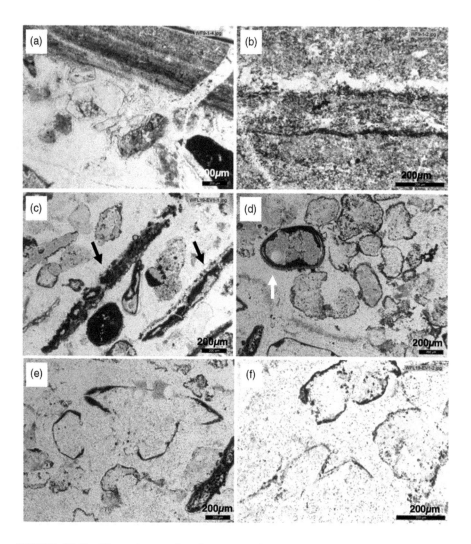

FIGURE 10.11 Photomicrographs of representative lithologies of the uppermost cherty unit of the Strelley Pool Formation in the Goldsworthy greenstone belt. Transmitted light. (a) The matrix of the basal pebbly sandstone (Figure 10.10c), containing extensively silicified euhedral crystals. (b) Pyrite-rich carbonaceous laminae in chert clast in the basal sandstone. (c) Sandstone with subvertically to vertically stuck planer structures (Figure 10.10d), which are composed of oxides and voids (the arrows). (d) Rounded to irregularly shaped framework grains coated by oxides in the same specimen of (c). One grain (the arrow) has oolitic texture. (e, f) Extensively silicified euhedral crystals in the same sample of (c).

Light-colored cherts are generally composed of laminae composed of chalcedonic chert or megaquartz and those identified minute crystal ghosts or pseudomorphs, tiny particles of unknown origin, and/or very fine light yellowish brown particles [possibly organic matter (OM)], and particles of dolomite [$CaMg(CO_3)_2$], apatite

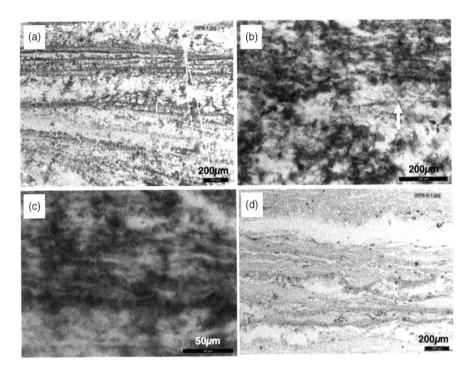

FIGURE 10.12 Photomicrographs of representative lithologies and facies of the uppermost cherty unit of the Strelley Pool Formation in the Goldsworthy greenstone belt. Transmitted light. (a) Finely laminated carbonaceous chert. (b) Rarely preserved filamentous, though ambiguous, structure (the arrow) observed in the finely laminated carbonaceous chert. (c) Blobby textured laminae of the finely laminated carbonaceous chert. (d) Varicolored laminated to banded carbonaceous chert with fossil-like microstructures, characterized by secondary chert fills.

$[Ca_{10}(PO_4)_6(OH,F,Cl)_2]$ and barite $(BaSO_4)$. Mineral particles occasionally comprise peloids. Relatively large hexagonal-shaped pseudomorphs now composed of micro- to mesocrystalline quartz are also found. Hematite (Fe_2O_3) pseudomorphs of lenticular crystals are often found to comprise laminae, although in other places they are just voids outlined by hematite or replaced by quartz.

10.4.3 Depositional Environment of the Uppermost Cherty Unit

The uppermost cherty unit is characterized by drastic vertical change of lithology as represented by the type section (Figure 10.9d). Sugitani et al. (2015) also described six sections measured from 150 m interval and demonstrated significant lateral variations, which are basically attributed to changes in thickness of each lithology, except for the section enriched in light-colored cherts (Figure 2 in Sugitani et al., 2015) as discussed above. Such drastic lateral and vertical changes in lithostratigraphy indicate that the depositional environment was unstable and

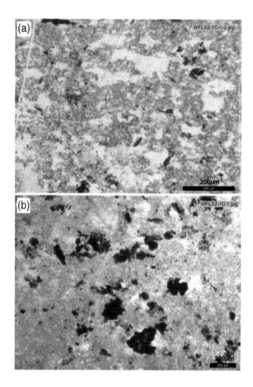

FIGURE 10.13 Photomicrographs of the carbpnaceous of the SPF in the Goldsworthy green-stone belt. Transmitted light. (a) Massive carbonaceous chert, with fossil-like microstructures. (b) Clustered sulfide grains in the massive carbonaceous chert.

even highly fluctuated. Quiet and detritus free depositional environment, repre-sented by laminated chert deposition, was interrupted repeatedly by depositional episodes of higher but wide range of energy. This had caused reworking of uncon-solidated sediments, inflow of sand-sized detritus, destruction of consolidated sediments, and transportation of large clasts. Such environmental fluctuations were at least partially attributed to volcanic activities, as evidenced by the pres-ence of volcaniclastic sandstone. Volcanisms were likely explosive, as suggested from angular chert boulder up to 50 cm across and abundant vesicular volcanic fragments in the volcaniclastic sandstone (Sugitani et al., 2015). In Sugitani et al. (2015), it was suggested that the site of deposition was coastal, having been sub-jected to evaporitic environment. This model seems to be strengthened by newly identified euhedral crystal pseudomorphs common in the different lithologies (e.g., Figures 10.11a, e, f). Although their original mineralogy is still unknown, their shape and extensive silicification suggest that their evaporitic origin (thus soluble) is plausible. The massive black cherts hosting fossil-like microstructures had not been precipitated from seawater. Contribution of hydrothermal fluids is evident from nonseawater-type REEs signatures and anomalously high heavy metal concentrations (Sugitani et al., 2015). Thus, one of the potential origins of the massive chert with conical structure is siliceous sinter (*Column*) formed in terrestrial hydrothermal systems (Sugitani et al., 2015).

10.4.4   FOSSIL-LIKE MICROSTRUCTURES IN THE MASSIVE BLACK CHERTS

Morphologically diverse fossil-like microstructures including spheroid, lens, filament, and film have been described from the massive black cherts and black-white laminated cherts (Sugitani et al., 2010, 2015). To date, microfossils have been identified from four localities (Figure 10.9b).

10.4.4.1   Spheroids

Spheroids are various in morphology (Figure 10.14). Their size ranges from 5 to 80 μm. Small (< 15 μm) spheroids tend to occur as colonies, occasionally in association with film-like structures or being enveloped. In general, their preservation state is poor. One colonial specimen is exceptionally well-preserved and is composed of more than 100 tightly packed spheroids (Figure 10.14a). Medium-sized (~20 μm) spheroids are less common; the colonial specimen shown in Figure 10.14b is associated with an envelope-like material. Large spheroids are either truly spherical to ovoid, the latter of which is less common (Figure 10.14c). Many of them are hollow, filled with pure microcrystalline quartz masses. They are solitary or colonial and occasionally enveloped (Figure 10.14c). Distinct type of spheroid is characterized by hyaline, thin, and transparent wall and by foam-like envelop. The specimen shown in Figure 10.14d has an attached small half-spheroid. Other specimens will be described and discussed in detail in the next chapter.

10.4.4.2   Lenses

Lenses are the most abundant. Morphological subtypes including asymmetric, symmetric, single- or biased-flanged, and nonflanged are identified (Figure 10.15). The specimen in Figure 10.15a represents the asymmetric type, characterized by flange attachment out of the major axis of the central body. Symmetric type shown in Figure 10.15b appears to be less common than asymmetric types. This specimen is characterized by thin-walled central body and displays smooth outline from the central boy to the flange. The sectioned specimen in Figure 10.15c represents the biased tailed type, with asymmetric attachment of flange. Hollow central body and tapered morphology of flange are clearly demonstrated. Nonflanged type was also identified (Figure 10.15d).

10.4.4.3   Filaments and Films

Filaments (filamentous structures) are not common in the Waterfall locality. One specimen worth to be breifly described here is a long (~800 μm) discontinuous filament with fluffy carbonaceous matter on the surface. Its width, excluding fluffy portion, is ca. 20 μm. It appears to be bent and has notched terminal. Films (film-like structures) are common in this locality and are morphologically diverse, like the Farrel Quartzite population. In addition to relatively simple single-sheeted films,

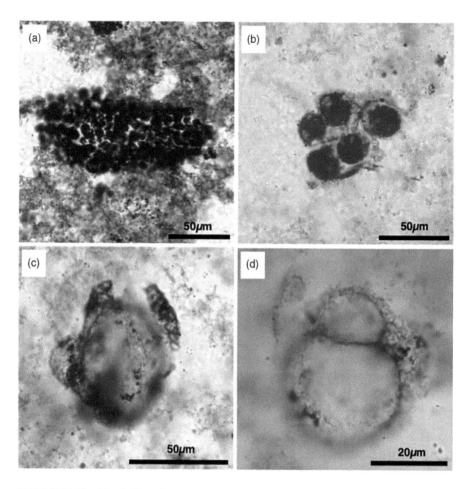

FIGURE 10.14 Fossil-like microstructures from the SPF massive cherts in the Goldsworthy greenstone belt. Transmitted light. (a) Ovoid colonial cluster composed of small spheroids. (b) Colonial cluster composed of several spheroids, with envelop-like substructure. (c) Solid-walled hollow large ovoidal spheroid with an envelope. (d) Thin- and smooth-walled hollow spheroid with an attached hemisphere, with a foam-like envelope.

films with attached small spheroids and those that appear to be branched are present. Some films are hyaline, whereas others are granular.

10.4.5 Minor Occurrences of Fossil-Like Microstructures

Minor occurrences of fossil-like lenticular microstructures have already been reported also from black-white laminated cherts (Figure 10.16a) (Sugitani et al., 2015). Further examination of petrographic thin sections has revealed an

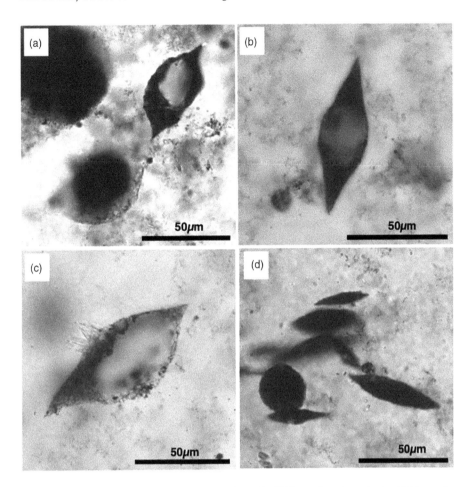

FIGURE 10.15 Fossil-like microstructures from the SPF massive cherts in the Goldsworthy greenstone belt. Transmitted light. (a) Two lenticular structures. The white arrow shows the equatorial view (asymmetric), whereas the solid arrow shows the inclined polar view. (b) Lenticular structure of symmetric type. (c) Lenticular structure characterized by uneven width of flange. Note that the flange attachment is out of the major axis of the central body. (d) Loose cluster composed of nonflanged lenses.

occurrence of colonial small spheroids in this lithology, though poorly preserved (Figure 10.16b) and minor occurrence of lenses and spheroids in other lithologies. Poorly preserved small spheroids and lenses are also identified in chert clasts in the basal sandstone (Figure 10.16c and d). Though very poorly preserved, clustered small spheroids and lenses have been newly identified also from carbonaceous chert laminae interbedded within the light-colored chert (Figure 10.16e and f).

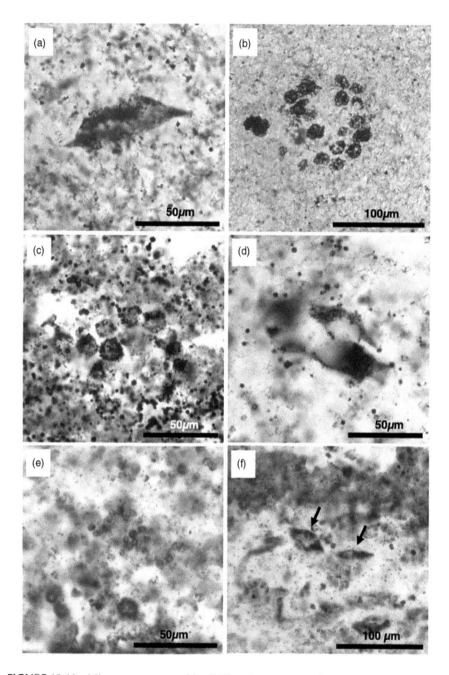

FIGURE 10.16 Minor occurrences of fossil-like microstructures from the uppermost unit of the Strelley Pool Formation in the Goldsworthy greenstone belt. Specimens identified lithologies other than the massive chert. All are poorly preserved. Transmitted light. (a) and (b) Lens and colonial small spheroids, respectively, in laminated to banded chert. (c) and (d) Colonial small spheroids and clustered lenses, respectively, in the tabular laminated chert clast in the basal sandstone. (e) and (f) Colonial cluster composed of small spheroids and lenses (the arrows) from carbonaceous laminae within the light-colored chert.

COLUMN: SILICEOUS SINTER

Siliceous sinter refers to siliceous deposits precipitated from hot springs or geysers at geothermal (terrestrial hydrothermal) systems. Primary mineral facies are mostly opaline silica ($SiO_2 \cdot nH_2O$) (opal-A), which could be recrystallized to microcrystalline quartz via midway phases such as opal-CT. Siliceous sinters exhibit various morphotypes and textures. They were once considered to be formed by purely abiogenic processes including temperature drop and evaporation of overflown hot spring waters and splashes (Walter, 1976). However, it has been demonstrated that facies of modern siliceous sinter and compositions of microbial communities are closely related with each other, basically depending on distance (i.e. temperature) from the hydrothermal vent (e.g., Jones et al., 1997; Cady and Farmer, 1996; Braunstein and Lowe, 2001). For example, spicular sinter forms close to an active vent or associated pool, dominantly through an inorganic process at high temperature (evaporation of splashes). In contrast, bedded sinter characterized occasionally by palisades and shrubs represents more distal facies, where the temperature may drop to around 30°C, and biomats of filamentous cyanobacteria could develop. Columnar and conical sinters represent intermediate facies, from splash zone to outflow system. Microbes and their activities are often involved in the formation of columnar and conical sinters. Rapid silicification of stromatolites has been reported from geothermal systems (Jones et al., 2002).

REFERENCES

Allwood, A.C., Burch, I., Walter, M.R. 2007. Stratigraphy and facies of the 3.43 Ga Strelley Pool Chert in the southwestern North Pole Dome, Pilbara Craton, Western Australia. *Geological Survey of Western Australia*, Record 2007/11.

Allwood, A.C., Walter, M.R., Kamber, B.S., Marshall, C.P., Burch, I.W. 2006. Stromatolite reef from the Early Archaean era of Australia. *Nature* 441, 714–718.

Allwood, A.C., Kamber, B.S., Walter, M.R., Burch, I.W., Kanik, I. 2010. Trace elements record depositional history of an Early Archean stromatolitic carbonate platform. *Chemical Geology* 270, 148–163.

Braunstein, D., Lowe, D.R. 2001. Relationship between spring and geyser activity and the deposition and morphology of high temperature (> 73°C) siliceous sinter, Yellowstone National Park, Wyoming, U.S.A. *Journal of Sedimentary Research* 71, 747–764.

Cady, S.L., Farmer, J.D. 1996. Fossilization processes in siliceous thermal springs: Trends in preservation along thermal gradients. *Evolution of Hydrothermal Ecosystems on Earth (and Mars?)*, Wiley, Chichester (Ciba Foundation Symposium 202), pp. 150–173.

Flügel, E. 2010. *Microfacies of Carbonate Rocks: Analysis, Interpretation and Application*. Springer-Verlag, Berlin, Heidelberg.

Hickman, A.H. 2008. Regional review of the 3426-3350 Ma Strelley Pool Formation, Pilbara Craton, Western Australia. *Geological Survey of Western Australia*, Record 2008/15.

Hickman, A.H. 2012a. Review of the Pilbara Craton and Fortescue Basin, Western Australia: Crustal evolution providing environments for early life. *Island Arc* 21, 1–31.

Hickman, A.H. 2012b. North Shaw, W.A. Sheet 2755 (2nd edition): Geological Survey of Western Australia, 1: 100 000 Geological Series.

Jones, B., Renaut, R.W., Rosen, M.R., Ansdell, K.M. 2002. Coniform stromatolites from geothermal systems, North Island, New Zealand. *PALAIOS* 17, 84–103.

Jones, B., Renaut, R.W., Rosen, M.R. 1997. Biogenicity of silica precipitation around geysers and hot-spring vents, North Island, New Zealand. *Journal of Sedimentary Research* 67, 88–104.

Lowe, D.R. 1980. Stromatolites 3,400-Myr old from the Archean of Western Australia. *Nature* 284, 441–443.

Lowe, D.R. 1983. Restricted shallow-water sedimentation of Early Archean stromatolitic and evaporitic strata of the Strelley Pool Chert, Pilbara Block, Western Australia. *Precambrian Research* 19, 239–283.

Smithies, R.H., Van Kranadonk, M.J., Hickman, A.H., 2004. De Grey, W.A. Sheet 2757 (version 2.0): Western Australia Geological Survey, 1:100 000 Geological Series.

Sugitani, K., Grey, K., Allwood, A.C., Nagaoka, T., Mimura, K., Mimura, M., Marshall, C.P., Van Kranendonk, M.J., Walter, M.R. 2007. Diverse microstructures from Archaean chert from the Mount Goldsworthy – Mount Grant area, Pilbara Craton, Western Australia: Microfossils, dubiomicrofossils, or pseudofossils? *Precambrian Research* 158, 228–262.

Sugitani, K., Kohama, T., Mimura, K., Takeuchi, M., Senda, R., Morimoto, H. 2018. Speciation of Paleoarchean life demonstrated by analysis of the morphological variation of lenticular microfossils from the Pilbara Craton, Australia. Astrobiology 18, 1057–1070. doi:10.1089/ast.2017.1799.

Sugitani, K., Lepot, K., Nagaoka, T., Mimura, K., Van Kranendonk, M., Oehler, D.Z., Walter, M.R. 2010. Biogenicity of morphologically diverse carbonaceous microstructures from the ca. 3400 Ma Strelley Pool Formation, in the Pilbara Craton, Western Australia. *Astrobiology* 10, 899–920.

Sugitani, K., Mimura, K., Nagaoka, T., Lepot, K., Takeuchi, M. 2013. Microfossil assemblage from the 3400 Ma Strelley Pool Formation in the Pilbara Craton, Western Australia: Results from a new locality. *Precambrian Research* 226, 59–74.

Sugitani, K., Mimura, K., Suzuki, K., Nagamine, K., Sugisaki, R. 2003. Stratigraphy and sedimentary petrology of an Archean volcanic-sedimentary succession at Mt. Goldsworthy in the Pilbara Block, Western Australia: Implications of evaporite (nahcolite) and barite deposition. *Precambrian Research* 120, 55–79.

Sugitani, K., Mimura, K., Takeuchi, M., Yamaguchi, T., Suzuki, K., Senda, R., Asahara, Y., Wallis, S., Van Kranendonk, M.J. 2015. A Paleoarchean coastal hydrothermal field inhabited by diverse microbial communities: The Strelley Pool Formation, Pilbara Craton, Western Australia. *Geobiology* 13, 522–545.

Van Kranendonk, M.J. 2000. Geology of the North Shaw 1:100 000 sheet: Geological Survey of Western Australia, 1:100 000 Geological Series Explanatory Notes, ISSN 1321-229X; sheet 2755.

Van Kranendonk, M.J. 2004. Carlindie, W.A. Sheet 2756: Western Australia Geological Survey, 1: 100 000 Geological Series.

Van Kranendonk, M.J., Smithies, R.H., Hickman, A.H., Bagas, L., Williams, I.R., Farrell, T.R. 2004. Event stratigraphy applied to 700 million years of Archaean crustal evolution, Pilbara Craton, Western Australia. *Geological Survey of Western Australia Annual Review*. 2003–04, pp. 49–61.

Walsh, M.M. 1992. Microfossils and possible microfossils from Early Archean Onverwacht Group, Barberton Mountain Land, South Africa. *Precambrian Research* 54, 271–293.

Walter, M.R. 1976. Geyserites of Yellowstone National Park: An example of abiogenic "stromatolites" In M.R. Walter ed. Stromatolites, Developments in Sedimentology 20, 85–112. Elsevier, Amsterdam.

11 Biogenicity of the Pilbara Microstructures

11.1 INTRODUCTION

Readers might be bewildered with herein described assemblages of fossil-like microstructures from the Pilbara Craton, Western Australia, because some of them appear to be far from the general images of Archean organisms. General images of Archean organisms are tiny and simple in morphology; their preservation states should be poor. The Pilbara microstructures are large. Even structures classified as "small spheroids" range from 5 to 15 μm with an average of 10 μm in diameter, significantly larger than typical prokaryotic coccoid cells (0.1–5.0 μm). Microstructures classified into "large spheroids" range from 15 to 270 μm along the major dimension. Lenticular microstructures range from 20 to 100 μm across. As described in this chapter, some of the large spheroids and lenses can be extracted by acid maceration, retaining their morphologies. Thus, the Pilbara microstructures may represent the oldest assemblage of acritarch (*Column*). The microstructures often comprise composite structures correlative to stages of reproductions. Ironically these features of the Pilbara microstructures might have caused doubts on their credibility as microfossils in scientific society. To my knowledge, however, papers that criticize our interpretation in a direct manner have not yet been published, although some have suggested that lenticular microstructures represent colonies suspended in water column or microbially colonized volcanic vesicles but not cellular microfossils (Javaux, 2019; Wacey et al., 2018a, b).

In my opinion, the biogenicity of the Pilbara microstructures ever described, if not all, has been fully demonstrated by previous publications (Sugitani et al., 2007, 2009a, b, 2010, 2013, 2015a, b; Grey and Sugitani, 2009; Oehler et al., 2009, 2010, 2017; Sugahara et al., 2010; Schopf et al., 2010, 2017; House et al., 2013; Lepot et al., 2013; Delarue et al., 2020; Alleon et al., 2018). In this chapter, I discuss on the biogenicity of the Pilbara microstructures more comprehensively, based on the criteria proposed in Chapter 7.

11.2 GEOLOGIC CONTEXT 1: AGES OF ROCKS

The localities, from which fossil-like microstructures have been discovered, are within fully described Archean sedimentary successions in the Pilbara Craton. However, the exact ages of the sedimentary successions are not always known. In this section, I would like to tackle this issue.

11.2.1 The Farrel Quartzite

As described in the chapter 9, there have been some twist and turn in assignment of the clastic sedimentary unit at the southern flank of Mount Goldsworthy and Mount Grant to the established stratigraphic unit in the Pilbara Supergroup.

The reason for this confusion was the lack of enough and precise age data for this region, although according to Australian Stratigraphic Units Database, the Farrel Quartzite (FQ) had deposited between c. 3,200 and c. 2,940 Ma. Assignment of the clastic sedimentary unit at the Mount Grant and Mount Goldsworthy to the Farrel Quartzite may need to be regarded as still provisional. However, there is no doubt about its Archean age, considering the spatial relationship of this greenstone belt with the well-dated Neoarchean (ca. 2.7 Ga-old) Fortescue Group. On the other hand, the clastic sedimentary unit is likely younger than 3.4 Ga, considering stratigraphic relationship with the Strelley Pool Formation (SPF) at the western Mount Grant.

11.2.2 The Strelley Pool Formation

The Strelley Pool Formation is one of the most studied Archean sedimentary successions in the Pilbara Craton, and its lithostratigraphy has been extensively studied. Therefore, assignments of the studied successions to the Strelley Pool Formation are highly promising. This is particularly obvious for the Panorama localities, which are within the region of extensive mapping made by previous studies (e.g., Allwood et al., 2006, 2007). Suspicious readers may cast doubts on reliability of assignment of the other two localities to this formation. In the following reasons, however, the assignment of the studied sedimentary successions in the Warralong and the Goldsworthy greenstone belts to the Strelley Pool Formation is reasonable. First, sedimentary successions in the Warralong and the Goldsworthy greenstone belts contain an association of stromatolites and evaporites that characterizes the Strelley Pool Formation in the Panorama greenstone belt. Second, these sedimentary successions have distinct chert lithofacies identical to those in the Panorama greenstone belt. They include "chert" composed of sand-sized reworked carbonaceous grains and chert characterized by laminae composed of fine pyrite grains, in which fossil-like microstructures of the similar morphological variation have been identified.

11.3 GEOLOGIC CONTEXT 2: SEDIMENTARY ORIGIN OF HOST CHERTS

It is fundamental to demonstrate that the host carbonaceous cherts containing fossil-like microstructures are sedimentary in origin, in order to claim that the microstructures are genuine microfossils. Previous studies have described putative microfossils in Archean volcanic rocks, which were interpreted as fossilized endolithic microbes (e.g., Furnes et al., 2004). This means that the sedimentary origin of the host rock is not always essential for the biogenicity of contained microstructures. Additionally, veins could be hosts for microfossils, if they were precipitated from relatively "low (<120°C)" temperature hydrothermal fluids (Takai et al., 2008). Despite this, the biogenicity of ancient microstructures is undoubtedly supported by the sedimentary origin of their host rocks. In the present case, multiple lines of evidence support their sedimentary origin.

First, the host black cherts are laterally continuous. The FQ black chert in the Goldsworthy greenstone belt can be traced up to 7 km along the strike at the Mount Grant (Figure 9.2). The SPF black chert in the Goldsworthy greenstone belt can be traced over 100 m along the strike, with intermittent missing due to erosional contact with the overlying volcanic unit (Figure 10.9). The SPF black chert in the Warralong greenstone belt is also traced over 200m along the strike (Figure 10.7), with again intermittent missing due to erosional contact with the overlying conglomerate. Lateral traceability of the black cherts in the Panorama greenstone belt has not been examined by myself, which is however compensated by the detailed mapping (Allwood et al., 2006). Second, the black cherts at every localities occur as part of sedimentary successions composed of terrigenous clastic rocks such as sandstone, volcaniclastic rocks, sandstone, massive chert of evaporite origin, jasper, banded iron formation (BIF), and/or carbonates. Third, the black chert itself contains clastic layers. The FQ black chert contains a thin layer composed of volcaniclastics and possible evaporite minerals, whereas the SPF black cherts contain a layer composed of reworked carbonaceous grains. Fourth, both the SPF and FQ black cherts display cross lamination, though locally (Figure 11.1). These lines of evidence indicate the sedimentary origin of the black cherts, which is consistent with the absence of large-scale cross cutting veins composed of black cherts in the vicinity of the black cherts. Trace element {rare earth elements (REEs)} support sedimentary origin of the FQ and the SPF black cherts (Allwood et al., 2010; Sugitani et al., 2007, 2013, 2015; Sugahara et al., 2010; 2013, 2015b).

11.4 GEOLOGIC CONTEXT 3: PRIMARY ORIGIN OF HOST CHERTS

In this section, the primary origin of the carbonaceous black cherts is discussed: the term "primary" here is used for when the cherts were originated from silica- and organic-rich sediments or formed by early silicification of organic-rich sediments. If the cherts are secondary but not primary, one may suspect the possibility that the later chertification process was associated with remobilization of organic matter (OM), resulting in formation of microfossil-like structures. This possibility may need to be once considered, because secondary cavity-fill portions are often carbonaceous (Figure 11.2a).

In Sugitani et al. (2007), the primary origin of the FQ black cherts containing fossil-like microstructures were argued by the presence of silica spherulites, which were variable in diameter but usually range from 5 to 10 μm, outlined by finely dispersed minute carbonaceous particles. Oehler and Schopf (1971) and Oehler (1976) experimentally demonstrated that chalcedonic quartz spherulite can form from colloidal

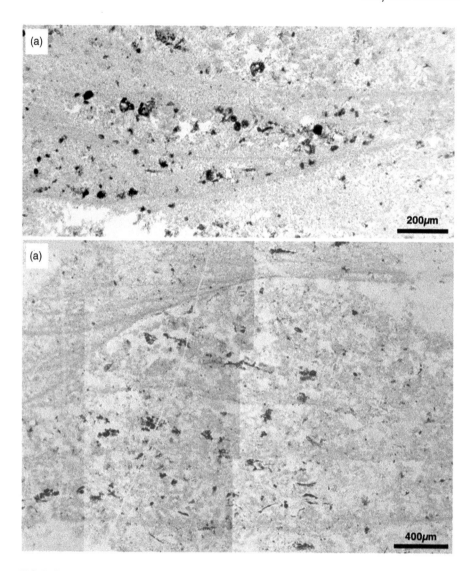

FIGURE 11.1 Cross-lamination locally present in carbonaceous cherts containing microfossil-like objects. The FQ specimen (a) and the SPF specimen (b) from the Goldsworthy greenstone belt. Transmitted light.

gel. Due to spherulites, the matrix of the Bitter Springs Formation chert has a lacey and reticulate appearance (Schopf, 1968). Similar texture was described for the FQ black cherts (Figure 7 in Sugitani et al., 2007). While this previous observation was more or less ambiguous, the texture is now better reproduced from newly prepared thin sections (Figure 11.2b). Similar spherulites are identified also in the SPF black cherts (Figure 11.2c and d). Furthermore, the additional two lines of evidence are consistent with primary origin of the black cherts. One is the presence of sand-sized

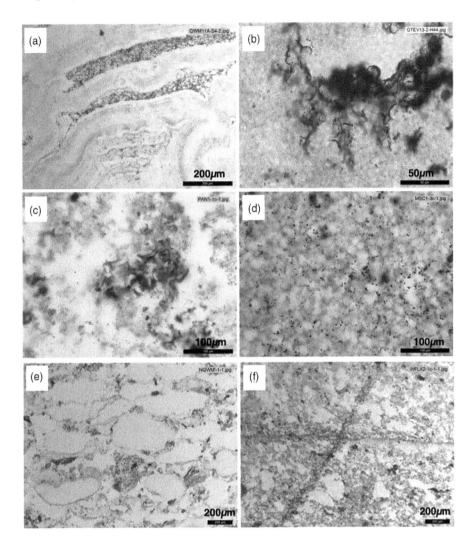

FIGURE 11.2 Fabrics of primary origin of the chert studied here {except for (a)}. Transmitted light. (a) Involvement of carbonaceous matter in colloform-textured portion, indicating redistribution of organic matter during early diagenesis. (b) Silica spherulites in FQ black chert. (c) Silica spherulite in SPF black chert (the Panorama locality 2). (d) Silica spherulite in SPF black chert (the Warralong locality 1). (e) Silica granules, indicative of silica precipitation in water column. FQ black chert from Mt. Goldsworthy. (f) Possible silica granules in SPF black chert (the Goldsworthy locality 1, massive facies).

granules composed of pure chert. Such silica granules were first described from the Barberton greenstone belt by Stefurak et al. (2014) (the Chapter 5). The authors interpreted them as primarily precipitates within water columns. Such silica granules have been identified in the FQ black cherts (Figure 11.2e). Similar structures can be seen also in the SPF black cherts (Figure 11.2f). These lines of evidence point to the formation of the examined black cherts by combined processes of deposition of silica granules, in-situ precipitation of silica-gel, and early silicification, the latter two of which possibly played a major role.

In addition to primary silica precipitation and early silicification, later pervasive silicification is evidenced by the presence of silicified mafic to volcaniclastic fragments and probable evaporitic minerals in thin layers interbedded within the black cherts (Figures 9.9d and 10.11). However, this process was not associated with addition of carbonaceous matter to siliciclastic layer and thus cannot explain the formation of the black cherts. As described in the next section, this process removed OM and in some cases was associated with ferruginization. It is also unlikely that the black cherts originated from carbonate and/or siliciclastics. Although carbonate minerals and fine silicates are found, their occurrence is very minor in the FQ cherts and the SPF black cherts in the Warralong and the Goldsworthy greenstone belts. In the SPF black cherts in the Panorama greenstone belt, carbonate minerals are common, but they occur as euhedral grains in a close association with lenticular structures (Figure 10.6b). There have been identified no dispersed carbonate minerals or patches as can be seen in some laminated cherts of the Cleaverville Formation that can be interpreted as silicified carbonates (Figure 5.11).

11.5 SYNGENICITY

The black cherts were subjected to silicification events related to invasion of silica-rich fluids. This event appears to have occurred repeatedly from the early stage when the precursors of black cherts were not yet consolidated to the later stage after its complete lithification. This phenomenon could place constraint on timing when fossil-like microstructures were involved in their host sediments. An example is shown in Figure 11.3a, which represents the earlier stage of silicification before consolidation, as evidenced by irregularly shaped margin that faces quartz mass composed of inner megaquartz mosaic and outer microquartz mass showing chalcedonic extinction figure. Margin of the black chert was bleached and reddish in color, indicating metasomatic silicification associated with ferruginization. In this region, the fossil-like microstructures and the matrix have been subjected to the same alteration (Figure 11.3b$_{1-3}$). It may be also noted that such microstructures are often cut by quartz vein (Figure 11.3b$_2$, c, d), indicating invasion of silica-rich fluids after complete lithification. These features can be taken as petrographic evidence for syngenicity. This is confirmed by Raman spectral data showing the same thermal maturity for the microstructures as the matrix OM (Sugitani et al., 2007, 2010).

11.6 BIOGENICITY

In this section, biogenicity of the FQ and SPF fossil-like microstructures are discussed comprehensively based on size and size distribution, shape, occurrence, taphonomy, and chemical and isotopic compositions.

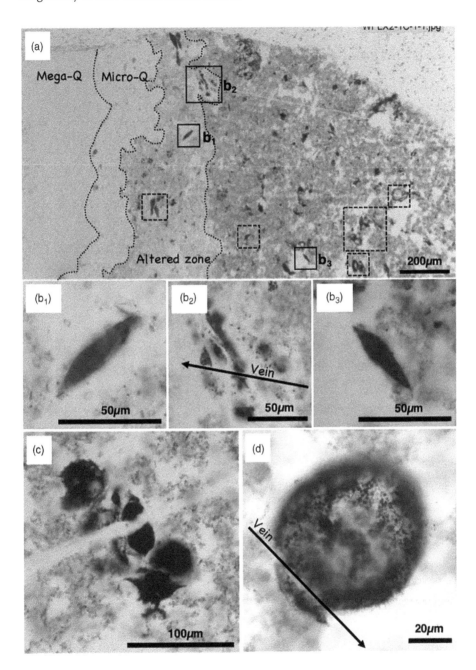

FIGURE 11.3 Occurrences of fossil-like microstructures (lenses) suggesting their syngenicity. Transmitted light. (a) Silicification and ferruginization of the SPF black chert from the Goldsworthy greenstone belt. Insets show the microstructures. (b₁) Ferruginized lens. (b₂) Ferruginized lenses cut by thin quartz vein. (b₃) Unaltered lens. (c) Lenticular microstructure of the Farrel Quartzite cut by quartz vein. (d) Large spheroid of the Strelley Pool Formation cut by quartz vein (the arrow).

11.6.1 Films

Films (film-like structures) are found from all the localities, although abundance and morphological diversity are extensive in the Farrel Quartzite. Thus, FQ films are discussed here. The biogenicity of the SPF films has been argued from carbonaceous composition and morphological similarity to modern microbial films (Sugitani et al., 2010, 2013) and detection of nitrogen- and oxygen- rich organic molecules in them based on X-ray absorption near edge structures (XANES) (Alleon et al., 2018).

Films occur throughout the FQ black cherts, and their distribution is random and disperse (Figure 11.4a). Carbonaceous compositions and taphonomic features are consistent with biogenicity. Taphonomic features include bent, notched edges, and hyaline to granular nature. Membranous preservation of some specimens has been confirmed by extraction of this type of morphology by acid maceration (not shown here). Among numerous specimens, elaborate films have been identified, including single-sheeted film with regularly spaced short ridges (10 μm step and 2–10 μm height) (Figure 11.4b), complicatedly branched film (Figure 11.4c), and branched film on which small spheroids are attached (Figure 11.4d). Though biogenic, it is unlikely

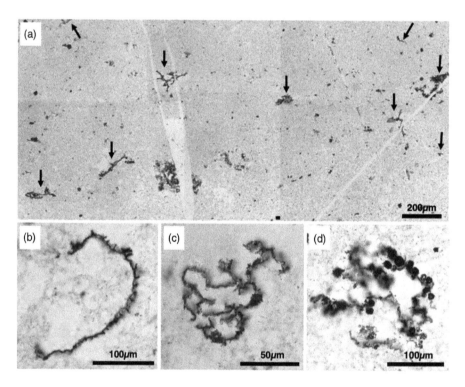

FIGURE 11.4 Film-like structures, from FQ. Transmitted light. (a) Random distribution of abundant film-like structures (arrows) in the carbonaceous black chert from FQ. (b) Single-sheeted film (sectioned) with regularly spaced ridges. (c) Branched film. (d) Degraded branched film with small spheroids.

that the film-like structures are fossilized cellular life. As Sugitani et al. (2007) suggested, it is the best interpretation that they represent reworked and fragmented biofilm. Textural and morphological variations may correspond to species diversity that had constructed biofilms.

11.6.2 FILAMENTS

Carbonaceous filamentous structures have been found both from the Farrel Quartzite and the Strelley Pool Formation except for the Warralong greenstone belt. Variations in morphology and occurrence are recognized including (1) broad (~20 μm) filament, (2) thin (~3μm) filament, and (3) thread-like filaments less than 1 μm in width.

The biogenicity of filamentous structures needs to be assessed in the following basic traits of morphology, including hollowness, constant width, and taphonomy in addition to carbonaceous composition. Furthermore, colonial occurrence and association with mat-like fabrics should be favored for biogenicity, because many of filamentous microbes are basically considered to be benthic and often comprise biomats. Broad filaments identified from the Farrel Quartzite (Figures 9.14f) and the Strelley Pool Formation satisfy the compositional and morphological criteria and thus can be called possible microfossils. They show fragmented appearance and occur dispersedly but do not comprise colonial cluster and/or mat-like fabric. It is likely that they comprised biomats elsewhere and were reworked and transported to the site of deposition. Biogenicity of the SPF broad filaments from the Panorama locality 2 is further supported by their successful extraction from host rock by acid maceration (Delarue et al., 2020). Nano-scale Secondary Ion Mass Spectroscopy (Nano-SIMS) elemental mapping indicates their membranous nature associated with essential nutrients (N and P) (Figure 11.5a–c). The SPF thin filaments from the Panorama locality 1 satisfies morphological traits. Hollowness was demonstrated by an enlarged view of different specimens comprising a loose cluster (Figure 11.5d) and by images constructed using a confocal laser scanning microscopy (CLSM) (Figure 11.5e and f) (Schopf et al., 2017). They are locally identified in facies displaying mat-like fabric. Thus, blurred discontinuous filamentous structures identified as arrangement of fine carbonaceous particles can be interpreted as degraded such the thin filaments comprising the mat itself (Figure 10.2c).

Thread-like filaments clustered in irregularly shaped pure chert masses of the FQ black cherts show dense populations within a specific thin (~1 mm) layer (Figure 9.14d) and are also probably biogenic (Figure 11.6a), considering carbonaceous compositions and taphonomic features. However, their cellularity is uncertain, due to their very thin nature. It is possible that the threads represent fibrillar extracellular polymeric substances (EPS) (Leppard, 1986).

Thin filaments from the FQ black chert (Figure 9.14e) have not yet been confirmed for their carbonaceous compositions, although their appearance is the same as carbonaceous materials in the matrix. Most of them are blurred or thread-like, which could be attributed to taphonomic degradation, because one specimen is found to have preserved its solid shape (Figure 11.6b), whereas the other specimen, probably slightly degraded, suggests its hollow interior (Figure 11.6c). They occur as loose cluster (Figure 9.14e), which are restricted to fractures filled

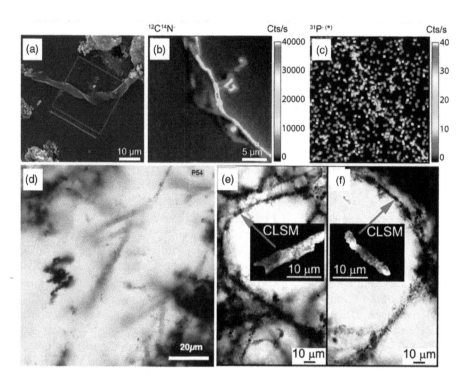

FIGURE 11.5 Photomicrographs of filaments from SPF at the Panorama localities 1 and 2. (a) Broad filament extracted by acid maceration, from the Panorama Locality 2. (b) and (c) show results of elemental mapping using Nano-SIMS for the specimen shown in (a). (a)–(c) are reproduced from Delarue et al. (2020), with permission from Elsevier. (d) Loose cluster of cylindrical filaments from the Panorama Locality 1. (e) and (f) Cylindrical cellularity and kerogenous composition of the filaments equivalent to those in (d) are confirmed by CLSM imaging. (e) and (f) are reproduced from Schopf et al. (2017), with permission from Elsevier.

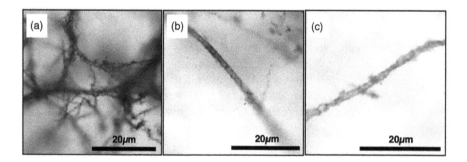

FIGURE 11.6 Enlarged views of thin filaments from FQ shown in Figure 9.14d and e. (a) Bundled and entangled filaments (Figure 9.14d). Each filament is less than 0.5 μm in width. (b) and (c) show filaments that occur within early formed fractures (Figure 9.14e). (b) The best-preserved specimen (composite photograph). (c) Slightly degraded specimen, demonstrating the cylindrical cellularity. Transmitted light.

with microcrystalline quartz. The fractures locally have jagged outline, and some appear to be closed systems. Additionally, the fractures are cut by quartz vein. These features suggest that the fractures had formed before complete lithification of sediments. Their biogenicity is possible but requires further evidence such as concentrations of essential elements (N, S, and P).

11.6.3 SMALL SPHEROIDS

Small spheroids are defined for spherical structures from 5 to 15 μm in diameter, identified both from the Strelley Pool Formation and the Farrel Quartzite (Figure 9.13a and Figure 10.14a). This size range is set based on the results of measurements of spheroid structures (n = 185 for the SPF and n = 480 for the FQ) (Sugitani et al., 2007, 2010). The structures, particularly those around 10 μm in diameter, are very

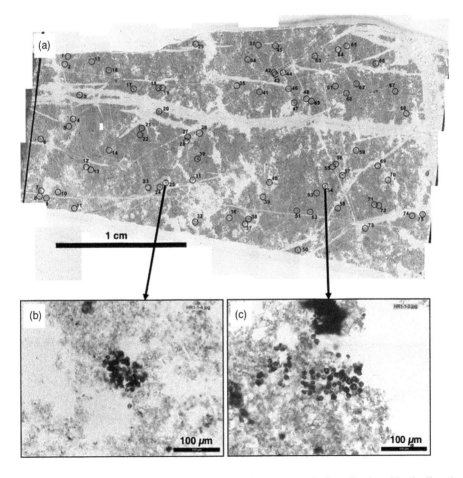

FIGURE 11.7 (a) Abundant distribution of colonies composed of small spheroids, the Farrel Quartzite. (b) and (c) show examples of individual colonies. Transmitted light.

common in the Farrel Quartzite. As shown in Figure 11.7a, more than 70 colonies composed of 5–100 individuals can be occasionally identified in a petrographic thin section whose area and thickness are 8.7 cm^2 and 30 μm, respectively. The colonies are completely embedded within the matrix composed of microcrystalline quartz and cloudy carbonaceous materials (Figure 11.7b and c); no signs for invasions can be seen. Abundance of colonial small spheroids is much lesser in the SPF black cherts, although they are commonly found from the three greenstone belts.

Their biogenicity is consistent with carbonaceous compositions, colonial occurrences, and taphonomic features. Taphonomic features include partial

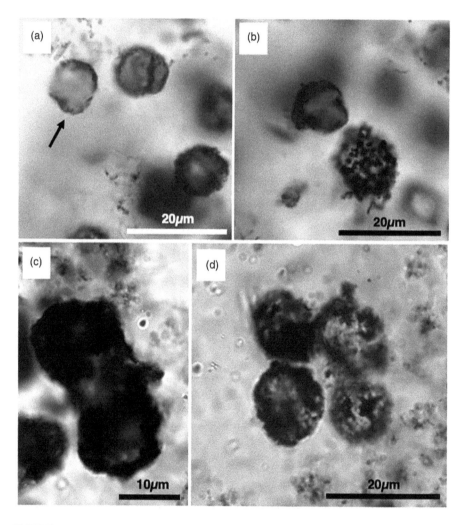

FIGURE 11.8 Small spheroids from the Farrel Quartzite. (a) Small spheroid with a partial breakage (the thin arrow). (b) Wrinkled and granulated specimens of small spheroid. (c) Specimen possibly representing an intermediate stage of binary fission. (d) Tetrad arrangement, as a possible result of successive binary fission. Transmitted light.

breakage, wrinkle, and granulation of their walls (Figure 11.8a and b). Schopf et al. (2010) measured two colonies of the FQ small spheroids, which range from 11.5 to 13.5 μm (n = 15) and from 7 to 10.5 μm (n = 9) in diameter, respectively. Calculated divisional dispersion indices (DDI) are less than 2. DDI for 15 of 20 colonies from the Panorama locality are also less than 3 (Sugitani et al., 2013). The measured DDI strengthen biogenicity of the small spheroids and suggest that they reproduced simply by binary fissions. A specimen that can be interpreted as an intermediate stage of binary fission (Figure 11.8c) and a tetrad arrangement (Figure 11.8d) as a possible result of successive binary fission support their biogenicity.

These morphological features suggestive of biogenicity are consistent with geochemical data. Nano-SIMS elemental mapping revealed that carbonaceous composition ($^{12}C^-$) of the FQ small spheroids are associated with N ($^{12}C^{14}N^-$) and S ($^{32}S^-$), well known as essential elements (Oehler et al., 2010) (Figure 11.9). Enhancement of silicon ($^{28}Si^-$) and oxygen ($^{16}O^-$) associated with distribution of C, N, and S has been already reported from the genuine microfossil assemblage of the Neoproterozoic Bitter Springs Group and has been interpreted as a result of silica permineralization of OM (Oehler et al., 2006, 2009). Carbon isotopic compositions of individual small spheroids were measured by House et al. (2013) for the FQ population (Figures 11.10) and Lepot et al. (2013) for the SPF one. Both studies demonstrated their significantly light carbon isotopic compositions ($\delta^{13}C_{PDB} < -30‰$), distinct from carbonaceous clots and disseminated carbonaceous matter in the matrix and veins.

It should be here noted that some of the above discussions (the narrow size distribution and evidence for binary fission) cannot be applied to small spheroids comprising colonies together with large spheroids and/or lenses. The biogenicity of these types need to be considered in the context of biogenicity of associated larger spheroids and lenses, as discussed later.

11.6.4 LARGE SPHEROIDS

Large spheroids are defined for spheroid structures larger than 15 μm along the major dimension, automatically set by the range for the small spheroids described above. This broad morphotype is classified into (1) flexible-walled large spheroid, (2) robust and thick-walled large spherical spheroids, (3) robust-walled large oblate spheroids, and (4) robust and thin-walled large spherical spheroids. Here, the term "flexible" is used for spheroids that appear to be compressed or tightly folded in petrographic thin sections and in macerates. The term "robust" is used for other spheroids that do not exhibit such habits.

11.6.4.1 Flexible-Walled Large Spheroids

Flexible-walled large spheroids are common in the Farrel Quartzite, whereas to date, only one possible specimen has been identified from the Strelley Pool Formation at the Panorama locality 2. Thus, the biogenicity of the SPF specimen is not discussed here. In the FQ black cherts, up to more than 20 specimens can be occasionally identified in a petrographic thin section with an area of 8.7 cm^2 and a thickness of 30 μm. I have long been puzzled by the presence of a significant

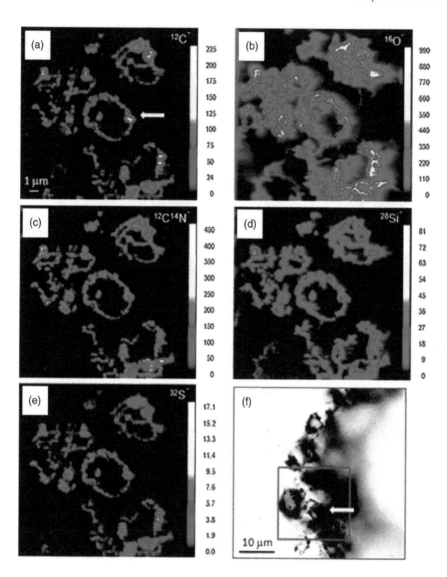

FIGURE 11.9 Nano-SIMS elemental maps of FQ small spheroids. (a–e) Nano-SIMS maps for carbon ($^{12}C^-$) (a), oxygen ($^{16}O^-$) (b), nitrogen ($^{12}C^{14}N^-$) (c), silicon ($^{28}Si^-$) (d), sulfur ($^{32}S^-$) (e). Yields of these ions (intensity) are shown by color bars and scales. (f) Optical microphotograph of the small spheroids subjected to the nano-SIMS analyses (the inset). (Reproduced from Oehler et al. (2010), with permission from Mary Ann Liebert.)

number of "film-like" structures with a smooth outline and relatively narrow size between 100 and 300 μm along the major dimension. These features are inconsistent with the interpretation that they are fragmented biofilms. I have re-interpreted such structures as flexible-walled large spheroids based on identification of compressed large spheroids in acid maceration and their appearance in petrographic

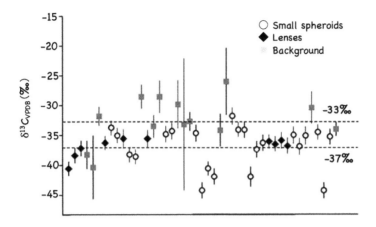

FIGURE 11.10 $\delta^{13}C$ composition (± 1 σ, ‰ VPDB) of the Farrel Quartzite microstructures and OM by SIMS. The analyzed spots include 21 small spheroids, 10 lenses, and 14 background organic carbon (in order of analysis). Dashed lines show weighted mean values for the background carbon (−33‰) and for both lenses and small spheroids (−37‰). (Image courtesy by House et al. (2013).)

thin sections quite similar to spheroid microfossils with flexible wall from the 2.5 Ga-old Gamohaan Formation, South Africa (Czaja et al., 2016). Carbonaceous compositions and syngenicity were confirmed by Raman spectral analyses of specimens in macerates and petrographic thin sections (not shown here). Taphonomic features of micron to submicron scale can be identified in petrographic thin sections. Degree of compression is various (Figure 11.11a and c), and the surface texture ranges from hyaline to granular (Figure 11.11b and d). Variation of the surface texture, likely attributed to different degrees of taphonomic degradation, can be seen also for extracted specimens (Figure 11.11e and f).

This type of large spheroids occurs solitarily and never comprise colony-like cluster, meaning lack of biological occurrence and sign of vital activity (but see Section 12.6.1). Additional data such as individual carbon isotopic data and distribution of essential elements such as N, S, and P may be required to assert their biogenicity. However, it may be emphasized that this type of spheroid is similar to sphaeromorph acritarch, the most common morphotype of organic-walled microfossils reported from the wide age interval from the Paleoproterozoic to the Paleozoic. Thus the biogenicity of this morphotype is likely. Though the term "sphaeromorph acritarchs" was not used, spheroidal microfossils has also been reported from the ca. 3.2 Ga-old siliciclastic rocks in the Barberton greenstone belt, South Africa (Javaux et al., 2010).

11.6.4.2 Robust and Thick-Walled Large Spherical Spheroid

Robust and thick-walled large spherical spheroids are here defined for spheroids without compressive occurrence and with solid appearance. This morphotype has been identified both from the Farrel Quartzite and the Strelley Pool Formation.

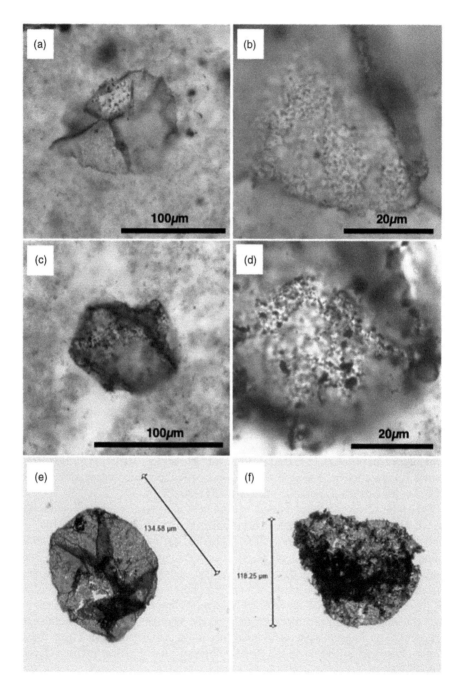

FIGURE 11.11 Photomicrographs of flexible large spheroids from the Farrel Quartzite. (a) Highly compressed specimen in petrographic thin section. (b) Enlarged view of the specimen in (a), characterized by smooth and hyaline surface. (c) Weakly compressed specimen in petrographic thin section. (d) Enlarged view of the specimen in (b), characterized by granular surface. (e) and (f) show specimens extracted by acid maceration, showing different degrees of degradation.Transmitted light.

They range from ~20 to 50 µm along the major dimension and thus are significantly smaller than the flexible-walled large spheroids. This category involves specimens showing significant variation in morphology and texturally. Here, a major population identified from the FQ black cherts is discussed for its biogenicity. Though their solid appearance at glance, they, if not all, are likely hollow, being filled with microcrystalline quartz as suggested by the presence of multiple specimens with irregularly shaped openings (Figure 9.13e) and partially transparent wall (Figure 11.12a). Although many of this type occur as solitarily in petrographic thin sections, the presence of paired specimens (Figure 11.12b) and colonial clusters of various arrangements (Figure 11.12c and d) cannot be overlooked. Considering robust

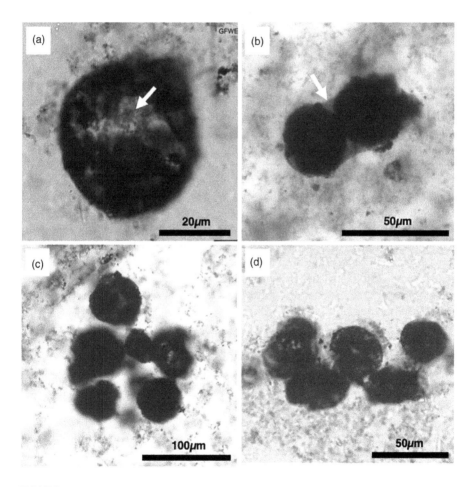

FIGURE 11.12 Photomicrographs of robust and thick-walled spheroids, from the Farrel Quartzite. Transmitted light. (a) Specimen with partially transparent (the arrow) wall. (b) Paired specimens connected by racy material (the arrow). (c, d) Clusters of thick-walled spheroids of different arrangements.

and thick-walled spheroids are found in macerates (not shown here), they are not aggregations of carbonaceous particles. It seems reasonable to suggest that some specimens of this morphotype are probable microfossils, although geochemical data including isotopic compositions are required for confirmation.

11.6.4.3 Robust-Walled Large Oblate Spheroids

Robust-walled large oblate spheroids, which are defined by their elliptic shape and rigid appearance, have been identified both from Farrel Quartzite and the Strelley Pool Formation (Figures 9.13f, and 10.14c). Number of this morphotype is much smaller than the other spheroid morphotypes and cannot always be discriminated from the robust and thin-walled large spherical spheroids discussed below.

The biogenicity of this morphotype is consistent with hollowness and taphonomy, in addition to carbonaceous composition. Hollowness, which is often uncertain for thick-walled specimens, is confirmed by specimens with thin and transparent walls. Some thick-walled specimens are also found to be hollow, due to partial breakage. Variation in wall appearance from hyaline to granular can also be attributed to taphonomy. A few specimens have envelopes or are associated with film-like structures. More importantly, this morphotype occurs as pairs potentially corresponding to various stages of binary fissions (Figure 11.13a and b) and as colonial clusters of various types including detached cluster and chained cluster in which individuals are connected by thin film (Figure 11.13c and d). These consistent lines of evidence indicate that this type of spheroid is a probable microfossil. The colonial specimen from the SPF black chert (the Waterfall locality) is worth to be mentioned in some details (Figure 11.13e). This specimen is composed of several individuals, ranging from 30–50 µm across. As suggested by the arrowed individual, their morphology may be more likely elliptic. Alternatively, it represents some compression. Thin envelop is locally found, and walls of individuals have tiny carbonaceous clots. The structure that can be interpreted as ruptured spheroid is associated with this cluster (Figure 11.13f). Such occurrence and taphonomy are consistent with their biogenicity, although geochemical and isotopic data would be required for confirmation.

11.6.4.4 Robust and Thin-Walled Large Spherical Spheroids: The FQ Assemblage

Robust and thin-walled large spherical spheroids from the Farrel Quartzite range from 40 to 70 µm in diameter and are characterized by wrinkled or dimpled carbonaceous wall (Figure 9.13c and d). In addition to such the complex wall architecture, there can be seen remarkable occurrences: some specimens are associated with multiple small spheroids (Figure 11.14a and b) or relatively large single spheroid (Figure 11.14c and d). Taphonomic features such as partial breakage and granulation of their walls are also identified. Although paired specimens suggestive of reproduction are absent and this morphotype seems to be not acid-resistant, these lines of evidence, particularly morphological elaboration, are consistent with their biogenicity (Sugitani et al., 2007, 2009b). Again, geochemical and isotopic data would be required for confirmation.

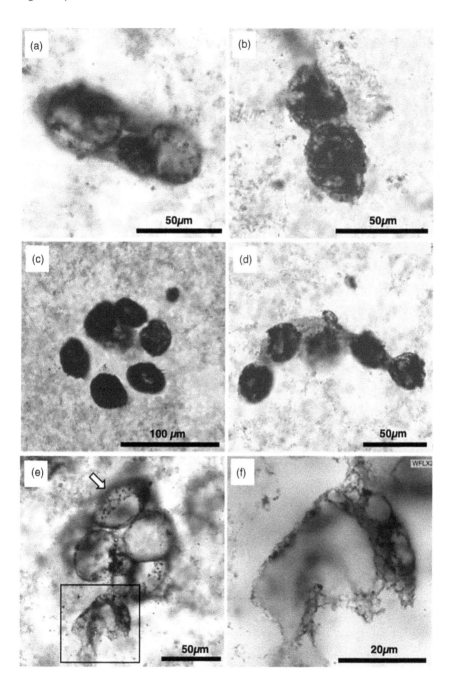

FIGURE 11.13 Photomicrograph of robust-walled large oblate spheroids from the Farrel Quartzite (a–d) and the Strelley Pool Formation (e, f). Transmitted light. (a) Paired and not-fused specimens that appear to be thin-enveloped. (b) Paired and fused specimens. (c) Colonial cluster of ellipsoids. Note its regular arrangement. (d) Chained ellipsoids that appear to be connected by thin film. (e) Colonial cluster of thin-walled elliptic spheroids, with a thin envelop. (f) Enlarged view of the inset in (e) representing raptured cell.

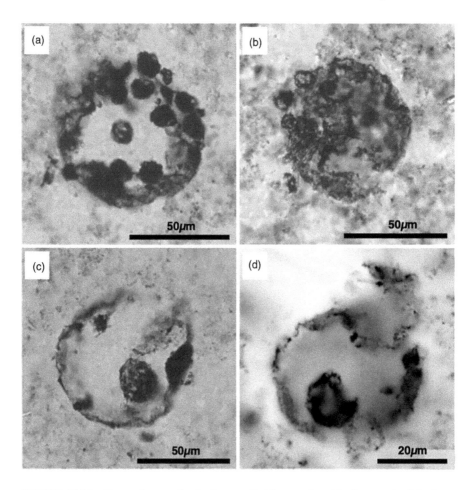

FIGURE 11.14 Photomicrograph of robust-walled large spheroids characterized by inner objects, from the Farrel Quartzite. Transmitted light. (a) Specimen with several small spheroids inside. (b) Possible equivalent but more degraded specimen to that in (a). (c) and (d) Specimens containing a single spheroid inside.

11.6.4.5 Robust and Thin-Walled Large Spherical Spheroids: The SPF Assemblage

Robust and thin-walled large spherical spheroids described from the Strelley Pool Formation are distinct from those from the FQ population, in lack of any architecture of wall reprsented by Figure 10.14d. Though only a few specimens have been identified from the Goldsworthy locality and the Panorama locality 2, the biogenicity of this morphotype can be argued based on carbonaceous composition, hollowness, colonial occurrence, taphonomy, and morphology suggestive of reproduction (Figure 11.15). Colonial specimen shown in Figure 11.15a are composed of individuals, ranging from 20 to 40 μm across in appearance, which are closely packed with each other, being represented by mutual compressed contacts. As can be seen in the specimen in Figure 10.14d, an attachment smaller hemispheric spheroid on larger spheroid, with

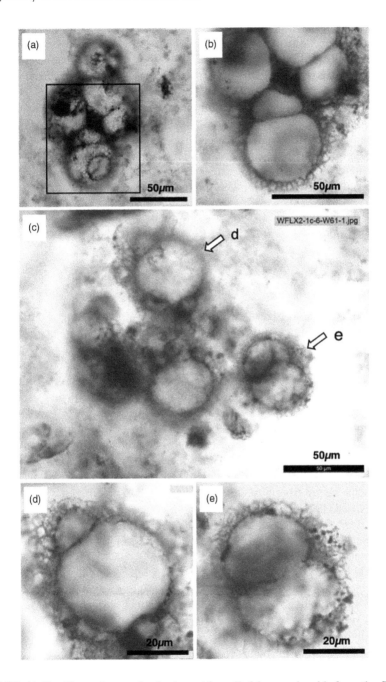

FIGURE 11.15 Photomicrograph of robust thin-walled large spheroid, from the Strelley Pool Formation. Transmitted light. (a) Colonial cluster of thin-walled spheroids. Surface reticulation represents foam-like envelop. (b) Enlarged view of the inset in (a), taken at deeper focal depth, showing completely hollow insides and compressed "cells". (c) Loose colony of thin-walled spherical spheroids with foam-like envelope, from SPF. (d) and (e) Enlarged views of two insets representing the reproduction stage.

a sharp straight contact in cross section is identified (Figure 11.15b). Loose cluster of thin-walled spheroids of various sizes (Figure 11.15c) with a foam-like envelop is likely equivalent to the specimens in Figure 10.14. This specimen also contains a sphere with an attached smaller hemispheric sphere (Figure 11.15d) and attached spheres of the same size (Figure 11.15e).

There can be seen no specimens in macerates. Analyses of carbon isotopic compositions and elemental distribution have not yet been performed. However, such disadvantage toward biogenicity is compensated by other features. Compressed mutual contacts in this specimen can be attributed to the wall flexibility. It may be noted that the attached hemispheric spheroid may represent budding or divisional process, one of the reproduction styles of microorganisms as discussed in the next chapter. Enveloped clusters exhibit remarkable complexity of occurrence. These consistent lines of evidence point to the biogenicity of this morphotype.

11.6.5 LENSES

Lenses (lenticular structures), as well as small spheroids, are the most abundant morphotype and characterize the Pilbara microstructures. They have been identified from all the localities of the Farrel Quartzite and the Strelley Pool Formation. This morphotype was previously described as "spindle". However, reconstruction of 3d-images and careful examination of numerous specimens in petrographic thin sections and in macerates have revealed that spindles merely represent superficial equatorial view of lenticular structures. The biogenicity of the lenticular structures is supported by multiple and convergent lines of evidence, including (1) carbonaceous composition, (2) narrow size distribution from 20 to 100 μm across, (3) abundant occurrence (up to 100 individuals in petrographic thin section with an area of 8.8 cm^2 and thickness of 30 μm, (4) nonsolid nature of central body, (5) taphonomic features, (6) presence of structures corresponding to reproduction, (7) colonial occurrence, (8) extractable nature from host chert by acid (HF-HCl) maceration, (9) cytological complexity, and (10) biological geochemistry (Sugitani et al., 2007, 2009a; Grey and Sugitani, 2009; Sugitani et al., 2015a; Oehler et al., 2010, 2017; House et al., 2013; Lepot et al., 2013; Delarue et al., 2020; Alleon et al., 2018). Evidence numbered from 4 to 9 is explained in some detail below.

Nonsolid nature is an important feature, which means that lenticular structures are not mineral artifacts. Some appear to be clearly hollow, filled with pure chert masses as shown in, e.g., Figures 9.14a and 10.6a and otherwise have inner substructures such as spheroids and irregularly shaped clots (Figure 11.16a). Others occasionally appear to be solid, which is however attributed to thick (~2 μm or more) wall or presence of inner dense alveolar structure or rarely closely packed spheroids, as described later. Various taphonomic features are observed, including partial breakage and associated release of inner materials (Figure 11.16a and b), partially wrinkled wall (Figure 11.16c and d) independent of crystal boundaries of the matrix chert, and degradation of the whole body into granulation and vesiculation (Figure 11.16e and f). Paired specimens are relatively common (Figure 11.17a and b). The relationship between counterparts is various from being fused to being

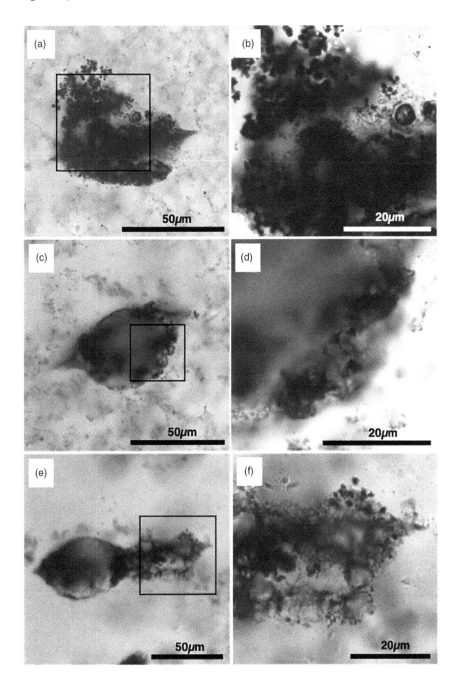

FIGURE 11.16 Lenticular microstructures showing partial taphonomy. Transmitted light. (a) Translucent specimen the Farrel Quartzite, containing a small spheroid inside. (b) Enlarged view of the inset in (a) showing partial breakage. (c) Sectioned specimen from the Farrel Quartzite, having partial wrinkled wall (the inset). (d) Enlarged view of the inset in (c). (e) Paired specimen from the Strelley Pool Formation. One of the two is highly degraded (the inset). (f) Enlarged view of the inset in (e).

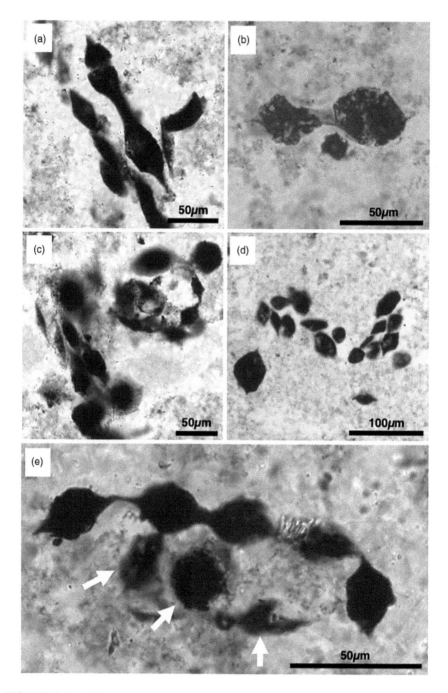

FIGURE 11.17 Paired lenticular microstructures and colonial clusters, from the Goldsworthy localities. Transmitted light. (a) Paired lenticular structures from the Strelley Pool Formation. (b) Paired lenticular structures from the Farrel Quartzite. Note that one of them appears to be shrunk. (c) and (d) show randomly oriented colonial clusters from the Strelley Pool Formation and the Farrel Quartzite, respectively. (e) Chained colony from the Strelley Pool Formation. Note that the arrowed objects out of focus are also lenses.

connected by flange, corresponding to various stages of binary fissions. Colonial cluster, as a consequence of successive reproductions, is also common. Many of them are composed of randomly oriented individuals (Figure 11.17c and d). Though rare, chained cluster is also present (Figure 11.17e). Colonial clusters composed of lenticular structures and other morphotypes such as spheroids and ellipsoids are also present. Such colonies have important implications of life cycle, which will be discussed in Chapter 12.

Lenticular structures can be extracted by acid maceration (Figure 11.18a), which means that they are not composed of loosely aggregated particles but membranous material. In acid macerates, delicate structures represented by chained cluster are also identified (Figure 11.18b). Secondary electron microscopy (SEM) and Focused Ion Beam-Transmitted Electron Microscopy (FIB-TEM) analyses of extracted specimens revealed delicate structures and textures that cannot be seen by a petrographic microscope (Figure 11.18c–e) (Sugitani et al., 2015b). The central body contains abundant dense globules around a few microns across and is characterized by inner

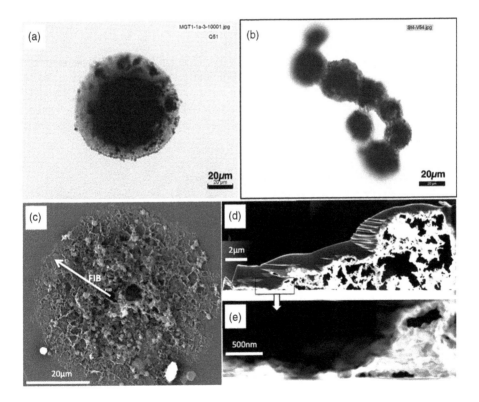

FIGURE 11.18 (a) Lenticular structure extracted by acid maceration, from the Farrel Quartzite. Transmitted light. (b) Chained lenses extracted by acid macereation, from the Goldsworthy locality of the Strelley Pool Formation. Transmitted light. (c) SEM image of the SPF lenticular structure extracted by acid maceration, from the Goldsworthy locality. The arrow shows the FIB sectioning line. (d) TEM image of foil made by FIB. (e) Enlarged image of the inset in (d).

alveolar structure and the flange is reticulated or exhibits fibrillar network in a sub-micron scale.

Geochemical aspects of lenticular structures, like small spheroids described above, also point to their biogenicity. Oehler et al. (2010) performed Nano-SIMS analyses on the FQ lenticular structures and demonstrated that their carbonaceous walls ($^{12}C^-$) are closely associated with distribution of N ($^{12}C^{14}N^-$) and S ($^{32}S^-$). The authors also reported enhanced association of O ($^{16}O^-$) and Si ($^{28}Si^-$) with these essential elements. Carbon isotopic compositions of individual lenses were measured by House et al. (2013) and Oehler et al. (2017) (Figure 11.10) for the FQ population and Lepot et al. (2013) and Oehler et al. (2017) for the SPF population. Both studies clearly demonstrated their significantly light carbon isotopic compositions ($\delta^{13}C_{PDB} < -30‰$), distinct from carbonaceous clots and disseminated carbonaceous matter in the matrix and veins. As to the SPF lens, Alleon et al. (2018) detected nitrogen- and oxygen-rich organic compounds such as amide functional groups on specimen exposed on freshly crushed rock surface, comparative to the ca. 1.9 Ga-old Gunflint microfossils. These lines of evidence are all consistent with the biogenicity of the lenticular structures.

11.7   REFUTING OBJECTIONS

While biogenicity of the Pilbara microfossils, if not all morphotypes, has been established through extensive and multidisciplinary studies as described above, a few skepticisms against biogenicity or cellularity of lenticular structures still remain. Also, their inhabitation in soils has been proposed despite our detailed descriptions of lithostratigraphy and petrography of fossil-bearing cherts. In this section, I will contest against these interpretations.

11.7.1   ARE LENTICULAR MICROSTRUCTURES VOLCANIC VESICLES OR THOSE MICROBIALLY COLONIZED?

Some skepticisms to the cellularity of lenticular microfossils and to even their biogenicity still persist. I guess that such skepticism is the inside out of peculiarity of lenticular microfossils in their size and morphology. Here I would like to pick up the study of Wacey et al. (2018a, b), in which the authors suggested that volcanic vesicles could mimic lenticular microfossils, mentioning to the possibility that the inner walls of volcanic vesicles could be colonized by microbes.

11.7.1.1   Descriptions of Wacey's Volcanic Vesicles Mimicking Microfossils

Wacey et al. (2018a) described an assemblage of tephra (scoria, tubular pumice, plus vesicular and nonvesicular volcanic glass shards) from clasts in sandstone and conglomerate of the Strelley Pool Formation and stated that they include "morpho-types that closely resemble previously described microfossils from this unit and elsewhere". Numerous spheroidal vesicles and subordinate eye- and lens-shaped morphotypes mostly ranging 10–50 μm in diameter were described from clasts of scoria, whereas tubes 5–15 μm in diameter from pumice. Additionally, vesicular and non-vesicular subangular shard particles have been identified. The key and common feature of these volcanic structures is being lined by anatase (TiO_2), with or without OM. These volcanic materials, if not all, have been extensively silicified to be composed of microcrystalline quartz. Based on this observation, the authors reinterpreted spheroid microfossils described by themselves from the Strelley Pool Formation (Wacey et al., 2011) as pseudofossils. In another paper (Wacey et al. 2018b), the authors described spheroidal to lenticular microstructures lined by organic C and/or pyrite from the ~3.48 Ga Dresser Formation in the Pilbara Craton, Western Australia. Based on size distribution, wall ultrastructure, and chemistry of some specimens, the microstructures were interpreted as volcanic vesicles. While vesicles in volcanic glass would hardly be misinterpreted as microfossils, the authors suggested the possibility that reworked, isolated, and silicified vesicular volcanic glasses, adsorbing OM onto their surfaces, would mimic microfossils. Thus, it is implied that some of the previously described FQ and SPF lenticular microfossils (e.g., Sugitani et al., 2007, 2010) may be volcanic vesicles (Wacey et al., 2018b).

11.7.1.2 Lenticular Microfossils Are Not Originated from Volcanic Vesicles

Wacey et al. (2018a, b) do not directly deny the biogenicity of flanged lenticular microstructures. However, their works may have given unfounded negative impressions on the biogenicity of this and the other morphotypes. Since I have realized from some personal communications that the volcanic vesicle origin of lenticular microfossils has attracted not a few researchers, it is indispensable for me to make it clear that any morphotype of the Pilbara microfossils did not originated from volcanic vesicles. Their vesicular origin can be rejected simply by their occurrences. If the structures were contained in volcanic fragments, some of them should be "fragmented". It is unusual that volcanic rock would be fragmented without breaking vesicles, because vesicles represent physical weakness. When we go back to the occurrence of the Pilbara microstructures, the problem will immediately be resolved. As described before, the microstructures, except for large spheroids and thick filaments, often occur as colonial clusters. While hundreds of colonial clusters containing lenses have been identified, none of them contain fragmented individuals located on their outer edges.

11.7.1.3 Dresser Vesicles May Be Lenticular Microfossils

In my view, at least some of lenticular structures described from the Dresser Formation (Wacey et al., 2018b) are likely microfossils. I have never seen the materials, and the following arguments are not beyond my suggestions. Some of the Dresser lenses are walled with carbon and free from oxides. Blurred appearance of carbonaceous wall is similar to that of the Pilbara lenticular microfossils that have been subjected to later silicification (Figure 11.3). The specimen shown in Figure 5b of Wacey et al. (2018b) is not a simple lens but appears to have protrusions, which could be equivalent to two-dimensional appearance of flange. Additionally, these two protrusions are out of the major axis of the central body. Such asymmetricity has also been described as a submorphotype of the Pilbara lenticular microfossils

(Sugitani et al., 2009a) (also see Figure 11.16c). As discussed in the next chapter, morphometric studies of the extracted SPF lenticular microfossils revealed that their morphological variations are systematic and likely related to different environments of their habitats. Namely, the morphological variations can be interpreted in the context of diversification from the common ancestor. The Strelley Pool Formation is aged ~3.4 Ga, which is 0.1 Ga younger than the Dresser Formation. It is not unrealistic to assume that the similar-shaped microbes had already been present as ancestoral species at the Dresser time.

11.7.2 DOES THE FARREL QUARTZITE MICROFOSSIL ASSEMBLAGE REPRESENT SOIL COMMUNITIES?

Retallack et al. (2016) proposed the hypothesis that the FQ microfossil-bearing carbonaceous cherts represent permineralized "paleosols". Furthermore, they suggested that the described microfossils were actinobacteria, purple sulfur bacteria, and methanogenic Archaea in terrestrial habitat. Their assertions are not only inconsistent with ours at all but also based on incorrect interpretations and citations. My colloborators and I promptly published a comment paper (Sugitani et al., 2017), which was replied (Retallack et al., 2017). As the sedimentary and chemical precipitate origin of the carbonaceous cherts has been discussed in detail in sections 11.3 and 11.4, I do not repeat this matter here. Instead, I would like to focus on the issue whether the section measured by Retallack et al. (2016) exactly corresponds to the CE2 or not. This is very critical, because Retallack et al. (2016) states "Attention focused on detailed sampling of a short geologic section (Fig. 2) corresponding to unit CE2 of Figure 2 of Sugitani et al. (2007) in the uppermost Farrel Quartzite…" and "Putative Archean microfossils from the very same beds of the Farrel Quartzite sampled here (Fig.5) passed a variety of critical tests for biogenicity …". I am worrying if readers would accept these statements and reinterpretation of the described microfossils as soil microbial communities, as the environment of habitat is very critical to considering about their metabolisms. Many of our previous studies on the Farrel Quartzite and the overlying Cleaverville Formation have been repeatedly cited in Retallack et al. (2016). I sincerely acknowledge this but would like to claim that some were incorrectly cited.

11.7.2.1 Ambiguity of the Examined Locality

Arguments between Retallack et al. and us (Sugitani et al., 2017; Retallack et al., 2017) were not productive. This is largely attributed to the ambiguity in the locality examined by Retallack et al. (2016). Furthermore, the ambiguity was multilayered. Judging from the photograph provided by Figure 2 of Retallack et al. (2016), Retallack's measured section ~6 m thick "probably" corresponds to the lower chertevaporite unit (CE1) but does not correspond to CE2 at the boundary between the Farrel Quartzite and the Cleaverville Formation. The caption says, "Overview of contact between Farrel Quartzite and site studied here (arrow) and by Sugitani et al. (2003, 2007) immediately below the Cleaverville Formation…". The photograph showing the flank of Mt. Grant was taken from the southern foot of Mt. Grant, from which the contact between the Farrel Quartzite and the Cleaverville Formation should not be visible, based on my experience in field trips. The contact which can be traced over 7 km is always located along the main ridgeline of the Mt. Grant. Also, the locality traced based on the GPS data shown in Retallack et al. (2016) is far away from our type locality and the point shown in Figure 2 of Retallack et al. (2016) (Sugitani et al., 2017). Such ambiguity has complicated the problems. Nevertheless, if the GPS data correctively show the measured section, I cannot entirely exclude the possibility that the section corresponds to the CE2.

11.7.2.2 Ambiguity of the Examined Materials

Descriptions of the section measured by Retallack et al. (2016) were detail. However, the occurrence of fossiliferous carbonaceous chert is ambiguous. The measured section is 6 m thick, including ~2.5 m unit that are interpreted to have several paleosol horizons. Retallack et al. (2016) say "Paleosols of the Farrel Quartzite are developed on quartzo-feldspathic sand with small amounts of clay and rock fragments (Figure 11.6), including volcanic rocks, schist, and chert". There was no description of layered chert ~15 cm thick that characterizes the upper chert-evaporite unit (CE2) (Sugitani et al., 2007). Retallack et al. (2016) described that massive fossiliferous and pyritic chert labelled R4336 occurred at A horizon of "Jurnpa clay loam", the lowermost "paleosol" unit. As far as looking at microphotographs provided in Figure 9A of Retallack et al. (2016), the chert (R4336) at A horizon is similar to carbonaceous cherts we have collected from several horizons of the Farrel Quartzite. However, in detailed figure for "Jurnpa clay loam" ca. 30 cm thick, no chert was described. Instead, tuff or claystone corresponds to A horizon and R4336. In either case, the lithostratigraphy of the "paleosol" unit is quite distinct from that of CE2. One solution for the lithostratigraphic discrepancy between CE2 and the "paleosol unit" is that these two represent the same strata, but different facies such as center vs. margin in the same basin. In other words, I do not deny the possibility that the Rettalack's section contains paleosol horizons. Nevertheless, the paleosol origin of the fossiliferous chert (R4336) is unlikely.

11.7.2.3 Misunderstanding of Previous Studies and Incorrect Citations

Setting aside the scientific side of things, the study by Retallack et al. (2016) involves some problems in presentation and citation, which are addressed here. My intention is not to argue out disadvantages in Retallack et al. (2016), but to inform our previous observations correctly to readers.

Retallack et al. (2016), for example, stated "Microfossils of the Farrel Quartzite have been considered marine (House et al., 2013), but their matrix lacks lamination, stromatolites, or soft sediment deformation". This is incorrect, because lamination and soft sediment deformation were already described in Sugitani et al., (2007) as key features of sedimentary origin of fossiliferous cherts. It is also obvious that the lack of stromatolites does not provide evidence for subaerial deposition. In my first paper of the Farrel Quartzite (Sugitani et al., 2003), we described several types of extensively silicified crystals and identified vertical to subvertically grown giant (~30 cm) crystals as nahcolite and clustered prismatic crystals and vertically oriented bladed crystals as barite. We have interpreted nahcolite as evaporitic origin, whereas barite as hydrothermal origin. However, Retallack et al. (2016) say "Evaporite sand crystals of nahcolite and barite… have been interpreted as evaporites of coastal salina (Sugitani et al., 2007)". We had never specified coastal salina as a depositional environment for these crystals. Also, citing our work on trace element geochemistry on the fossil-bearing cherts (Sugahara et al., 2010), Retallack et al. say "The East Pilbara lower Cleaverville Formation and underlying Farrel Quartzite, however, share slightly super-chondritic Y/Ho rations and slight Eu anomalies, characteristic of freshwater, and distinct from modern ocean values and the marine stromatolitic, 3.4 Ga Strelley Pool Chert". This is incorrect at all. Slightly super-chondritic Y/Ho ratios and slight europium anomalies merely suggest that the cherts are not purely marine. Indeed, we carefully stated in the abstract of Sugahara et al. (2010) "The black cherts were thus precipitated from a water mass influenced significantly by for example continental run-off, ground water, and/or geothermal water but not from high-T hydrothermal solution…". It is unfortunate to find these misunderstandings and misquotations of my works.

COLUMN: ACRITARCHS

Acritarchs refer to organic-walled microfossils that can be extracted by acid maceration (generally HCl-HF decomposition of rocks under room temperature) and cannot be classified into any established taxa. They are abundantly extracted from rocks younger than 2.0 Ga. Though being unassigned to any specific taxonomic groups, it seems to be generally considered that acritarchs possibly represent eukaryotic or cyanobacterial microfossils, including egg cases of protists, resting cysts of *chlorophyta* and *dinoflagellates*, and sheaths of cyanobacteria, because only eukaryotic and cyanobacterial organisms have been thought to have ability to produce acid-resistant vesicles. In this context, a successful extraction of organic-walled microfossils from the Archean rocks back to 3.4 Ga (Javaux et al., 2010; Sugitani et al., 2015a) may pose a problem to this prerequisite. As discussed in Chapter 13, the Archean microfossils composed of acid-resistant organic walls may represent early eukaryotic or cyanobacterial lineages. Alternatively, unknown prokaryotic lineages could have produced vesicles with similar physicochemical property. Furthermore, we may need to consider the possibility that the acid-resistant property could have been obtained through diagenetic processes. In either case, the extracted structures are not aggregation of carbonaceous particles but are composed of continuous organic walls, providing evidence for biogenicity. Additionally, the extracted specimens can be subjected to various analyses, including precisely targeted FIB-TEM analyses, elemental mapping, and isotopic analyses, and morphometry, providing further evidence for their biogenicity.

REFERENCES

Alleon, J., Bernard, S., Le Guillou, C., Beyssac, O., Sugitani, K., Robert, F. 2018. Chemical nature of the 3.4 Ga Strelley Pool microfossils. *Geochemical Perspectives Letters* 7, 37–42.
Allwood, A.C., Burch, I.W., Walter, M.R. 2007. Stratigraphy and facies of the 3.43 Ga Strelley Pool Chert in the southwestern part of the North pole Dome, Pilbara Craton, Western Australia. *Geological Survey of Western Australia Record* 2007–11.

Allwood, A.C., Kamber, B.S., Walter, M.R., Burch, I.W., Kanik, I. 2010. Trace elements record depositional history of an early Archean stromatolitic carbonate platform. *Chemical Geology* 270, 148–163.

Allwood, A.C., Walter, M.R., Kamber, B.S., Marshall, C.P., Burch, I.W. 2006. Stromatolite from the Early Archaean era of Australia. *Nature* 441, 714–718.

Czaja, A.D., Beukes, N.J., Osterhout, J.T. 2016. Sulfur-oxidizing bacteria prior to the Great Oxidation Event from the 2.52 Gamohaan Formation of South Africa. *Geology* 44, 983–986.

Delarue, F., Robert, F., Derenne, S., Tartèse, R., Jauvion, C., Bernard, S., Pont, S., Gonzalez-Cano, A., Duhamel, R., Sugitani, K. 2020. Out of rock: A new look at the morphological and geochemical preservation of microfossils from the 3.46 Gyr-old Strelley Pool Formation. *Precambrian Research* 336, 105472.

Furnes, H., Banerjee, N.R., Muehlenbachs, Staudigel, H., de Wit, M.J. 2004. Early life recorded in Archean Pillow Lavas. *Science* 304, 578–581.

Grey, K., Sugitani, K. 2009. Palynology of Archean microfossils (c. 3.0Ga) from the Mount Grant, Pilbara Craton, Western Australia: Further evidence of biogenicity. *Precambrian Research* 173, 60–69.

House, C.H., Oehler, D.Z., Sugitani, K., Mimura, K. 2013. Carbon isotopic analyses of ca. 3.0 Ga microstructures imply planktonic autotrophs inhabited Earth's early oceans. *Geology* 41, 651–654.

Javaux, E.J. 2019. Challenges in evidencing the earliest traces of life. *Nature* 572, 451–460.

Javaux, E.J., Marshall, C.P., Bekker, A. 2010. Organic-walled microfossils in 3.2-billion-year-old shallow-marine siliciclastic deposits. *Nature* 463, 934–938.

Lepot, K., Williford, K.H., Ushikubo, T., Sugitani, K., Mimura, K., Spicuzza, M.J., Valley, J.W. 2013. Texture-specific isotopic compositions in 3.4 Gyr old organic matter support selective preservation in cell-like structures. *Geochimica et Cosmochimica Acta* 112, 66–86.

Leppard, G.G. 1986. The fibrillar matrix component of lacustrine biofilms. *Water Research* 20, 697–702.

Oehler, J.H. 1976. Hydrothermal crystallization of slica gel. *Geological Society of America Bulletin* 87, 1143–1152.

Oehler, D.Z., Robert, F., Mostefaoui, S., Meibom, A., Selo, M., McKay, D.S. 2006. Chemical mapping of Proterozoic organic matter at sub-micron spatial resolution. *Astrobiology* 6, 838–850.

Oehler, J.H. Robert, F., Walter, M.R., Sugitani, K., Allwood, A., Meibom, A., Mostefaoui, S., Selo, M., Thomen, A., Gibson, E.K. 2009. NanoSIMS: Insights to biogenicity and syngeneity of Archaean carbonaceous structures. *Precambrian Research* 173, 70–78.

Oehler, D.Z., Robert, F., Walter, M.R., Sugitani, K., Meibom, A., Mostefaoui, S., Gibson, E.K. 2010. Diversity in the Archean biosphere: New insights from NanoSIMS. *Astrobiology* 10, 413–424.

Oehler, D.Z., Walsh, M.M., Sugitani, K., Liu, M.-C., House, C.H. 2017. Large and robust lenticular microorganisms on the young Earth. *Precambrian Research* 296, 112–119.

Oehler, J.H., Schopf, J.W. 1971. Artificial microfossils: Experimental studies of permineralization of blue-green algae in silica. *Science* 174, 1229–1231.

Retallack, G.J., Krinsley, D.H., Fischer, R., Razink, J.J., Langworthy, K.A. 2016. Archean coastal- plain paleosols and life on land. *Gondwana Research* 40, 1–20.

Retallack, G.J., Krinsley, D.H., Rischer, R., Raznik, J.J., Langworthy, K.A. 2017. Reply to comment by K. Sugitani et al. (2016) on the article "Archean coastal plain paleosols and life on land" by G.J. Retallack et al. (2016), Gondwana Research 40, 1–20. *Gondwana Research* 44, 270–271.

Schopf, J.W. 1968. Microflora of the Bitter Springs Formation, Late Precambrian, Central Australia. *Journal of Paleontology* 42, 651–688.

Schopf, J.W., Kudryavtsev, A.B., Osterhout, J.T., Williford, K.H., Kitajima, K., Valley, J.W., Sugitani, K. 2017. An anaerobic ~3400 Ma shallow-water microbial consortium: Presumptive evidence of Earth's Paleoarchean anoxic atmosphere. *Precambrian Research* 299, 309–318.

Schopf, J.W., Kudryavtsev, A.B., Sugitani, K., Walter, M.R. 2010. Precambrian microbe-like pseudofossils: A promising solution to the problem. *Precambrian Research* 179, 191–205.

Stefurak, E.J.T., Lowe, D.R., Zentner, D., Fischer, W.W. 2014. Primary silica granules – A new mode of Paleoarchean sedimentation. *Geology* 42, 283–286.

Sugahara, H., Sugitani, K., Mimura, K., Yamashita, F., Yamamoto, K. 2010. A systematic rare-earth elements and yttrium study of Archean cherts at the Mount Goldsworthy greenstone belt in the Pilbara Craton: Implications for the origin of microfossil-bearing black cherts. *Precambrian Research* 177, 73–87.

Sugitani, K., Grey, K., Allwood, A.C., Nagaoka, T., Mimura, K., Minami, M., Marshall, C.P., Van Kranendonk, M.J., Walter, M.R. 2007. Diverse microstructures from Archaean chert from the Mount Goldsworthy – Mount Grant area, Pilbara Craton, Western Australia: microfossils, dubiomicrofossils or pseudofossils?, *Precambrian Research* 158, 228–262.

Sugitani, K., Grey, K., Nagaoka, T., Mimura, K. 2009a. Three-dimensional morphological and textural complexity of Archean putative microfossils form the northeastern Pilbara Craton: Indications of biogenicity of large (>15μm) spheroidal and spindle-like structures. *Astrobiology* 9, 603–615.

Sugitani, K., Grey, K., Nagaoka, T., Mimura, K., Walter, M.R. 2009b. Taxonomy and Biogenicity of Archaean spheroidal microfossils (ca. 3.0 Ga) from the Mount Goldsworthy-Mount Grant area in the northeastern Pilbara Craton, Western Australia. *Precambrian Research* 173, 50–59.

Sugitani, K., Lept, K., Nagaoka, T., Mimura, K., Van Kranendonk, M., Oehler, D.Z., Walter, M.R. 2010. Biogenicity of morphologically diverse carbonaceous microstructures from the ca. 3400 Ma Strelley Pool Formation, in the Pilbara Craton, Western Australia. *Astrobiology* 10, 899–920.

Sugitani, K., Mimura, K., Nagaoka, T., Lepot, K., Takeuchi, M. 2013. Microfossil assemblage from the 3400 Ma Strelley Pool Formation in the Pilbara Craton, Western Australia: Results from a new locality. *Precambrian Research* 226, 59–74.

Sugitani, K., Mimura, K., Suzuki, K., Nagamine, K., Sugisaki, R. 2003. Stratigraphy and sedimentary petrology of an Archean volcanic-sedimentary succession at Mt. Goldsworthy in the Pilbara Block, Western Australia: implications of evaporite (nahcolite) and barite deposition. *Precambrian Research* 120, 55–79.

Sugitani, K., Mimura, K., Takeuchi, M., Lepot, K., Ito, S., Javaux, E.J. 2015a. Early evolution of large micro-organisms with cytological complexity revealed by microanalyses of 3.4 Ga organic-walled microfossils. *Geobiology* 13, 507–521.

Sugitani, K., Mimura, K., Takeuchi, M., Yamaguchi, T., Suzuki, K., Senda, R., Asahara, Y., Wallis, S., Van Kranendonk, M.J. 2015b. A Paleoarchean coastal hydrothermal field inhabited by diverse microbial communities: The Strelley Pool Formation, Pilbara Craton, Western Australia. *Geobiology* 13, 522–545.

Sugitani, K., Van Kranendonk, M.J., Oehler, D.Z., House, C.H., Walter, M.R. 2017. Comment: Archean coastal-plain paleosols and life on land. *Gondwana Research* 44, 265–269.

Takai, K., Nakamura, K., Toki, T, Tsunogai, U., Miyazaki, M., Miyazaki, J., Hirayama, H., Nakagawa, S., Nunoura, T., Horikoshi, K. 2008. Cell proliferation at 122°C and isotopically heavy CH_4 production by a hyperthermophilic methanogen under high pressure cultivation. *Proceedings of National Academy of Sciences USA* 105, 10949–10954.

Wacey, D., Kilburn, M.R., Saunders, M., Cliff, J., Brasier, M.D. 2011. Microfossils of sulphur-metabolizing cells in 3.4-billion-year-old rocks of Western Australia. *Nature Geoscience* 4, 698–702.

Wacey, D., Noffke, N., Saunders, M., Guagliardo, P., Pyle, D.M. 2018a. Volcanogenic pseudo-fossils from the ~3.48 Ga Dresser Formation, Pilbara, Western Australia. *Astrobiology* 18, 539–555.

Wacey, D., Saunders, M., Kong, C. 2018b. Remarkably preserved tephra from the 3430 Ma Strelley Pool Formation, Western Australia: Implications for the interpretation of Precambrian microfossils. *Earth and Planetary Science Letteres* 487, 33–43.

12 Lifecycle and Mode of Life of the Pilbara Microfossils

12.1 INTRODUCTION

Based on the established biogenicity, we go to the next step, namely proposal of possible lifecycle, style of reproduction, and mode of life of the Pilbara microfossils, both from the ca. 3.0 Ga-old Farrel Quartzite (FQ) and the ca. 3.4 Ga-old Strelley Pool Formation (SPF). The hypotheses are made largely based on morphology, morphology related to reproduction, and occurrence of colony. Although signs of reproduction have already been mentioned for discussions on biogenicity in Chapter 11, I would like to propose more comprehensive and in-depth interpretations here.

It is well expected that readers would have concerns about my (over?) interpretations of very ancient microfossils. Although the Pilbara microfossils are exceptionally well preserved, their preservation statuses are poorer compared with some Proterozoic microfossils, which allowed the proposal of biological affinity and even lifecycles (e.g., Agić et al., 2015 and references therein). However, it seems worth to propose provisional views of life stages and reproduction styles of the Pilbara microfossils, which could be basis for future studies. Comparisons with modern microorganisms here are not intended to imply biological affinities, which would be discussed separately in the next chapter.

12.2 REPRODUCTION STYLES, LIFE CYCLE, AND COLONY OF MODERN MICROBES

Before interpreting the Pilbara microfossils, reproduction styles and life cycles of modern unicellular microorganisms and morphological variations of their colonies are summarized. They are expected to provide clues to interpret variations in morphology and colonial occurrences of the Pilbara microfossils.

12.2.1 ALTERNATIVE REPRODUCTION STYLES TO BINARY FISSIONS

It may be generally considered that unicellular microbes have a simple life cycle such as vegetative cell growth and binary fission to produce off-springs. However, even unicellular microbes have various life stages and related life cycle variants. Alternatives to binary fissions such as asymmetric division and multiple fissions have been identified for a wide range of unicellular microorganisms of the three domains

DOI: 10.1201/9780367855208-12

(e.g., Angert, 2005). A few taxa of cyanobacteria are known to reproduce by multiple fissions. Genus *Stanieria* does not reproduce by binary fission. The starting
cell, called baeocyte, is generally small, approximately less than 2 μm. Baeocyte
vegetatively enlarges its size and eventually can be a large (up to 30 μm) cell with
thick sheath (extracellular envelope), in which a number of off-springs, again called
baeocytes, are produced. This sporangium-like "mother" cell tears open to expel
the baeocytes. Thus, the colony of *Stanieria* can be composed of spheroids of various sizes with broken sheaths (Waterbury and Stanier, 1978) (Figure 12.1a). Two
other cyanobacterial genera *Dermocarpella* and *Pleurocapsalean* reproduce in a
more complex fashion combined with binary and asymmetric divisions. Multiple
fissions are not restricted to cyanobacteria and eukaryotes. Some spheroid sulfur
bacteria may reproduce by multiple fissions and symbiotic *Metabacterium polyspora* as well (Angert and Losick, 1998). Budding, a type of asymmetric divisions,
has been extensively known for eukaryotic organisms such as yeasts, although some
bacteria including coccoid unicellular cyanobacteria (*Hyphomonas polymorpha*,
genus *Chamaesiphon*) and possibly large spheroid sulfur bacteria "*Candidatus
Thiomargarita nelsonii*" reproduce by budding (Waterbury and Stanier, 1978;
Salman et al., 2013) (Figure 12.1b).

12.2.2 Morphological Change of Cells

Many of unicellular microorganisms may have resting or dormant stage, generally
in the consequence of adaptation to unfavorable conditions for vegetative growth,
such as deficient in nutrients and other resources and desiccation. The resting-stage
cell is called "cyst" or "spore". Figure 12.1c shows a cyst of eukaryotic unicellular
alga, which is quite different in morphology of its flagellated motile vegetative stage.
The genus *Pterosperma* (Prasinophyceae), a non-colony forming unicellular green
alga, has a unique life cycle (Figure 12.1d), including (1) small motile unicell with
flagella that can reproduce by binary fission, (2) initial phase of phycoma formation
starting by loss of flagella, (3) rounding of cell, with wall thickening and development of flange-like ala, (4) increase in volume of phycoma, (5) production of zoo-
spores in inner vesicle, (6) release of the inner vesicle, and (7) rupture of the vesicle
and release of motile cells (Parke et al., 1978; Tappan, 1980). It is also well known
that in some filamentous cyanobacteria, vegetative cells could differentiate into heterocyst specialized for nitrogen fixation and into thick walled, dormant cells called
akinete (Figure 12.1e).

12.2.3 Morphology of Colonies

Unicellular microorganisms tend to reproduce in a short period at an order of minutes under favorable environmental conditions, which often results in mass of cells
called colony. When microorganisms are adhesive, their colonies are often associated with formation of biofilms. Nonadhesive microorganisms also comprise
colonies, which are often enveloped mucilage sheath like coccoid cyanobacterial
Microcystis (Figure 12.2a) or rigid sheath like *Candidatus Thiomargarita nelsonii*
(Figure 12.2b). When dividing cells are not liberated and retain with each other,

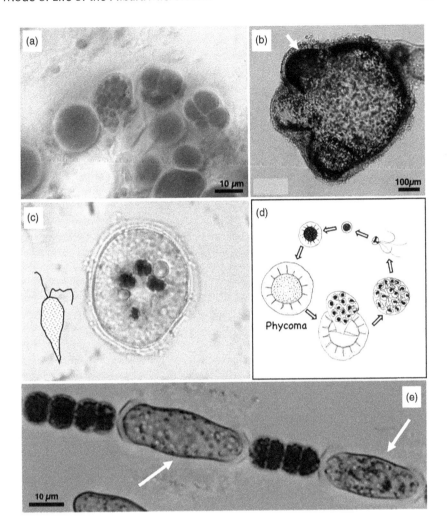

FIGURE 12.1 (a) Multiple fissions of cyanobacterial genus *Stanieria*. Rupture of the mother cell and spouting of baeocytes. (Photo courtesy by Sergei Shalygin.) (b) A large sulfur bacterium "Candidatus Thiomargarita nelsonii" with possible buds (the arrows). (Reproduced from Salman et al. (2013), with permission from Springer.) (c) Cyst of eukaryotic unicellular flagellated motile alga, genus *Chattonella* (ca. 30 μm across). (Photo courtesy of Mineo Yamaguchi.) Illustration shows a morphology of cell at the vegetative stage. (d) Life cycle of green alga genus *Pterosperma*. (e) Modern akinete (the white arrows) bearing nostocalean cyanobacteria (Dolichospermum hangangense sp.). (Reproduced from Choi et al. (2018)., with permission from The Korean Society of Phycology.)

connected cells comprise filamentous or chain-like colony, which is called trichome and occasionally ensheathed, as represented by some cyanobacteria (Figures 12.1e and 12.2c). A large sulfur bacterium, *Thiomargarita namibiensis* (*Column*), also comprises trichome, although they can be solitarily or comprise randomly clustered colony (Figure 12.2d).

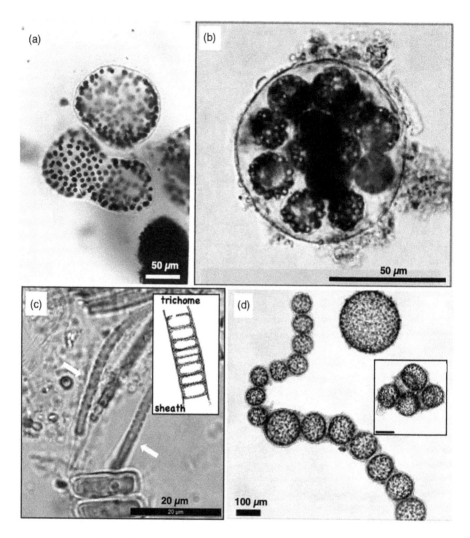

FIGURE 12.2 (a) Planktonic cyanobacteria, genus *Microcystis*. (Photo courtesy by Ho-Dong Park.) Note that the colonies are enclosed by mucilaginous sheath. (b) Colony of sulfur bacterium "*Candidatus Thiomargarita nelsonii*" enclosed by rigid sheath. {Reproduced from Salman et al. (2011), with permission from Elsevier.} (c) Trichome-forming cyanobacteria (d) Photomicrograph of sulfur-oxidizing bacteria, *Thiomargarita namibiensis*. The chained colony is enclosed by mucilaginous sheath. The inset is a randomly clustered colony. {Reproduced from Salman et al. (2011), with permission from Elsevier.}

12.3 FILMS

Films, some of which are associated with small spheroids, are commonly abundant in the Farrel Quartzite and the Strelley Pool Formation, although variations in abundance between localities can be seen. They are best interpreted as reworked and fragmented biofilms composed largely of extracellular polymeric substances (EPS)(*Column* in Chapter 7). Film-like structures are various in morphology and texture. They are either single-sheeted, blanched, and rarely layered or honeycomb-structured (Figures 9.14c, 11.4). Textural variations including hyaline, granular, or wrinkled are identified. These morphological and textural variations may be attributed not only to taphonomy but also to species and/or community diversity that secrete EPS. Reworked biofilms are well documented from the Barberton greenstone belt, South Africa and are relatively thick and composed of stacked multilayered structures (Tice and Lowe, 2006), suggesting benthic mat communities. On the other hand, it is equivocal if the FQ and SPF films were derived from benthic mat communities. Their derivation from adhesive communities cannot be ruled out. Rather, I favor this. In either case, they are allochthonous. Considering that the host cherts were likely precipitated from water masses influenced by hydrothermal fluids (though their contribution was various), the supposed adhesive communities might have inhabited around hydrothermal vent sites. It is possible that intermittent strong outflow of hydrothermal fluids or spouting of geyser had peeled biofilms off from substrates and transported them to the site of deposition.

12.4 FILAMENTS

The Pilbara filamentous microfossils include (1) sinuous tubular thin filaments comprising biomat-like fabrics from the Strelley Pool Formation at the Panorama locality 1 (Figure 10.2c), (2) curved possibly tubular thin filaments that occur as loose cluster in fractures before consolidation of sediments from the Farrel Quartzite (Figure 9.14e, 11.6b,c), (3) relatively broad hollow filaments occurring solitarily and show fragmented and reworked habit from both the Farrel Quartzite (Figure 9.14f) and the Strelley Pool Formation, (4) clusters of bundles of carbonaceous threads from the Farrel Quartzite (Figure 9.14d, 11.6a), and (5) poorly preserved thread comprising siliceous stromatolites from the Strelley Pool Formation (Figure 7.8b). While the FQ carbonaceous threads can be interpreted as fibrillar EPS, the others may represent fossilized cellular or related structures such as trichome, sheath, or hyphae (Figure 12.2c and d). Though not conclusive, thin tubular filaments in fractures and biomats and threads comprising stromatolite may represent trichome. On the other hand, relatively broad hollow filaments are likely sheath enclosing trichome. Geochemistry, if combined with variations in occurrence and width, could provide clues to provide further taxonomic constraints on these filaments. To date, such approach has been performed only for on the SPF filaments at the Panorama locality 1 and those at the Goldsworthy locality (Schopf et al., 2017, Delarue et al., 2020, Sugitani et al., 2015b). As described above, these filaments comprise biomats of different morphologies and fabrics. Although examined specimens are limited, the former is flat layered, whereas the latter comprises columnar structures (stromatolite in the strict sense). Carbon isotopic values ($\delta^{13}C_{PDB}$) of the former are lighter than −30‰ (Sugitani et al., 2010), whereas the latter have $\delta^{13}C_{PDB}$ values around −27‰ (Sugitani et al., 2015b). The former is associated with pyrite (FeS_2)

and silicified gypsum ($CaSO_4 \cdot 2H_2O$)/anhydrite ($CaSO_4$). The filaments were interpreted to be sulfur-metabolizing bacteria (Schopf et al., 2017), based on sulfur isotopic compositions of associated pyrite and similarity in occurrence to younger equivalents (Schopf et al., 2015). On the other hand, the latter filaments are not associated with sulfides. It is unlikely that they were sulfur metabolizers. Filamentous microbes, most likely prokaryotic, comprising these two types of biomats likely represent different taxa.

12.5   SMALL SPHEROIDS

Small spheroids (less than 15 μm in diameter) occur both from the FQ and the SPF localities, although abundance and preservation states are much higher in the former. Though simple in morphology, occurrence of small spheroids is various, including (1) colony exclusively composed of small spheroids, (2) colony attached with film-like structures, and (3) colony composed of small spheroids and other morphologies such as larger spheroids and lenses, which could be further classified into subtypes. Among these three types, the former two have been identified both from the Strelley Pool Formation and the Farrel Quartzite, whereas the third type appears to be restricted to the latter. Occurrence of colonies composed of small spheroids and larger spheroids and lenses suggests that these different morphotypes represent different life stages of the same taxa, as discussed later.

Small spheroid populations comprising colonies without large spheroids and lenses are believed to had simple life cycle such as vegetive cell growth and binary fission to produce two daughter cells. Scarcity of morphology of the midway stage of fission can be interpreted in the consequence of rapid reproduction. Two types of mode of life are suggested, including adhesive and planktonic. Small spheroids attached on films (EPS) likely had adhesive mode of life (Figure 11.4d), although such occurrence does not immediately indicate that the EPS was secreted by the attached small spheroid organisms. Small spheroids comprising simple spherical to irregularly-shaped colonies (Figure 12.3a), on the other hand, likely had planktonic mode of life. Some spherical colonies appear to have been once enclosed by envelop or sheath as shown in Figure 12.3a,b, like colonies of cyanobacterial *Microcystis* enclosed by mucilaginous sheaths (Figure 12.2a).

Autotrophic and planktonic mode of life for some of small spheroids has also been suggested from individual carbon isotopic values (House et al., 2013). The values of small spheroids ($\delta^{13}C_{PDB} = -36.9 \pm 0.3$ ‰) are more depleted than disseminated organic matter (OM) in the matrix ($\delta^{13}C_{PDB} = -32.7 \pm 0.6$ ‰) (Figure 11.10). This excludes the possibility that the spheroid microfossils were heterotrophs, which would have heavier isotopic signatures than the substrates represented by the matrix OM. It was also implied that such light carbon isotopic compositions suggest planktonic lifestyle, as microorganisms in this lifestyle could utilize carbon source (CO_2) without limitation, contrastive to benthic lifestyle (House et al., 2013; Kaufman and Xiao, 2003).

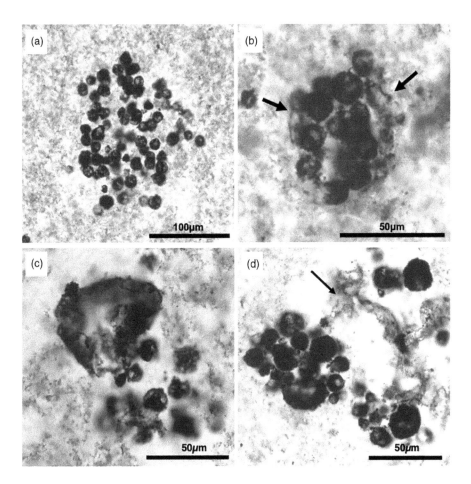

FIGURE 12.3 Small spheroids, from the Farrel Quartzite (FQ). Transmitted light. (a) Loosely packed spherical colony of small spheroids. (b) Tightly packed spherical colony of small spheroids that appear to be enclosed with thin film (the arrows). (c) Broken large spheroid associated with multiple small spheroids, from the Farrel Quartzite. (d) Colony composed of spheroids of various sizes with degraded film-like structure (the arrow).

12.6 LARGE SPHEROIDS

Large spheroids are here defined for spheroid and more or less ellipsoidal objects larger than 15 μm across. They can be classified into five types (see Chapter 11). Here, two morphotypes are discussed.

12.6.1 FLEXIBLE-WALLED LARGE SPHEROIDS

Almost all flexible large spheroids are identified from the Farrel Quartzite (Figures 9.13b and 11.11). From the Strelley Pool Formation, only one possible equivalent specimen has been found from the Panorama locality 2. As described earlier, flexible large spheroids occur as compressed or folded. Though not robust in appearance, they can be extracted by acid maceration. Measurements of flexible large spheroids (n = 256, unpublished data) from the Farrel Quartzite show that the major dimension ranges from 81.7 µm to 270.1 with an average of 147.5 µm, whereas the minor dimension ranges from 54.6 µm to 218.5 with an average of 115.5 µm. The aspect ratio ranges from 0.43 to 1.0, with an average of 0.80. Flexible large spheroids occur dispersedly and never comprise colony-like clusters. Clearly identifiable pairs have never been seen. It seems more reasonable to assume that they represent resting-stage but not vegetative one.

12.6.2 ROBUST AND THIN-WALLED LARGE SPHERICAL SPHEROIDS AND ROBUST-WALLED LARGE OBLATE SPHEROID

These morphotypes have been identified both from the Farrel Quartzite (relatively common) and the Strelley Pool Formation (rare). Several representative specimens are addressed here.

Robust and thin-walled large spherical spheroids from the Farrel Quartzite range from 40 to 60 µm in major dimension and is generally characterized by wrinkled wall. Occurrence is various, including solitary, an association with multiple small spheroids, and an association with an inner solitary spheroid (Figures 9.13c, d, f and 11.14). However, paired specimens have not yet been discovered, which excludes the possibility of vegetative stage for this type of spheroid. Key features for elucidating life cycle of this type of spheroid include (1) specimens containing multiple small spheroids inside and (2) colonies composed of multiple small spheroids and a single large spheroid, which often appears to be broken habit (Figure 12.3c and d). Namely, small spheroids may include taxa that reproduce by multiple fissions, which is known for wide clades of bacteria including cyanobacterium *Stanieria* (Figure 12.1a), and large sulfur bacteria (Salman et al., 2011) (Figures 12.1b and 12.2b). In Chapter 9, I have described that large solitary spheroids often contain a single sphere inside. Such specimen may represent the resting stage of this type of spheroid microbes.

Robust and thin-walled large spherical spheroids from the Strelley Pool Formation are not wrinkled and tend to comprise colonial clusters, and thus are distinct from those identified in the Farrel Quartzite. The SPF specimens of thin-walled large spherical spheroids are characterized by smaller hemispheres attached on larger spheres (Figures 10.14d and 11.15). Colonies and individuals are commonly surrounded by foam-like texture that can be interpreted as envelop. This type of asymmetric division may be called budding.

12.7 LENSES

Morphology, texture, and architecture of colony of lenses (lenticular microfossils), particularly those in the FQ assemblage, are diverse. The diversity likely reflects taxonomic variations of lenticular microbes, though potentially overprinted by taphonomy and life cycle variants. Here, these variations are described.

Morphometry-based taxonomy of the SPF lenticular microfossils will be discussed separately in Section 12.7.2.

12.7.1 Variations in Morphology, Texture, and Colony

12.7.1.1 Area, Oblateness, and Flange Width

Lenticular microfossils are composed of central spheroid body and surrounding sheet-like flange. However, their shape is various both in polar and equatorial views. In polar view, this morphotype is various in area, flange width, and oblateness. These variations had already been noticed in the early stage of this study (e.g., Sugitani et al., 2007, 2009a, b) and have later been confirmed by examinations of the structures extracted by acid maceration, which enabled us to observe the shape of microfossils at the polar view. For example, specimens shown in Figure 12.4a and b tell us that the ratio of flange width against the radius (relative flange width) is various. Furthermore, significant variation in oblateness has been identified (Figure 12.4c and d). Although such variations in area, oblateness, and relative flange width are expected to reflect taxonomic diversity of lenticular microfossils, statistical and qualitative analyses have been made only for the SPF specimens (Sugitani et al., 2018). The results are summarized in Section 12.6.3.

12.7.1.2 Symmetricity vs. Asymmetricity and Distortion

Most of the lenticular microfossils are of symmetric shape at polar view. Namely, the central body is surrounded by flange of nearly constant width (see. Figure 12.4). However, there can be seen specimens in which flange does not surround the central body but extends from one side of the body (Figure 12.5a), or flange width is distinct between one side and the other along the major dimension (Figure 12.5b). These submorphotypes can be depicted as having asymmetric flange.

Another type of asymmetry is observed for the equatorial view. It may be unknowingly imaged that representative and majority of lenticular microfossils are a symmetric spheroid body surrounded by flange exactly around its equator, like a specimen shown in Figure 12.6a. However, many of lenticular microfossils have flanges that do not extend from the major axis of the central body (Sugitani et al., 2009a). Namely the central bodies of many of the specimens tend to be asymmetric at the equatorial view; flanges extend from the widest portion from the main bodies. In addition to such the clearly definable asymmetricity, many specimens exhibit undefinable asymmetricity and can be designated as being distorted. It should be noted that such distortion and even asymmetricity of the central body do not always represent primary morphology, although this is not the case for the specimen in Figure 12.6b. This specimen is characterized by asymmetric central body and peripheral flange has an open space (void) inside. Such open space may represent vacuole, which can be formed by heterotrophic gas production. However, one third of the central body (the arrow in the figure) is not vacuolated but exhibits reticulation of carbonaceous matter that characterizes the inner structure of some of lenticular microfossils (Oehler et al., 2009; Sugitani et al., 2015a). Therefore, the asymmetric submorphotype of lenticular microfossil (Sugitani et al., 2009a) is still retained. Although the specimen

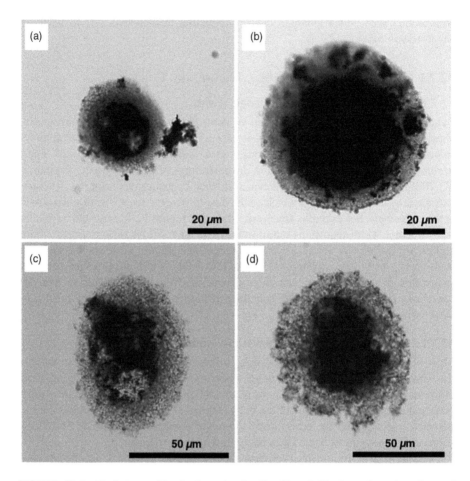

FIGURE 12.4 Variations of lenticular microfossils. (a) and (b) show the polar views of spherical lenses with different sizes and different ratios of flange width to the radius of the whole body, from the Farrel Quartzite. (c) and (d) show the polar views of spherical and oblate lenses from the Strelley Pool Formation. Transmitted light.

in Figure 12.6c has symmetrically attached flange, its central body is distorted and has low-relief cobbs. These cobbs do not correspond to inner voids. At least one of them (the upper cobb) appears to be attributed to pushing out by inner object. Such objects, if they are composed of OM, could represent daughter cells, spores or colony of other potentially heterotrophic microbes. The specimen in Figure 12.6d shows overall distortion, which cannot be described qualitatively.

Whatever the causes are, distortion of lenticular microfossils makes it difficult to classify lenticular microfossils further, if based solely on individual morphology. At this stage, therefore, morphological classification of lenticular microfossils, namely identification of submorphotypes, is vague and fragmentary, except for the case that the classification was made focusing polar view of extracted specimens from the

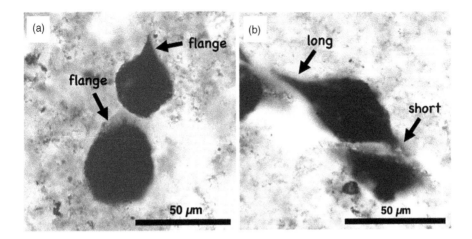

FIGURE 12.5 Lenticular microfossils with asymmetric flange, from the Farrel Quartzite. Transmitted light. (a) Two adjacent specimens with flange extended only from one side of the central body. Note that the central body of the upper specimen is fat and nearly spherical. (b) Lens with flange of uneven width.

Strelley Pool Formation, as discussed later. Distortion and breakage of lenticular microfossils, however, may provide us with invaluable supporting evidence for biogenicity, because such irregularity and incompleteness cannot be generally expected for solid materials including minerals and rock fragments (also see arguments against volcanic vesicle model of lenticular microfossils at Section 11.7.1).

12.7.1.3 Flange Fabrics

In most of the examined specimens, flange, a thin tapered appendage surrounding central body, is granular- or spongy-textured. This textural feature could be attributed to taphonomy. However, several distinct fabrics can be identified. One is translucent and hyaline, although only several specimens have been identified to date, both in petrographic thin sections and macerates (Figures 12.7a and 12.8a). Such flange tends to be narrow in width, less than one fourth of the radius of the whole body. Reticulated flange is the most common, identified again both in petrographic thin sections and macerates (Figures 12.7b and 12.8b). The most conservative interpretation of reticulation is crystal imprint, namely, redistribution and segregation of OM due to crystallization of the matrix microquartz, although the boundaries and sizes of the present matrix quartz do not appear to correspond to the reticulation. This mineral imprint model also cannot explain why such reticulation can be seen "only" for some of the lenticular microfossils but not for spheroids and films. The primary origin and thus taxonomic trait of the reticulation are still retained. Striated flange may be the next common type (Figures 12.7c and 12.8c). Striations are identified as portions poorer in carbonaceous matter than the surroundings and are identified for specimens in macerates in addition to those in petrographic thin

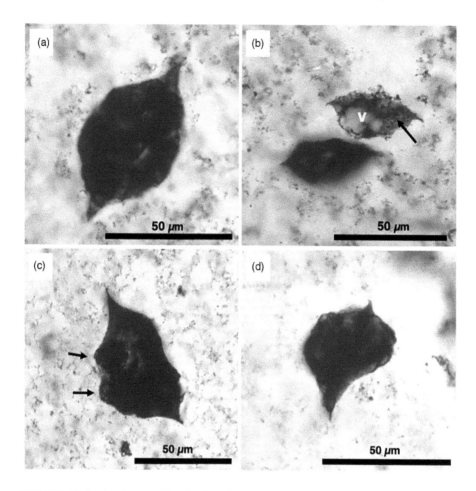

FIGURE 12.6 Lenticular microfossils of various shapes, from the Farrel Quartzite. Transmitted light. (a) Symmetric type, with flange extending exactly from the equator of the symmetric central body. (b) Sectional equatorial view of asymmetric type (the above) having open space inside (v). Flange extends from one side of the body. (c) and (d) Distorted lenticular microfossils. The specimen in (c) is characterized by low-relief cobbs (the arrows).

sections. The striations are a few microns in width and are lined up at regular intervals. Radiated striation is also identified: the supposed terminal is located out of the center of the microfossil bodies, and striations appear to penetrate into the central body but disappear there and reappear in the flange. Fibrillar structured flange is as rare as hyaline type, but clearly identifiable as fine splits of flange in a few extracted specimens (Figure 12.8d). Potential equivalent in petrographic thin section show some different appearance. Like radiated striation, fibrils appear to be radiated and penetrate into the central body (Figure 12.7d). One may consider that the striations and fibrils represent silicified mineral crystal grown within microfossils. Again, however, such structures can be seen only for lenticular microfossils.

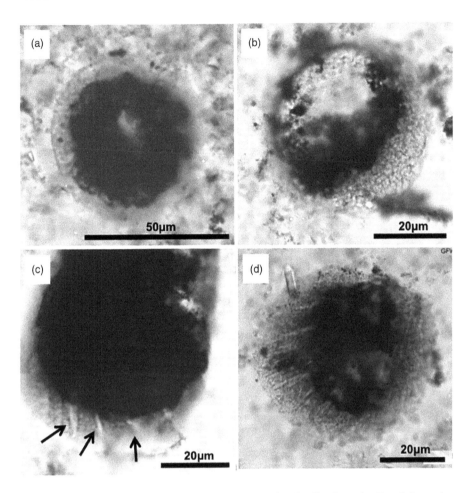

FIGURE 12.7 Various flange fabrics of lenticular microfossils, from the Farrel Quartzite. Transmitted light. (a) Translucent hyaline flange. (b) Reticulated flange. (c) Striated flange. Striations are indicated by arrows. (d) Fibrillar flange.

12.7.1.4 Architectures of Colony

Several types of colonies have been identified, including (1) Type 1: colony composed solely of detached lenses (Figure 12.9a), (2) Type 2: colony composed of lenses and spheroids (Figures 12.9b and c), (3) Type 3: colony composed of connected lenses (Figure 12.9d), and (4) Type 4: colony composed of lenses and film. Here, Types 1–3 are described, as Type 4 is very rare, only in the FQ assemblage. Type 1 colony can be further classified into several subtypes based on arrangement of individual lenses. Lenses are either randomly or regularly oriented in colonies. Regular orientation is represented by parallel arrangement, which has been common in the SPF population (Sugitani et al., 2013). Type 2 colonies can be subdivided into two subtypes based on spheroid size. One is characterized by small spheroids of almost the same size

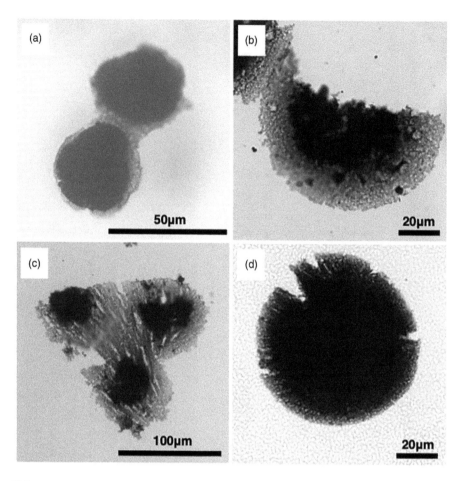

FIGURE 12.8 Various flange fabrics of the FQ lenticular microfossils extracted by acid maceration. Transmitted light. (a) Translucent hyaline flange. (b) Reticulated flange. (c) Striated flange. (d) Fibrillar flange.

(~10 μm) (Figure 12.9b), whereas the other by various sizes of spheroids (Figure 12.9c). The latter type appears to dominate colonies containing lenticular microfossils. Type 3 colonies are composed of lenticular microfossils that are attached with each other to comprise chained or sheet-like colonies. It has previously been suggested that some of the chained colonies represent merely sectional view of sheeted colonies, based on focusing pattern of individuals of "chain-like" colony at different focal depths (Sugitani et al., 2011). The presence of sheet-like colonies has been confirmed by their identification in macerates (Figure 12.8c). In both chained and sheet-like colonies, individuals of lens are connected with each other by sharing flange. It may be noted that the SPF colonies are dominated by Type 1, whereas the FQ ones by Type 2, particularly composed of lenses and spheroids of various sizes. Type 3 has been identified both from the Farrel Quartzite and the Strelley Pool Formation.

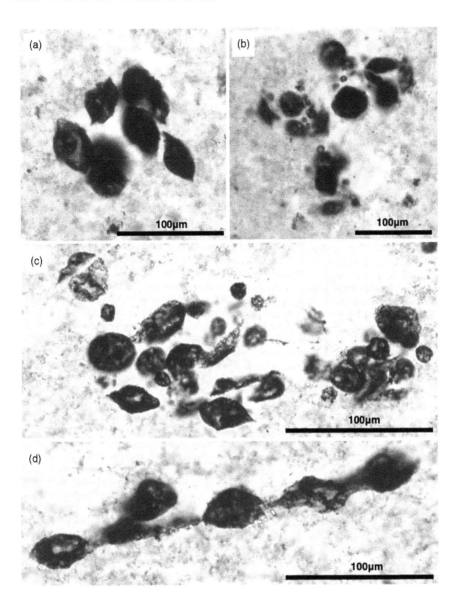

FIGURE 12.9 Various types of colony of lenticular microfossils. All from the Farrel Quartzite. Transmitted light. (a) Colony composed exclusively of lenses. (b) Colony composed of lenses and small spheroids. (c) Colony composed of lenses and spheroids of various sizes. (d) Chained (or sheeted) colony.

12.7.2 LIFECYCLES AND REPRODUCTION STYLES OF LENTICULAR MICROBES

Diversity of lenticular microfossils in morphology and colony reflects at least partially their taxonomic diversity. In addition, such diversity particularly in compositions and morphologies of their colonies could provide us with invaluable information about life cycle and reproduction styles of lenticular microbes. In this section, provisional life cycles of lenticular microbes with some implications for reproduction styles are discussed.

12.7.2.1 Simple Life Cycle with Binary Fission

Simple binary fission is one of the reproduction styles of lenticular microbes, which is fully evidenced by the presence of colonies composed exclusively of lenticular structures and composite structures corresponding to various stages of reproduction (Figure 12.10). Complete binary fission and detachment of the two daughter cells would have resulted in the formation of colonies composed of separated individuals. Chained and sheet-like colonies likely formed as a result of continuous binary fissions without liberation of individuals. The presence of colony composed of separated individuals and chained colony in the SPF population suggests that the chain-forming lenticular microfossils and non-forming ones are not taxonomically discernable (Figures 11.17c and e).

12.7.2.2 Multiple Fission with Baeocyte Formation

It is possible that the FQ lenticular microbes include taxa reproduced by multiple fissions with baeocyte formation, which is inferred from the following observations. Though only one specimen has ever been discovered, lenticular microfossil containing small spheroids inside has been identified (Figure 12.11a and b). As described above, lenticular microfossils comprise colonies together with spheroids (Figures 12.9b and c) and irregular film-like objects are often associated with each other (Figure 12.11c). My interpretation for this is that the film-like objects are walls of fragmented lenticular microfossils and that spheroids represent expelled baeocytes. Spheroids in the mixed colony are all small or are various in size but tend to be smaller than associated lenses. The former may represent the situation right after the rupturing of mother cells, whereas the latter represents the later stage in which the released baeocytes had already grown to various degrees (Figure 12.9c). Unevenness of spheroid size is not inconsistent with my interpretation. Such unevenness of cells in colony produced by multiple fissions has been found for cyanobacterial *Stanieria* (e.g., Wu et al., 2016). It is suggested that released spheroid baeocytes had grown up vegetatively and eventually become to be flanged lenses, in which multiple baeocytes would be produced (Figure 12.11d). This hypothesized life cycle of lenticular microbes was inspired from life cycles of a genus of green algae, *Pterosperma* and a genus of cyanobacteria, *Stanieria*.

In addition to this type of multiple fissions characterized by baeocyte formation, lenticular microbes might had reproduced by multiple fissions characterized by production of larger and non-spherical baeocytes. This inference comes from identification of some large specimens of lenticular microfossils in which several inner objects appear to be contained (Figure 12.12a). The presence of inner objects is inferred also from specimens characterized by somewhat irregular morphology with local convex or low-relief cobb. The colony composed of a small number (3–4) of lenticular microfossils is not rare and may represent clustered daughter cells (Figure 12.12b). Some lenticular microfossils from the Strelley Pool Formation are also found to be partitioned to multiple compartments, although the compartmentation

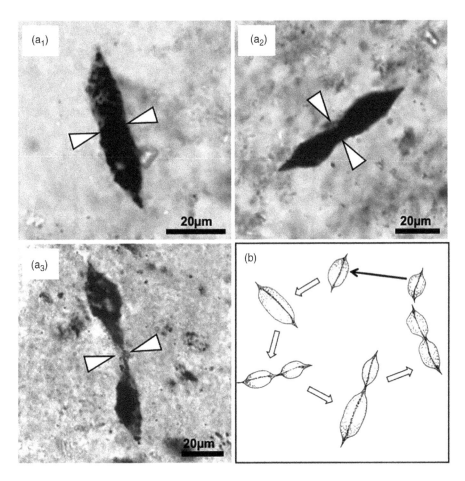

FIGURE 12.10 Simple binary fission of lenticular microfossil. Inferred from the SPF specimens. (a1–a3) show various stages of binary fissions. Transmitted light. (b) Schematic model of reproduction of lenticular microfossils by binary fissions.

is irregular (Figure 12.12c). Such compartmentation may be produced by crystal grains. Namely, due to growth of crystals insides the lens, carbonaceous matter was swept up to the grain boundaries to have formed apparent partitions. However, this model seems unlikely at least for this specimen in the following reasons. First, carbonaceous partitions appear to be independent of grain boundaries of the matrix quartz. Second, grain size and extinction habit of microquartz do not differ between the inside and the outside of the lens, excluding the possibility that the inside of the lens was once composed of different minerals such as carbonate and then replaced by microquartz. Third, one of the compartments is filled with carbonaceous matter. Thus, the partitioning is not artifact, but more likely considered to be primary texture, although further identification of similar specimens and detailed examination are indispensable.

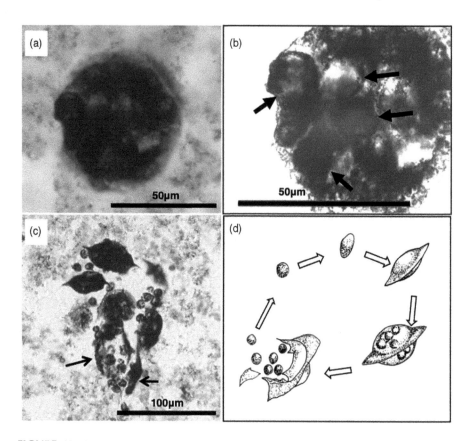

FIGURE 12.11 Hypothetical reproduction style of lenticular microbes, constructed from the FQ specimens. (a) Specimen containing multiple small spheroids inside. (b) The same specimen of (a), under stronger transmitted light. The arrows indicate internal spheroids. (c) Colony composed of lenses and small spheroids. The arrows indicate possible fragmented walls of lens. (d) Schematic model of reproduction of lenticular microfossils by multiple fissions. Transmitted light for (a–c).

12.7.2.3 Possible Asymmetric Division

Lenticular microfossils often comprise chain-like structure. Chain-like structure was previously interpreted to have been formed by continual binary fissions without liberation. Additionally or alternatively, similar composite structure can be formed by sequential budding as observed for some yeast, producing pseudomycelium (Figure 2 in Dayo-Owoyemi et al., 2014). Thus, asymmetric division (budding) is worth to be considered as a possible reproduction strategy of lenticular microbes. This is not unrealistic, because some chain-like structures are composed of individuals of various sizes. The extracted specimen from the Strelley Pool Formation appears to be a paired specimen at glance, but another much smaller than others is attached (Figure 12.12d). Another specimen suggestive of asymmetric division has

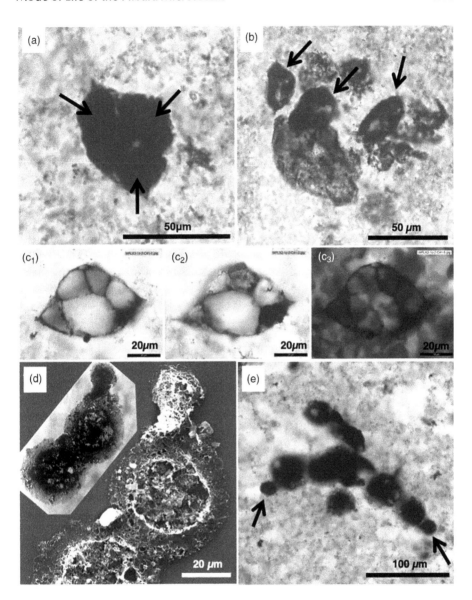

FIGURE 12.12 Photomicrographs of specimens of lenticular microfossils suggestive of various types of reproduction. Specimens for (a), (b), and (e) are from the Farrel Quartzite, whereas those of (c) and (d) are from the Strelley Pool Formation. Transmitted light, except for c_3, which was taken under crossed nicols. (a) Specimen that appears to have three inner objects (the arrows). (b) Three lenses and associated film, possible remnant of cell wall of the mother cell. (c) Chambered lens taken under different focal depths. Note that one of the chambers (the lower right) is filled with opaque materials, possibly carbonaceous matter (C_3). (d) Two paired lenses with an attached small one. Optical and SEM images for the extracted specimen. (e) Chain-like colony of lenses. Note that the two terminal lenses (the arrows) are smaller than the others.

been identified in petrographic thin section (Figure 12.12e). It is a chain composed of seven lenses; the size of individual lens tends to decrease from the one end to the other. Such the feature might be more likely interpreted in the context of sequential budding.

12.7.3 MORPHOMETRICS OF LENTICULAR MICROFOSSILS

A series of study of the Pilbara and the Barberton microfossils have revealed that diverse microbes had already evolved during the Archean. This fact itself evidences the early speciation, because all the extant organisms had likely evolved and diversified from the common ancestor. A morphometric approach to Archean microfossil taxonomy has been made by, for example, Schopf (1993), who measured cell sizes and shapes of the ca. 3.5 Ga-old trichome-like filamentous microstructures in petrographic thin sections and classified them into 11 taxa. However, some inaccuracies due to orientations of measured specimens in thin sections are unavoidable. In this context, lenticular microfossils have huge advantage, because they can be extracted from the host rocks by acid maceration. By placing the extracted specimens on a glass slide, we can directly examine polar views of lenticular microfossils, which means that we can correctly measure the diameter and the width of flange. Correct size data enable us to compare different populations in the context of morphological variation. In this section, the results of morphometric study performed by Sugitani et al. (2018) are summarized.

As described in Chapter 10, lenticular microfossils have been identified from four localities of the 3.4 Ga-old Strelley Pool Formation. Two of them, the Waterfall locality and the Panorama locality 2, were found to be prolific by examination of petrographic thin sections and had been subjected to acid macerations. Soon after starting examinations of palynomorphs from these two localities, I was aware that many of lenticular microfossils from the Panorama locality were oblate in the equatorial view. My students and I vigorously collected palynomorphic specimens using micropipette, under of a high-magnification binocular microscope. With great help of my students, more than 1,000 specimens have been collected.

It may be noted that the validity of this morphometric analyses requires the prerequisites that the lenticular microbes do not have lifecycle morphological variants but reproduce simply by binary fission. Lenticular microfossils often comprise colonies and occur as pairs and rarely chains. Paired specimens are various in style of attachment, from point attachment to fused attachment. This variation can be interpreted in the context of a sequence of binary fission (Figure 12.10). Specimens and colonies potentially suggestive of alternative reproductions (e.g., multiple fissions and asymmetric divisions) to binary fission are present. However, such specimens are rare. We examined well-preserved colonies of the Waterfall locality (n = 230) and the Panorama locality 2 (n = 65). Less than ten colonies appear to be composed of a mixture of spheroids and lenses. Therefore, it is acceptable to assume that the prerequisite is met by the SPF lenticular microbes. Compared with the SPF lenticular populations, the FQ ones are far diverse in morphology and occurrence of colony as described above, which could be at least partially attributed to life cycle variants and diverse life cycles. This means that the simple morphometrics could not be applied to the FQ populations.

12.7.3.1 Methodology

Morphometry was performed on the polar view of extracted lenticular microfossils mounted on a glass slide. Size data were obtained in this direction. Measurement was made for image taken by a digital camera DFC280 equipped to Leica DMRP microscope, using the software LAS (v.3.8). Extracted lenticular microfossils do not always retain their complete shapes, which is mainly attributed to incompletely preserved flange. This is inevitable, because the flanges are thin and fragile. As completely preserved specimens are minority, specimens preserving more than 3/4 of the whole shape were measured by extrapolation, resulting in a measurement of 415 specimens from the Waterfall locality and 433 from the Panorama locality 2. Three parameters such as whole area "S", oblateness "Ob", and flange ratio "Fr" were calculated using major and minor dimensions of the whole body (a and b) and the central body (a' and b') (Equations 12.1–12.3).

$$S = \pi \times a \times b \text{ (for the central body, } S' = \pi \times a' \times b') \qquad (12.1)$$

$$Ob = (a-b)/a \text{ \{for the central body, } Ob' = (a'-b')/a'\} \qquad (12.2)$$

$$Fr = (S-S')/S \qquad (12.3)$$

In order to omit outliers, the obtained parameters were examined, using the interquartile range method, giving rise to the final dataset (n = 395 for the Goldsworthy population and n = 414 for Panorama population). These data sets were subjected to analyses.

12.7.3.2 Results of Morphometric Analyses

Histograms for oblateness of the extracted lenticular microfossils indicate that our expectation was correct (Figure 12.13a$_1$, a$_2$). Circular type dominates the Waterfall population, whereas oblate one dominates the Panorama population. Furthermore, cluster analyses (Ward's method) using oblateness, whole area, and flange ratio as explanatory variables revealed that each population was composed of subpopulations. The Waterfall population was composed of circular subpopulation (n = 240) and slightly oblate one (n = 155): the former was further classified into two by areas. The Panorama population was also classified into three subpopulations, two of which were oblate but had distinct areas (n = 222 and n = 65, respectively), whereas the remaining one was relatively circular (n = 127). These classifications were statistically proved to be valid (Sugitani et al., 2018).

12.7.3.3 Morphological Variations – Cell Growth and Taphonomy?

It seems now clear that the SPF lenticular microfossils are composed of statistically defined several submorphotypes. In order to consider this variation in the context of speciation, we need to address the possibility that it could alternatively be caused by cell growth and taphonomy. As lenticular microbes were reproduced by simply

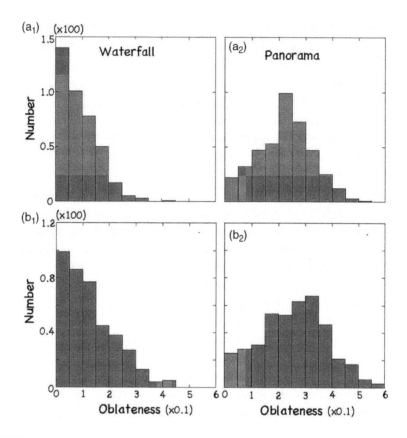

FIGURE 12.13 Histograms for oblateness of the whole bodies ($a_{1,2}$) and the central bodies ($b_{1,2}$) for the Goldsworthy populations and the Panorama populations. (Drawn from original data for Sugitani et al. (2018).)

binary fission as discussed earlier, vegetative cell growth result in an increase in oblateness, which could result in significant positive relationship between the oblateness and the major dimension. However, as shown in Figure 12.14, there can be seen no relationship between the two variables. Firstly, it seems unrealistic that only the Panorama population is rich in vegetatively enlarged individuals. Additionally, the observed morphological variations cannot be explained by either taphonomy or post-depositional deformation, because the central bodies of the two populations are also different in oblateness (Figure 12.13$b_{1,2}$). It is also unlikely that the original morphology of microfossils was modified by deformation of host rocks and sediment compaction. The host cherts are composed of equant microquartz, without preferred orientation of extinction. Lenticular microbes are thought to have been permineralized and/or encapsulated by silica before sediment compaction. There are no signs for redistribution of OM that comprises cells due to recrystallization from amorphous silica to microquartz.

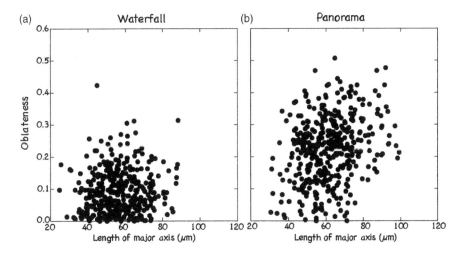

FIGURE 12.14   Relationship between oblateness and length of the major dimension for the Goldsworthy population (a) and the Panorama population (b). (Drawn from original data for Sugitani et al. (2018).)

12.7.3.4   Environmental Adaptation and Speciation of Lenticular Microbes

Our morphometric study demonstrated that the 3.4 Ga-old lenticular microbes could be classified into several submorphotypes at the equatorial view, which were defined by variations in oblateness and area. As discussed above, the morphological variations cannot be explained by vegetative cell growth and artifacts including taphonomy and deformation. So, how can such morphological variations of Archean lenticular life be explained?

In order to answer this question, we need to go back to the depositional environments of the host cherts. Lithostratigraphy and geochemistry of the Strelley Pool Formation at the Waterfall locality in the Goldsworthy greenstone belt and at the Panorama locality 2 in the Panorama greenstone belt had already been described in Chapter 10. Here, depositional environments of microfossil-bearing cherts from these two localities are briefly summarized. The Strelley Pool Formation at the Panorama locality 2 occurs as a thin (~5 m) unit composed dominantly of silici-fied fine-grained volcaniclastic rocks (ash) and sandstone composed dominantly of extensively silicified reworked carbonaceous grains, with minor carbonate and black chert. Sugitani et al. (2013) suggested based on their own observations and previ-ous detailed studies by Allwood et al. (2006, 2007) that the microfossil-bearing chert had deposited at intertidal to supratidal zone. The marine origin is consistent with seawater signatures of rare earth elements (REEs) (heavy REE-enrichments and super-chondritic Y/Ho values), although contribution of hydrothermal fluids was unambiguous. The Strelley Pool Formation at the Waterfall locality is much thicker and can be up to 80m thick. It is dominated by immature sandstones composed of volcanic and sedimentary fragments and quartz. The main body of microfossil-bear-ing black chert is unique in the presence of conical structures a few cm high. The structures were once suggested to be stromatolites (Sugitani et al., 2010), but later reinterpreted as "possible" siliceous sinters (Sugitani et al., 2015b). Although this interpretation is still provisional, there is no doubt that the cherts are geochemically distinct from the SPF fossiliferous cherts at the Panorama locality 2. The Waterfall cherts are characterized by various patterns of PAAS-normalized REEs including light- and middle-REE enrichment and nearly chondritic Y/Ho ratios and extreme enrichment of heavy metals. Thus, the two aqueous habitats were likely distinct with each other in physicochemical conditions such as concentrations of nutrients, salinity and pH.

Finally, we address the possibility that the morphological variations merely represent temporal and reversible alteration of cellular morphology. In response to environmental fluctuations, many organisms, particularly microorganisms, have the ability to change their morphology without genetic mutation. For example, *Caulobacter crescentus* changes the morphology of its stalk depending on the availability of nutrients, such as phosphate (Wagner et al., 2006). Therefore, the morphological variations of lenticular microfossils observed between the Goldsworthy populations and the Panorama ones could be temporal and reversible. However, we argue that the variations could be more reasonably interpreted as the result of allopatric (geographic) and adaptive speciation. The reasons for this include (1) spatial and environmental isolation between the two populations, ideal condition for speciation, (2) identification of morphological subpopulations in each population, and (3) the presence of a possible common type.

12.7.4 PLANKTONIC MODE OF LIFE OF LENTICULAR MICROBES

Many of lenticular microfossils are characterized by sheet-like appendage surrounding the central body. Already in the early stage of this research project, the presence of this appendage called "flange" had reminded us of their planktonic mode of life. This seems to be consistent with random distributions of colonies of lenticular microfossils in host cherts and the fact that their distribution is not controlled by laminae and layers. Furthermore, the assumed planktonic mode of life is consistent with carbon isotopic compositions of individual specimens (House et al., 2013). To test this hypothesis from different point of view, we performed sedimentation and floating-up simulations for virtual lenticular cells (Kozawa et al., 2019). Here, the methodology and results of only sedimentation simulation of this study are summarized.

12.7.4.1 Methodology

It was postulated that virtual fluid for simulation is incompressible and viscous, obeying the Navier-Stokes equation and that virtual cells were rigid. The latter postulation is consistent with the fact that the extracted lenses appear to be solid, contrastive to the flexible nature of films and large spheroids extracted by the same procedure. Also, films and large spheroids would be compressed by resin mounting with a cover glass. On the other hand, lenses retain their morphology in the same procedure. Thus, the rigid nature of lenses is likely a primary feature. Two types of motion are known for such rigid objects. One is translation motion of center of gravity with three degrees of freedom, being assumed to follow Newton's equation of motion. Another motion is rotation around the center of gravity, which has three degrees of freedom. This motion is assumed to follow Euler's equations.

This study (Kozawa et al., 2019) employed the following methods. The finite difference method on the staggered grids was used for the Eulerian dynamics of the virtual fluid, whereas Newton's equation of motion for the Lagrangian dynamics of the virtual cells. The volume penalization (VP) method (Kolomenskiy and Schneider, 2009) was employed in order to compute the force between the fluids and the embedded cells and fluid. By using discrete element method (Cundall and Strack, 1979), forces between cells to prevent overlapping were introduced.

12.7.4.2 Parameters of Virtual Cells

The defined parameters of virtual cells mimicking lenticular microfossils are shown in Figure 12.15. Virtual cell is constructed by sphere for the central body and oblate spheroid for flange. Whole sizes, oblateness, and flange ratios of the cells were introduced optionally from data obtained from the lenticular microfossils (n > 800) from the Strelley Pool Formation (Sugitani et al., 2018).

12.7.4.3 Sedimentation Simulation

A cubic box with 1 cm length was employed as the computational domain, which was discretized at the 400 grid points in each direction. The fluid filling this virtual space was assumed to be water with a density of 1.0 g/cm³ and a viscosity of 0.01 g/cm·s. Seventeen five virtual cells of an identical shape were set as $5 \times 5 \times 3$ with equal distance in the space and sedimented freely under the gravitational acceleration of 981 cm/s². Note that the initial orientations of the virtual cells were random, excluding the bias due to sedimentation velocity that depends on the initial degree of cell inclination. Therefore, sedimentation velocity was calculated as an average value of these 75 cells.

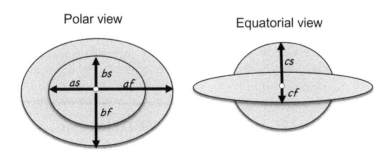

Ws (whole size) = *bf* × *af* × Pi
Bs (central bpdy size) = *bs* × *as* × Pi
Ob (oblateness) = *af* / *bf*
Rf (relative flange thickness) = *cf* / *cs*
Fr (flange ratio) (Ws – Bs) / Ws

FIGURE 12.15 Virtual cell for the sedimentation simulation.

12.7.4.4 Effect of Flange and Flange Thickness

The sedimentation velocity was calculated for the virtual cells of two sizes ($W_s = 1,000$ and $4,000$ μm^2) with a fixed flange ratio (F_r) of 0.5 and oblateness of 0 and varied relative flange thickness (R_f) (0.3, 0.4, 0.5). This was intended to examine the effect of flange on sedimentation velocity. We compared sedimentation velocities of the virtual flanged cells (V_{fsp}) with the spheres with the same mass (V_{esp}) (the equivalent sphere). It has been revealed that V_{fsp} is around 7% smaller than V_{esp}, and that thinner flange was more effective for reducing sedimentation velocity. Also, flange appears to work more effectively for larger cells. The cells with a projected area of $1,000$ μm^2 and those with $4,000$ μm^2 have a V_{fsp}/V_{esp} of 0.931 and of 0.924, respectively.

12.7.4.5 Effect of Oblateness

The effect of oblateness on sedimentation velocity was also examined, employing four patters of oblateness ($O_b = 0.0$, 0.1, 0.2, 0.3) that are applied to three combinations of size and flange thickness $\{(W_s, R_f) = (1,000$ μm^2, 0.3), $(4,000$ μm^2, 0.3), and $(4,000$ μm^2, 0.5)$\}$, with the fixed flange ratio ($F_r = 0.5$).

For the two cases of $(W_s, R_f) = (1,000$ μm^2, 0.3; $4,000$ μm^2, 0.5), the sedimentation velocity (V_{fsp}) decreases straightforwardly as O_b increases. For the case of $(W_s, R_f) = (4,000$ μm^2, 0.3), on the other hand, the variation in sedimentation velocity is complex: V_{fsp} $(O_b = 0.3) > V_{fsp}$ $(0.0) > V_{fsp}$ $(0.1) > V_{fsp}$ (0.2). As such, the effects of the oblateness, flange thickness, and the whole size on the sedimentation velocity are quite complicated, although the cells with oblateness of 0.2 tend to sediment more slowly than those with zero oblateness.

12.7.4.6 Summary

In order to obtain a better understanding about the traits of lenticular microbes such as large size (20–100 μm across, robust wall, and flanged), sedimentation simulations were performed. Analyzed virtual cells were of various sizes, oblateness, and relative flange thickness. Sedimentation velocities were compared with the mass-equivalent spheres. The results demonstrated that sedimentation velocities were reduced by flange: this function of flange was more effective for larger cells. Another implication from the simulations was that reduction of sedimentation velocity was most effectively obtained by the modest oblateness. Previously suggested planktonic mode of life for lenticular microbes has been successfully confirmed by this sedimentation simulations.

COLUMN: THIOMARGARITA AND BIG BACTERIA

Thiomargarita namibiensis, gram-negative proteobacterium, has been discovered from the Namibian continental shelf in 1997 (Schulz et al., 1999), where sediments were anoxic and sulfidic. This bacterium is chemolithoautotrophic. It utilizes H_2S as an electron donor and NO_3^- or O_2 as a terminal electron acceptor to produce organic matter from CO_2 and H_2O. Oxidation of H_2S results in the formation of elemental sulfur granules, which are stored in its periplasm. The sulfur granules beautifully look like pearls, which gave this bacterium genus name of *Thiomargarita*.

Like many of other bacteria, they are simply spheroid in morphology but are unusually large. In general, they range from 100 to 300 μm in diameter but can be up to 750 μm. Though not the largest bacterium, their cell size appears to be unusual as coccoid prokaryotic cells, which are generally less than 5 μm in diameter. As mentioned in the main text, such small size means higher surface area/volume ratio, effective for diffusive nutrient uptake. Thus, *Thiomargarita namibiensis* has much smaller surface area/volume ratio. So how does this giant bacterium overcome this advantage? Actually, their cells are not filled with cytoplasm, but largely occupied with a membrane-enclosed compartment called vacuole. Namely, their cytoplasm is only peripherally distributed. *Thiomargarita namibiensis* stores NO_3^- in vacuoles, which is utilized when this nutrient is not fully available from the environment. *Beggiatoa* is also sulfur-metabolizing bacteria, and cylindrical cells comprise trichome (filamentous body). Cylindrical cells of some marine species can be as large as 100 μm or more in diameter, and trichomes can be up to several cm long. They are also vacuolated and store nitrate inside. Whereas the unusually large cell size of some bacteria is thus closely associated with vacuolation, nonvacuolated large bacterium is also known. Heterotrophic *Epulopiscium* sp. can be up to 80×600 μm. They are not spherical and thus, surface area/volume ratios are not so low compared with equivalent spheroid cells. However, their size is thought to depend on nutrient-rich environment in gut of fishes. Some species of cyanobacteria such as, e.g., unicellular and spheroidal *Chroococcus giganteus* have cells 40–60 μm in diameter, although there can be seen no descriptions that they have vacuoles (Schulz-Vogt et al., 2007).

REFERENCES

Agić, H., Moczydłowska, M., Yin, L.-M. 2015. Affinity, life cycle, and intracellular complexity of organic-walled microfossils from the Mesoproterozoic of Shanxi, China. *Journal of Paleontology* 89, 28–50.

Allwood, A.C., Burch, I., Walter, M.R. 2007. Stratigraphy and facies of the 3.43 Ga Strelley Pool Chert in the southwestern North Pole Dome, Pilbara Craton, Western Australia. *Geological Survey of Western Australia, Record* 2007/11.

Allwood, A.C., Walter, M.R., Kamber, B.S., Marshall, C.P., Burch, I.W. 2006. Stromatolite reef from the Early Archaean era of Australia. *Nature* 441, 714–718.

Angert, E.R. 2005. Alternatives to binary fission in bacteria. *Nature Reviews Microbiology* 3, 214–224.

Angert, E.R., Losick, R.M. 1998. Propagation by sporulation in the guinea pig symbiont *Metabacterium polyspora*. *Proceedings of the National Academy of Sciences of the United States of America* 95, 10218–10223.

Choi, H.J., Joo, J.-H., Kim, J.-H., Wang, P., Ki, J.-S., Han, M.-S. 2018. Morphological characterization and molecular phylogenetic analysis of *Dolichospermum hangangense* (Nostocales, Cyanobacteria) sp. nov. from Han River, Korea. *Algae* 33, 143–156.

Cundall, P.A., Strack, O.D.L. 1979. A discrete numerical model for granular assemblies. *Géotechnique* 29, 47–65.

Dayo-Owoyemi, I., Rosa, C.A., Rodrigues, A., Pagnocca, F.C. 2014. *Wickerhamiella kiyanii* f.a., sp. nov. and *Wickerhamiella fructicola* f.a., sp., nov., two yeasts isolated from native plants of Atlantic rainforest in Brazil. *International Journal of Systematic and Evolutionary Microbiology* 64, 2152–2158.

Delarue, F., Robert, F., Derenne, S., Tartèse, S., Jauvion, C., Bernard, S., Pont, S., Gonzalez-Cano, A., Duhamel, R., Sugitani, K. 2020. Out of rock: A new look at the morphological and geochemical preservation of microfossils from the 3.46 Gyr-old Strelley Pool Formation. *Precambrian Research* 336, 105472.

House, C.H., Oehler, D.Z., Sugitani, K., Mimura, K. 2013. Carbon isotopic analyses of ca. 3.0 Ga microstructures imply planktonic autotrophs inhabited Earth's early oceans. *Geology* 41, 651–654.

Kaufman, A.J., Xiao, S. 2003. High CO_2 levels in the Proterozoic atmosphere estimated from analyses of individual microfossils. *Nature* 425, 279–282.

Kolomenskiy, D., Schneider, K. 2009. A Fourier spectral method for the Navier-Stokes equations with volume penalization for moving solid obstacles. *Journal of Computational Physics* 228, 5687–5709.

Kozawa, T., Sugitani, K., Oehler, D.Z., House, C.H., Saito, I., Watanabe, T., Gotoh, T. 2019. Early Archean planktonic mode of life: Implications from fluid dynamics of lenticular microfossils. *Geobiology* 17, 113–126.

Oehler, D.Z., Robert, F., Walter, M.R., Sugitani, K., Allwood, A., Meibom, A., Mostefaoui, S., Selo, M., Thomen, A., Gibson, E.K. 2009. NanoSIMS: Insights to biogenicity and syngeneity of Archaean carbonaceous structures. *Precambrian Research* 173, 70–78.

Parke, M., Boalch, G.T., Jowett, R., Harbour, D.S. 1978. The genus *Pterosperma* (Prasinophyceae): Species with a single equatorial ala. *Journal of the Marine Biological Association of the United Kingdom* 58, 239–276.

Salman, V., Amann, R., Girnth, A.-C., Polerecky, L., Bailey, J.V., Høgslund, S., Jessen, G., Pantoja, S., Schult-Vogt, H.N. 2011. A single-cell sequencing approach to the classification of large, vacuolated sulfur bacteria. *Systematic Applied Microbiology* 34, 243–259.

Salman, V., Bailey, J.V., Teske, A. 2013. Phylogenetic and morphologic complexity of giant sulphur bacteria. *Antonie van Leeuwenhoek* 104, 169–186.

Schopf, J.W. 1993. Microfossils of the early Archean Apex chert: new evidence of the antiquity of life. *Science* 260, 640–646.

Schopf, J.W., Kudryavtsev, A.B., Osterhout, J.T., Williford, K.H., Kitajima, K., Valley, J.W., Sugitani, K. 2017. An anaerobic ~3400 Ma shallow-water microbial consortium: presumptive evidence of Earth's Paleoarchean anoxic atmosphere. *Precambrian Research* 299, 309–318.

Schopf, J.W., Kudryavstev, A.B., Walter, M.R., Van Kranendonk, M.J., Williford, K.H., Kozdon, R., Valley, J.W., Gallardo, V.A., Espinoza, C., Flannery, D.T. 2015. Sulfur-cycling fossil bacteria from the 1.8-Ga Duck Creek Formation provide promising evidence of evolution's null hypothesis. *The Proceedings of the National Academy of Sciences* 112, 2087–2092.

Schulz, H.N., Brinkhoff, T., Ferdelman, T.G., Mariné, M.H., Teske, A., Jørgensen, B.B. 1999. Dense populations of a giant sulfur bacterium in Namibian shelf sediments. *Science* 284, 493–495.

Schulz-Vogt, H.N., Angert, E.R., Garcia-Pichel, F. 2007. Giant bacteria. *Encyclopedia of Life Sciences.*

Sugitani, K., Grey, K., Allwood, A., Nagaoka, T., Mimura, K., Minami, M., Marshall, C.P., Van Kranendonk, M.J., Walter, M.R. 2007. Diverse microstructures from Archaean chert from the Mount Goldsworthy – Mount Grant area, Pilbara Craton, Western Australia: Microfossils, dubiomicrofossils, or pseudofossils? *Precambrian Research* 158, 228–262.

Sugitani, K., Grey, K., Nagaoka, T., Mimura, K. 2009a. Three-dimensional morphological and textural complexity of Archean putative microfossils from the northeastern Pilbara Craton: Indications of biogenicity of large (>15 microm) spheroidal and spindle-like structures. *Astrobiology* 9, 603–615.

Sugitani, K., Grey, K., Nagaoka, T., Mimura, K., Walter, M.R. 2009b. Taxonomy and bioge-nicity of Archaean spheroidal microfossils (ca. 3.0 Ga) from the Mount Goldsworthy-Mount Grant area in the northeastern Pilbara Craton, Western Australia. *Precambrian Research* 173, 50–59.

Sugitani, K., Kohama, T., Mimura, K., Takeuchi, M., Senda, R., Morimoto, H. 2018. Speciation of Paleoarchean life demonstrated by analysis of the morphological variation of lenticu-lar microfossils from the Pilbara Craton, Australia. *Astrobiology* 18, 1057–1070.

Sugitani, K., Lepot, K., Nagaoka, T., Mimura, K., Van Kranendonk, M., Oehler, D.Z., Walter, M.R. 2010. Biogenicity of morphologically diverse carbonaceous microstructures from the ca. 3400 Ma Strelley Pool Formation, in the Pilbara Craton, Western Australia. *Astrobiology* 10, 899–920.

Sugitani, K., Mimura, K., Nagaoka, T., Lepot, K., Takeuchi, M. 2013. Microfossil assemblage from the 3400 Ma Strelley Pool Formation in the Pilbara Craton, Western Australia: Results from a new locality. *Precambrian Research* 226, 59–74.

Sugitani, K., Mimura, K., Takeuchi, M., Lepot, K., Ito, S., Javaux, E.J. 2015a. Early evolution of large micro-organisms with cytological complexity revealed by microanalyses of 3.4 Ga organic-walled microfossils. *Geobiology* 13, 507–521.

Sugitani, K., Mimura, K., Takeuchi, M., Yamaguchi, T., Suzuki, K., Senda, R., Asahara, Y., Wallis, S., Van Kranendonk, M.J. 2015b. A Paleoarchean coastal hydrothermal field inhabited by diverse microbial communities: the Strelley Pool Formation, Pilbara Craton, Western Australia. *Geobiology* 13, 522–545.

Sugitani, K., Mimura, K., Walter, M.R. 2011. Farrel Quartzite microfossils in the Goldsworthy greenstone belt, Pilbara Craton, Western Australia. Additional evidence for a diverse and evolved biota on the Archean Earth. In V. Tewari and J. Seckbach eds. *Stromatolites: Interaction of microbes with sediments.* pp. 115–132. Springer, Dordrecht.

Tappan, H. 1980. *The Paleobiology of Plant Protists.* p. 1028. W.H. Freeman and Co., San Francisco.

Tice, M. M., Lowe, D.R. 2006. The origin of carbonaceous matter in pre-3.0 Ga greenstone terrains: a review and new evidence from the 3.42 Ga Buck Reef Chert. *Earth-Science Reviews* 76, 259–300.

Wagner, J.K., Setayeshgar, S., Sharon, L.A., Reilly, J.P., Brun, Y.V. 2006. A nutrient uptake role for bacterial cell envelope extensions. *Proceedings of National Academy of Sciences of the United States of America* 103, 11772–11777.

Waterbury, J.B., Stanier, R.Y. 1978. Patterns of growth and development in pleurocapsalean cyanobacteria. *Microbiology and Molecular Biology Reviews* 42, 2–44.

Wu, Y.-S., Yu, G.-L., Jiang, H.-X., Liu, L.-J., Zhao, R. 2016. Role and lifestyle of calcified cya-nobacteria (*Stanieria*) in Permian-Triassic boundary microbialites. *Palaeogeography, Palaeoclimatology, Palaeoecology* 448, 39–47.

13 Facts and Problems of the Pilbara Microfossils and Related Issues

13.1 INTRODUCTION

I would like to address here the remaining problems regarding the Pilbara microfossil assemblage. The problems are their biological affinities. Key issues are whether the small spheroid and lenticular morphotypes are eukaryotic algae and/or cyanobacterial lineages, given that they were autotrophic (House et al., 2013; Lepot et al., 2013; also see Chapter 12). A few other issues to be resolved in the future studies are also addressed.

13.2 CYANOBACTERIAL MICROFOSSILS IN THE EARLY PRECAMBRIAN

13.2.1 RECORDS OF DESCRIBED POSSIBLE CYANOBACTERIAL MICROFOSSILS

As discussed in Chapter 6, the evolution of oxygenic photosynthesis was the most important turning point in the Earth's surface environment and biosphere. It is believed to have occurred at least 2.7 Ga and probably back to 3.0 Ga ago, based on stromatolite morphology, phylogenic study of modern cyanobacteria and other geochemical evidence, although direct evidence such as detection of cyanobacterial biomarker from the ca. 2.7 Ga-old Australian shales (Brocks et al., 1999) has later been discarded as contamination (French et al., 2015). On the other hand, cellularly preserved microfossils morphologically assignable to cyanobacteria have been reported from successions even older than 3.0 Ga. In this section, such microfossils would be reviewed, starting from descriptions of some Proterozoic records of cyanobacteria.

It is generally accepted that the microfossil assemblage described from the 1.89–1.84 Ga-old Belcher Supergroup, Hudson Bay, Canada (Hofmann, 1976) contains the certain oldest cyanobacterial microfossils (Demoulin et al., 2019 and reference therein). The microfossil assemblage there includes 18 genera and 24 species of filamentous and coccoid taxa, and have been identified from black chert layers, lenses, and nodules in stromatolitic dolostones. The depositional environment was interpreted to have been intertidal, shallow subtidal, and supratidal zones on a carbonate platform. The black cherts are thought to have been formed by very early silicification, which had enhanced remarkable preservation of microfossils.

Cyanobacterial microfossil in this assemblage has been denominated *Eoentophysalis belcherensis* (Figure 13.1a), after the modern cyanobacterium genus, *Entophysalis* (Golubic and Hofmann, 1976). Cell units of *Eoentophysalis belcherensis*

DOI: 10.1201/9780367855208-13

are spheroidal, ellipsoidal, or subpolyhedral and range from 2.5 to 9 μm across, often with tiny dark to opaque inclusions and pigmented lamellated envelopes. They occur as irregular clusters and layers and comprised pustular palmelloid colonies. *Entophysalis* is also known to comprise stromatolites in the Hamelin Pool of Shark Bay, Western Australia. Depositional environments of the Hamelin Pool stromatolites are also comparable to those of the *Eoentophysalis belcherensis*-bearing stromatolitic dolostones.

Less certain but promising fossilized cyanobacterial cells are structures that are morphologically comparable to akinete, dormant cell specialized from vegetative cell of trichome-forming cyanobacteria represented by nostocacean cyanobacteria (Figure 12.1e). An akinete, generally larger than vegetative cell, has an envelope and thick wall and can store essential nutrients, which allows itself to survive harsh

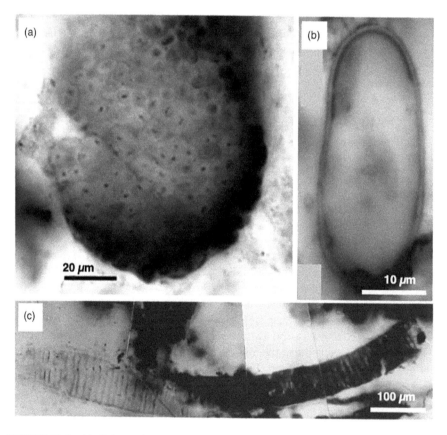

FIGURE 13.1 (a) Colony of cyanobacterial coccoid microfossils (*Eoentophysalis belcherensis*) from the ca. 1.9 Ga Belcher Supergroup, Canada (Hofmann, 1976). {Reproduced from Demoulin et al. (2019), with permission from Elsevier.} (b) *Archaeoellipsoides* from the >1.6 Ga Semri Group, India (Sharma and Shukla, 2019). (Photo courtesy by M. Sharma.) (c) *Oscillatoriopsis majuscule*, fossilized cyanobacteria-like trichomes from the 1.88 Ga Duck Creek Dolomite, Western Australia. {Reproduced from Knoll et al. (1988), with permission from Elsevier.}

environments such as desiccation, under which vegetative cells could not survive. When the environment becomes suitable for growth, the akinete germinates to be at vegetative stage. Possible fossilized akinetes, which are called *Archaeoellipsoides*, have been reported from the ~1.5 Ga-old Billyakh Group, Siberia, the 1.65 Ga-old McArthur Group, northern Australia, the >1.6 Ga-old Semri Group, India, and the 2.1 Ga-old Franceville Group, Gabon (Golubic et al., 1995; Amard and Bertrand-Sarfati, 1997; Tomitani et al., 2006; Sharma and Shukla, 2019). As Demoulin et al. (2019) cautiously note, cyanobacterial affinity of *Archaeoellipsoides* has not always been fully demonstrated, which is largely due to their simple morphology and solitary occurrence. Diagnostic occurrence is that such the differentiated cell structure, together with vegetative cell-like structures, comprises trichome-like filament (Pang et al., 2018). Among the above-referred older examples, double-walled ellipsoids reported from the >1.6 Ga-old Semri Group (Figure 13.1b) (Sharma and Shukla, 2019) seems the most plausible to me as fossilized akinetes.

From the ca. 1.8 Ga-old stromatolite-bearing Duck Creek Dolomite, Western Australia, two morphotypes of clearly septate broad filamentous microfossils (trichomes) have been described (Knoll et al., 1988) (Figure 13.1c). One specimen named *Oscillatoriopsis majuscula* sp. nov. is composed of cells approximately 63 µm wide and 6–11 long and is up to at least 700 µm long. The other one named *Oscillatoriopsis cuboides* sp. nov. is composed of cells 11–13 µm wide and long. Its length is 200 µm or more. Both types are not branched and do not have morphologically differentiated cells. Although they occur in discontinuous chert within thin bedded ferroan dolomite, but not in stromatolites, their morphological resemblance to trichomes of modern oscillatorian cyanobacteria is fascinating. However, the authors noted that the assignment to cyanobacterial affinity was conventional and mentioned the fact that species of *Beggiatoa*, the sulfur-utilizing heterotrophic bacteria, also fall within the morphological ranges of the *Oscillatoriopsis* species (Knoll et al., 1988).

From the Neoarchean successions of South Africa, possible cyanobacterial microfossils have also been reported. Altermann and Schopf (1995) described faint fluffy carbonaceous structures from silicified stromatolitic carbonates of the ~2.6 Ga-old Campbellrand Subgroup, which were interpreted as fossilized envelope (extracellular polymeric substance) of *Eoentophysalis* sp. (Figure 13.2a).

Kazmierczak and Altermann (2002) had etched polished slab of stromatolitic carbonate from the Campbellrand Subgroup using 5% formic acid (HCOOH) and identified the presence of two types of biostructures, including (1) spherical pitted bodies surrounded by rims and (2) smaller, clustered subglobular to irregularly polygonal pitted units. These structures were interpreted as fossilized mucilage capsules (outer sheath) of benthic coccoid cyanobacteria (*Entophysalis* and *Pleurocapsa*) based on morphological similarity and occurrence. They were mineralized by carbonates and Al-Fe silicates (Figure 13.2b), like mucilage capsules of modern coccoid cyanobacteria, *Entophysalis* (Figure 13.2c). In addition to these herein described coccoids and relatively narrow filaments, broad tubular filamentous microfossils have been reported also from the Gamohaan Formation of this subgroup (Klein et al., 1987). The Gamohaan filamentous microfossils, ranging from 15 to 25 µm in diameter, are unbranched and their length range from a few to many hundreds of microns (Figure 13.2d). They also occur as interwoven mass, displaying stromatolitic mat-like

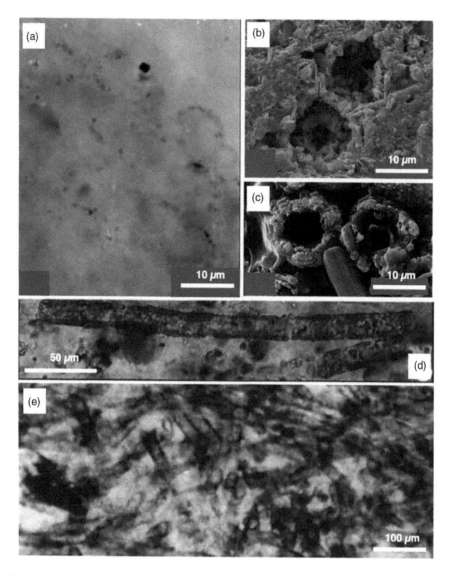

FIGURE 13.2 Microfossils from the ca. 2.5 Ga Campbellrand Subgroup, South Africa and related material. (a) Faint fluffy structure interpreted as remnants of fossilized envelope of benthic coccoids assigned to *Eoentophysalis sp.* (b) Structures interpreted as mucilage capsules of *Entophysalis* that were mineralized by calcium carbonates and Al-Fe silicates. (c) Modern equivalents to (d) from Lake Vai Si'i, Tonga {(a–c): Reproduced from Altermann and Kazmierczak (2003), with permission from Elsevier.} (d) Broad tubular filamentous microfossils (*Siphonophycus transvaalensis n. sp.*) potentially corresponding to sheaths encompassing trichomes of oscillatoriacean cyanobacteria, and (e) their interwoven mass, displaying stromatolitic mat-like fabric. ((d) and (e): Reproduced from Klein et al. (1987), with permission from Elsevier.)

fabric (Figure 13.2e). These well-preserved filamentous microfossils (*Siphonophycus transvaalensis* n. sp.) were interpreted to correspond to sheaths encompassing trichomes of oscillatorian cyanobacteria. Possible ensheathed trichomes, morphologically similar to oscillatorian cyanobacteria, have been described also from the ca. 2.7 Ga-old Tumbiana Formation of Western Australia (Schopf and Walter, 1983), from which tufted microbial mats supposedly built by cyanobacteira have been reported (Flannery and Walter, 2011).

Large (~300 μm across) organic-walled microfossils from the 3.2 Ga-old Moodies Group, South Africa may be cyanobacterial, considering their extraordinarily large size and the supposed environment of their habitat (photic zone and sulfide-poor), (Figure 8.7c, d). Colonial small carbonaceous spheroids from the ~3.4 Ga-old Kromberg Formation, South Africa, which has already been described in Chapter 8 (Figure 8.3e, f), also need to be mentioned here (Kremer and Kaźmierczak, 2017). The authors claimed that their morphological, colonial, and geochemical features were comparable to those described from younger microfossils convincingly interpreted as cyanobacteria. Although a cyanobacterial *Microcystis aeruginosa*-like benthic-planktonic life cycle was suggested, the authors did not exclude the possibility that the Kromberg spheroids represent "anoxygenic ancestors of cyanobacteria with the morphology of modern cyanobacteria but with a different physiology". In the follow-up studies (Kaźmierczak and Kremer, 2019), the authors suggested that the spheroid microfossils preserved cell cycle observed for modern cyanobacteria, which included vegetative cells, sporangia, and baeocyte-like spores. Based on identification of symmetrical and asymmetrical binary fissions and multiple fissions for these microfossils, their possible affinity to cyanobacterial Pleurocapsales was suggested (Figure 13.3).

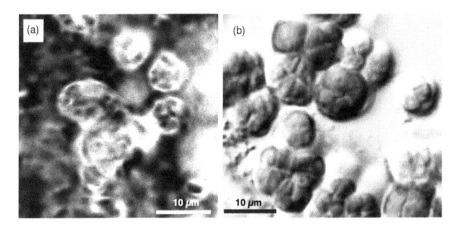

FIGURE 13.3 (a) Preserved multiple fissions of coccoid microfossils (?) in the 3.4 Ga Kromberg Formation, South Africa and (b) Modern *Pleurocapsa* sp. showing multiple fissions. (Reproduced from Kaźmierczak and Kremer (2019), with permission from Elsevier.)

Trichome-like microfossils from slightly older succession, the ca. 3.5 Ga Mount Ada Basalt, and the ca. 3.5 Ga Apex chert of Western Australia were already described and discussed in Chapter 8 (Figure 8.6d, e) (Awramik et al., 1983; Schopf, 1993). At the earliest stage of their studies, cyanobacterial affinity was suggested. While these microfossils have been subjected to severe skepticisms on their biogenicity and syngenicity, later analyses of carbon isotopic compositions of individual microfossils and confirmation of sample locality appear to push back the skepticisms. Nevertheless, the Apex filaments are unlikely cyanobacteria, because their habit was likely subsurface hydrothermal sytem. Unfortunately, habitat environment of the Mount Ada filament has not yet been well understood. As such, "potential" cyanobacterial microfossils have been reported from very deep time of the Earth's history.

13.2.2 CRITERIA FOR CYANOBACTERIAL MICROFOSSILS

As described above, assignment of microfossils to cyanobacterial affinity has been based largely on morphology and occurrence. Such approach is valid for microfossils younger than the Mesoproterozoic, because the preservation status of microfossils is, if not all, much better than that in older ages, and morphologies of cyanobacteria could plausibly be correlative to modern taxa. However, older microfossils, whose preservation status is generally poor and morphological similarity to modern taxa is less expected, should meet comprehensive criteria, in order to be assigned confidently to cyanobacteria. Recently, Demoulin et al. (2019) reviewed the fossil record of cyanobacteria and discussed criteria for identifying cyanobacterial microfossils in detail, including (1) morphology and division pattern, (2) ultrastructure, (3) paleoecology and behavior, (4) molecular fossils, (5) isotopic fractionation, and (6) intracellular biomineralization. Here, these criteria are simplified and introduced with my own considerations and modifications, including rearrangement of criteria entries.

1. Morphology: A few "multicellular" filamentous cyanobacteria have remarkable complexity represented by specialized cells known as heterocyst and akinetes that are morphologically distinct from vegetative cells. Such complexity, if identified for microfossils, could be diagnostic for cyanobacterial affinity.
2. Reproduction: Although not specific to cyanobacteria, multiple fissions for reproduction are the well described for cyanobacteria (Stanieria and Dermocarpa) (Waterbury and Stanier, 1978). Similar type of multiple fissions could be supporting evidence for cyanobacterial affinity.
3. Internal ultrastructure: Except for *Gloeobacter spp.*, all cyanobacteria have an internal membranous structures called thylakoids, which are not possessed by other prokaryotes. Surprisingly, such delicate structures could be fossilized and preserved over geologic time, particularly in association with clay minerals (e.g., Lepot et al., 2014).
4. Preservable sheath: Exopolysaccharidic envelope of cyanobacteria, generally called sheath, is less degraded than the cell itself and could be preservable as fossil. This preservation is enhanced in association with clay minerals and

aragonite. Many of acid-resistant organic walled microfossils from Proterozoic successions have been interpreted as cyanobacterial sheaths.

5. Paleoecology and behavior: Cyanobacteria are photoautotrophic and often comprise microbial mats such as stromatolites. Therefore, if fossils in question occur in stromatolites and/or sediments that represent photic zone environment, they possess the possibility of cyanobacterial lineage.

6. Molecular fossils: Biomolecules specific to cyanobacterial lineages and their derivatives, if detected from sedimentary rocks with morphologically cyanobacterial microfossils or directly from microfossils, could be strong evidence for cyanobacteria. Such biomolecules include, e.g., 2-methyl-hopanes, porphyrin, and scytonemin (Brocks et al., 2005; Summons et al., 1999). The oldest 2-methyl-hopanes and the oldest porphyrins have been recorded from 1.6 Ga and 1.1 Ga shales, respectively (Brocks et al., 2005; Gueneli et al., 2015)

7. Carbon isotopic signatures: Although different autotrophic metabolisms could produce organic matter (OM) of different carbon isotopic compositions, their values significantly overwrap with each other, except for those by methanogenesis characterized by extremely light carbon isotopic values (Figure 7.2). If microfossils have carbon isotopic compositions ($\delta^{13}C_{PDB}$) lighter than $-50‰$, their cyanobacterial lineages should be unlikely. If microfossils have carbon isotopic compositions from $-20‰$ to $-40‰$, their cyanobacterial affinities could be considered. Here, it should be, however, noted that the carbon isotopic compositions of autotrophically produced OM depend on not only the types of metabolism but also the isotopic compositions of inorganic carbon (CO_2 and HCO_3^-) as carbon source and their availability.

8. Intracellular biomineralization: Several types of active biomineralization have been reported for modern cyanobacterial cell interiors, including beads of intracellular Ca-carbonates and ferric phosphates (Couradeau et al., 2012; Brown et al., 2010). Such minerals, if identified in microfossils, could be taken as positive evidence for their cyanobacterial affinity.

13.3 EUKARYOTIC FOSSILS IN THE EARLY PRECAMBRIAN

13.3.1 RECORDS OF EUKARYOTIC MICROFOSSILS

Cells of eukaryotes are generally much larger and more complex than prokaryotes including cyanobacteria, which is thought to be products of endosymbiosis (*Column*). Therefore, identification of microfossils in ancient sediments as eukaryotes appears to be easier than that of cyanobacteria. The most conservative view is that the Mesoproterozoic organic-walled microfossils are the most reliable oldest microfossils of eukaryotes. According to Javaux and Lepot (2018), the oldest unambiguous eukaryotic microfossils are sphaeromorph acritarchs over 100 μm across characterized by concentric striation on their walls (*Valeria lophostriata*) (Figure 13.4a). This species has been described from the >1.65 Ga-old Mallapunyah Formation, Australia (Javaux, 2007). Similar sphaeromorphs characterized by concentric striations have

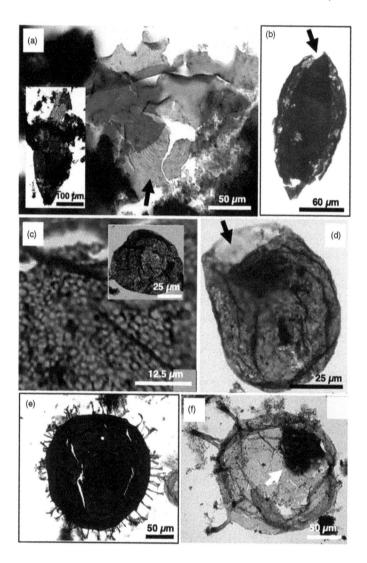

FIGURE 13.4 Proterozoic eukaryotic microfossils. (a) Concentric striation (the arrow) of *Valeria lophostriata*, from >1.65 Ga Mallapunyah Formation, Australia. Inset shows the whole image. (Reproduced from Javaux and Lepot (2018), with permission from Elsevier.) (b) Ovoidal acritarch from the Chuanlinggou Formation, northern China. The arrow shows the longitudinal rupture. (Reproduced from Peng et al. (2009), with permission from Elsevier.) (c) and (d) *Dictyosphaera delicata,* from the 1.4–1.7 Ga Ruyang Group, China. (c) shows interlocked polygonal plates of vesicle wall. Inset shows the whole view. (d) shows circular excystment opening. (Reproduced from Yin et al. (2005), with permission from Elsevier. e) *Shuiyousphaeridium macroreticulatum* from the Ruyang Group, northern China, with abundant regularly spaced cylindrical flared processes (Javaux et al., 2004). (Reproduced from Javaux and Lepot (2018), with permission from Elsevier.) (f) *Tappania plana* with unevenly distributed processes, some of which are branched (not shown) and bulbous protrusions (arrow) (Javaux et al., 2001). Note the intracellular inclusion (the white arrow). (Reproduced from Javaux and Lepot (2018), with permission from Elsevier.)

been reported from the Changzhougou Formation, China (Lamb et al., 2009). In Lamb et al. (2009), this formation was described as old as 1.8 Ga, although this age was later revised to be younger than $1,673 \pm 10$ Ma (Li et al., 2013). From the slightly younger (~1.65 Ga-old) Chuanlinggou Formation, China large ovoidal acritarchs have been reported (Peng et al., 2009). Their major axes range from 40 to 250 µm, with an average of 125 µm (n = 128) and a tightly clustered aspect ratio of 1.47. Namely, these microfossils display bipolar morphology. Additionally, they appear to be ruptured in a longitudinal manner (median split), and their walls are occasionally striated (Figure 13.4b). These features were interpreted to be consistent with a eukaryotic affinity (Peng et al., 2009). From the 1.7 to 1.4 Ga-old Ruyang Group, China, and the 1.5 Ga-old Roper Group, Australia, and others, different morphotypes of eukaryotic acritarchs have been reported. Large (normally > 40 µm) sphaeromorph, *Dictyosphaera delicata* is characterized by multilayered and reticulated wall (Figure 13.4c) and occasionally circular excystment opening (Figure 13.4d): these features are consistent with eukaryotic affinity (probably algae) (Yin et al., 2005). L. Yin and his colloborators have recently reported sphaeromorph acritarch having polygonal net-shaped ornamentation from the Paleoproterozoic Hutuo Group, North China that is assumed to have deposited around 2.0 Ga ago (Yin et al., 2020). Although this specimen was assigned to genus *Dictyosphaera*, only one specimen has ever been identified and lacks other diagnostic features described above. *Shuiyousphaeridium macroreticulatum* from the Mesoproterozoic Ruyang Group, China (Figure 13.4e) is characterized by multilayered vesicles, surface ornamentation identified by polygonal pattern, and furcated processes (e.g., Knoll et al., 2006) (also see below). *Tappania plana* (Figure 13.4f) from the Mesoproterozoic (~1.5 Ga ago) Roper Group, northern Australia has asymmetrically distributed processes, some of which are dichotomously branched and bulbous protrusions (e.g., Javaux et al., 2001). These large processed acritarchs are unambiguously eukaryotic, although further assignment to specific biological affinities (e.g., green algae vs. protists) has not yet be confirmed (Javaux and Lepot, 2018 and references therein).

13.3.2 Criteria for Eukaryotic Microfossils

To my knowledge, systematic consideration on how to identify microfossils as eukaryotic lineages has been proposed first by A. Knoll and his colloborators, which was based on examination of *Shuiyousphaeridium macroreticulatum* in the Mesoproterozoic Ruyang Group, China (Xiao et al., 1997). *Shuiyousphaeridium macroreticulatum* is an organic-walled spheroid vesicular microfossil characterized by numerous cylindrical hollow processes flaring outward and can be extracted by acid maceration of host rocks (Figure 13.4e). Specimens range from 50 to 300 µm in diameter. In addition to processes and large size, conspicuous features of this species identified under petrographic and electron microscopes are (1) multilayering and (2) ornamentation of vesicle walls. Vesicle wall *ca* 1.5 µm thick is multilayered, as expressed by different electron densities (Figure 13.5a). Vesicle outer walls have polygonal ridged reticulation ca 2 µm across on their surface. The same ornamentation is found also on the inner wall surface, but its relief is reverse, being composed of closely packed hexagonal plates (Figure 13.5b). In Knoll et al. (2006), it is stated that

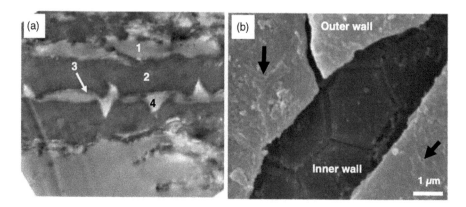

FIGURE 13.5 Ultrastructures of eukaryotic acritarch, *Shuiyousphaeridium macroreticulatum*, from the Ruyang Group, northern China. (a) TEM image of the two walls of compressed acritarch. 1: an outer layer composed of debris and processes, 2: an electron-dense layer corresponding to organic plates of inner wall, 3: an electron-tenuous thin layer that lines the inner side of the organic plates, 4: inner space of the vesicle. (b) SEM image of outer and inner walls characterized by surface ornamentation. The black arrows in the outer wall show ridges that delimit polygonal fields. The inner wall surface is composed of closely packed hexagonal plates. (Reproduced from Javaux et al. (2004), with permission from Wiley.)

> Prokaryotes can be large, they can have processes, and they can have preservable walls. But we do not know any prokaryote that combines the three characters, nor any that exhibit the complexity of form that light microscopy, SEM (a scanning electron microscope by KS), and TEM (a transmission electron microscope by KS) document in *Shuiyousphaeridium*. Many eukaryotes do exhibit these features in combination....

The formation process of surface ornamentation is suggested to be closely related to an endomembrane system and a cytoskeleton (structural protein filaments). Noteworthy is that the cytoskeleton originated and have been evolved in prokaryotes (e.g., Nath, 2010). In the last decade, homologs of elements of the eukaryotic cytoskeletal proteins have been identified in prokaryotes (e.g. Celler et al., 2013; Shih and Rothfield, 2006), and their involvement in cell growth, cell morphogenesis, cell division, DNA partitioning, and cell motility was implied. Furthermore, new elements specific to prokaryotes have been identified. Such facts may caution us against making an easy employment of criteria for eukaryotic affinity related to cytoskeletons. In either case, if this statement in Knoll et al. (2006) would be taken as criteria for eukaryotic cell recognition, it can be said that lenticular microfossils represent possible eukaryotic cells, because they possess two of the three features (large size and preservable wall). Arguments presented in Javaux and Lepot (2018) on identifying eukaryotic microfossils are also summarized below:

1. Size of microfossils is not be a good criterion, because eukaryotic cells can be very small down to 1–2 μm in diameter, whereas prokaryotic spheroid cells can be large up to 600 μm.

2. Wall ultrastructures such as "being multilayered" and specific wall ornamentation such as μm-scale polygonal reticulation could be reliable criteria.
3. Nonoccasional and nonaccidental opening structure correlative to excystment could be reliable criteria, although some cyanobacterial envelopes can
   be broken in a similar fashion to eukaryotic excystments.
4. Polyethylenic chain indicating preservation of algaenan produced by green
   algae, which can be detected by microanalyses using micro-FTIR (Fourier
   Transform Infrared Spectroscopy), is promising biomaker.
5. Sterane (cyclopentanoperhydrophenanthrenes) diagenetically derived from
   steroids or sterols of eukaryotes is promising only for Neoproterozoic and
   younger rocks and is difficult to be applied to older rocks, due to contamination and maturity problems.

It may be noted here that the above-described criteria and arguments do not mention to the presence/absence of fossilized organelles such as nuclei, chloroplast, and
mitochondria that characterize eukaryotic cells. This is attributed to difficulties in
identifying inner structures of fossil cells as those originated from specific organelles. In other words, similar structures could be produced by degraded and condensed cytoplasm. Objects identified inside coccoid microfossils have been called
intracellular inclusions (ICIs) (Figure 13.4f), whose origins have been extensively
debated. For example, ICIs in three-dimensionally preserved coccoid cells from the
~850 Ma-old Bitter Springs Formation, Australia were once interpreted as fossilized
nuclei (e.g., Schopf and Blacic, 1971), whereas this interpretation was later contested
by, e.g., Knoll and Barghoorn (1975). More recently, Pang et al. (2013) examined ICIs
of unambiguously eukaryotic ornamented microfossils (*Dictyosphaera delicata* and
Shuiyoushaeridium macroreticulatum). It was suggested that the ICIs were neither
fossilized nuclei nor condensed cytoplasm through taphonomy. The authors instead
suggested that the ICIs were formed by contradiction and consolidation of proto-
plasts of eukaryotic cells, leaving the cell wall as vesicles.

Finally, relatively recent works on microfossils of potential eukaryotic affinity (but
controversial) are introduced. Bengston et al. (2017) described filamentous structures
in vesicles and fractures within a basalt from the 2.4 Ga-old Ongeluk Formation,
South Africa. Based on similarity in occurrence and habits of the filaments to younger
mycelial fossils that are assigned to fungi, it was implied that the Ongeluk filamentous structures could put the origin of fungal clade to the Paleoproterozoic. Such
similarity-based approach has been taken also by Kaźmierczak et al. (2016) in order
to elucidate the origin of tubular structures from the 2.8–2.7 Ga-old Sodium Group,
South Africa. The authors claimed that the structures represented fossilized eukaryotic siphonalean organisms (green or yellow-green alga) (Figure 13.6a–c). Both
structures are not composed of carbonaceous matter; they are interpreted to have
been mineralized microfossils. As stressed by Bengston et al. (2017), the absence or
scarcity of OM should not be taken as strong counterevidence against biogenicity.
Unfortunately, however, this fact appears to lower the credibility of their interpretations such as eukaryotic microfossils (Javaux and Lepot, 2018). At this stage, it seems
to be premature to introduce these two Paleoproterozoic and Neoarchean microstructures as the earliest "genuine" eukaryotic microfossils.

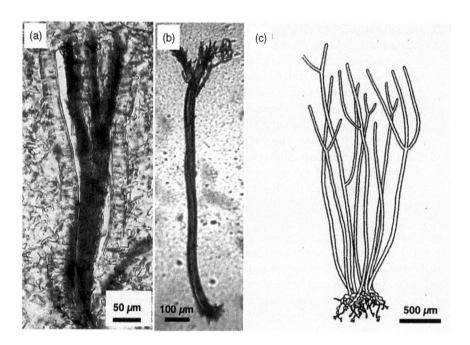

FIGURE 13.6 Tubular microfossils from ~2.8 to ~2.7 Ga-old lacustrine deposits of the Sodium Group of the Ventersdorp Supergroup, Kaapvaal Craton, South Africa and possible modern equivalent. (a) Optical micrograph of a thin section of branching nonseptate structures. The wall is composed of mineral, whereas the inside is kerogenous and dichotomous. (b) Fragments of tubular microfossils obtained from silicified stromatolites by maceration. (c) Drawings of modern marine siphonous green microalgae (*Pseudochlorodesmis furcellata*). (Reproduced from Kaźmierczak et al. (2016), with permission from Elsevier.)

13.3.3 RECORDS OF EUKARYOTIC MACROFOSSILS

In some textbooks, macroscopic fossil named *Grypania spiralis* is introduced as the possible oldest eukaryotic fossil. *Grypania spiralis* was discovered from the early Proterozoic (2.1 Ga-old, but later redated to 1.9 Ga-old) Negaunee Iron Formation, Michigan, North America (Han and Runnegar, 1992) (Figure 13.7a and b). *Grypania spiralis* is a string-like tubular fossil identified on bedding planes. They are 0.7–1.1 mm wide and 9 cm in maximum length. *Grypania spiralis* has been discovered from a wide range of the Proterozoic up to the Ediacaran age (Walter et al., 1976; Wang et al., 2016). Some species can be up to 2 mm in diameter and 50 cm in length (Figure 13.7c). Well-preserved *Grypania* indicates that they originally had a circular cross-section like spaghetti. The tip is rarely rounded, and those of India and China have countless streaks perpendicular to the direction of string growth. They appear to have spiral fibers inside and retains its coiled form during its life and death. *Grypania* from the Ediacaran Doushantuo Formation, South China is interpreted to be anchored or nestled into soft sediments at one end and suspended in the water column for photosynthesis (Wang et al., 2016). As noted above, it seems widely accepted that *Grypania* represents eukaryotic macroalga (e.g., Walter et al., 1990; Butterfield, 2009), although alternative interpretations have also been proposed, including e.g., cyanobacteria (Kumar, 1995).

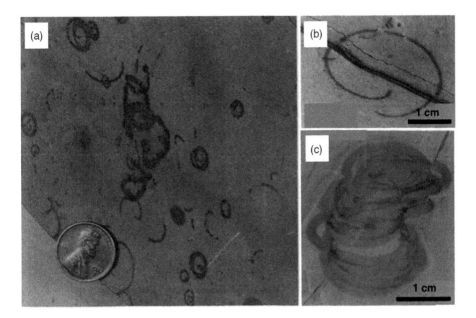

FIGURE 13.7 (a) Bedding surface of the 1.8 Ga Negaunee Iron Formation with fragments of *Grypania* and some thicker filaments. The coin for scale is 18.5 mm in diameter. (b) Enlarged view of Negaunee *Grypania*. (c) A ca. 1.1 Ga-old specimen of *Grypania spiralis* from the Vindhyan supergroup, central India. (Reproduced from Han and Runnegar (1992), with permission from AAAS.)

Different types of possible eukaryotic macrofossils have been described from the ca. 2.1 Ga-old sedimentary successions in Gabon, Western Africa. El Albani et al. (2010) discovered macroscopic fossil-like structures large enough to be visible to the naked eye (Figure 13.8a). The structures (n = 250) have a flat but three-dimensional shape and look like a jellyfish with a distorted shape, with a major axis of 7–120 mm, a minor axis of 5–70 mm, and a thickness of less than 1–10 mm. Most specimens have a streak pattern that extends radially toward the outer edge, with a central spheroid body. They are composed of pyrite, with a trace amount of OM. Based on morphological studies and isotopic compositions of C and S, the structures were interpreted as pyritized body fossils of colonial organisms. The authors further suggested the possibility that the fossilized organisms possessed multicellularity. Although the multicellularity in the wider sense can be seen also for bacteria (e.g., colonial cyanobacteria with differentiated cells), their centimeter size implicitly reminds us that the Gabon fossils are eukaryotic. The authors, on the other hand, cautiously said "The purported significance of the Gabon fossils is thus not multicellularity as such but the evidence they provide for a first appearance in the fossil record of macroscopic individuality" in their follow-up paper (El Albani et al., 2014), in which they provided further evidence for the biogenicity, describing new morphotypes, which include (1) apparently elongated morphotypes related to previously described lobate forms (Figure 13.8b), (2) circular disk-like forms (Figure 13.8c), (3) rounded aggregates of globules, and (4) spheroid acritarchs with compressed vesicle (Figure 13.8d). They were interpreted as fossils, based on the results of examinations of using microtomography,

FIGURE 13.8 Macrofossils from the 2.1 Ga-old black shales from the Francevillian Group, Gabon. (All reproduced from El Albani et al. (2014), with permission of PLOSone.) (a) Pyritize lobate macrofossils with radial fabric and inner pyrite concretions (the arrows). (b) Lobate macrofossil connected with an elongated structure. (c) Nonpyritized to weakly pyritized disk-like structures, composed of a core with radial striction, surrounded by a flange-like thin structure. (d) SEM image of spheroid microfossil (palynomorph) extracted by acid maceration.

isotopic geochemistry, and sedimentological analyses, excluding the possibility that they were formed through abiogenic processes such as nodular pyrite (FeS_2) formation and gas escape. These two studies have recently been followed by the argument that some of the described Gabon macrofossils, string-shaped structures up to 6 mm across and up to 170 mm long, represent multicellular or syncytial organisms with the ability of motility (El Albani et al., 2019). Notably, the authors do not suggest that the fossils represent animals. As a possible modern analog, amoeboid cells that would aggregate together into a migratory slug phase are proposed.

13.4   INTERPRETATION OF THE PILBARA MICROFOSSIL ASSEMBLAGES

In the above sections, we have reviewed records of cyanobacterial and eukaryotic microfossils and macrofossils. In the context of these records, how could the Pilbara microfossil assemblages be explained? I would like to discuss this issue in this section, focusing on spheroid and lenticular microfossils.

13.4.1 BIOLOGICAL AFFINITY OF SPHEROID MICROFOSSILS

As described in the earlier section, the oldest reliable fossil evidence of cyanobacteria has dated back to 1.8 Ga, and possible cyanobacterial microfossils have been identified from the 2.5 Ga-old Campbellrand Supergroup, South Africa, although it seems likely that oxygenic photosynthesis had evolved at least around 2.7 Ga ago, based on the morphological evidence of stromatolites. Molecular clock and geochemical studies also suggest that oxygenic photosynthesis occurred as early as ca. 3.0 Ga ago (Lyons et al., 2014; Planavsky et al., 2014) and ocean redox stratification developed ca. 3.2 Ga ago (Satkoski et al., 2015). Though controversial, the evolution of oxygenic photosynthesis can be dated back to 3.5 Ga (Falcon et al., 2010) and even 3.7 Ga (e.g., Rosing, 1999; Rosing and Frei, 2004). I am not sure if this idea that the oxygenic photosynthesis had evolved as earlier as 3.5 Ga is reliable. Conservatively it is reasonable to assume that this metabolic innovation had occurred around 3.0 Ga ago as discussed in Chapter 6. This is not necessarily in conflict with the widely accepted scenario that atmospheric oxygen began to increase around 2.4 Ga ago (the Great Oxidation Event). Molecular oxygen produced by oxygenic photosynthesis in aquatic environments was not necessarily released immediately to the atmosphere but could have been consumed by reductants such as CH_4, NH_4^+ and Fe^{2+} and possibly heterotrophs.

The depositional environment of the Farrel Quartzite (FQ) fossil-bearing black chert is assumed to have been shallow to subaerial, intermittently evaporitic, closed to semi-closed basin, where the sunlight could have reached the basin bottom (Chapter 9). The identified small and large spheroid microfossils and lenticular microfossils were likely planktonic, which gave them advantage to utilize the sunlight (Chapter 12). Based on highly heterogeneous nitrogen isotopic compositions of individual microfossils including significant positive $\delta^{15}N$ values up to 30 ‰, Delarue et al. (2018) argued that the habitat basin water contained free O_2 (Chapter 4). Such circumstantial evidence supports the hypothesis that the FQ assemblage of the Pilbara microfossils may include cyanobacterial affinities. If similarity-based assignment is allowed, colonial small spheroids morphologically can be correlative to *Microcystis* (Figure 12.3a): note that their carbon isotopic values are not inconsistent with this assignment (House et al., 2013). Though less promising, small spheroids associated with a large spheroid (Figure 12.3c) can be correlative to cyanobacterial *Dermocarpa*, although these two taxa are thought to have evolved from multicellular cyanobacteria later than 2.5 Ga ago (e.g., Schirrmeister et al., 2011, 2015). The cyanobacterial basal lineages include unicellular coccoid cyanobacteria, including planktonic marine or freshwater species of *Synechococcus* and freshwater mat-forming species of *Gloeobacter* (Sánchez-Baracaldo, 2015; Ponce-Toledo et al., 2017). *Gloeomargarita lithophora* is considered to be the present-day closet cultured relative of primary plastids (Ponce-Toledo et a., 2017). Blank and Sánchez-Baracaldo (2010) suggested that cyanobacterial first diversification had occurred in terrestrial, freshwater environments and that early cyanobacteria were planktonic or endolithic. Such implications are not inconsistent with the FQ microfossil assemblage and its assumed habitat environment.

If my assertion that the FQ assemblage includes cyanobacterial affinities could be allowed, how could small spheroids and robust-walled large spheroids from the

Strelley Pool Formation (SPF) be interpreted? The Strelley Pool Formation had deposited around 3.4 Ga-ago, more than 300 Ma older than the Farrel Quartzite, if its present stratigraphic assignment is correct (Chapter 11). Like the FQ cherts, the SPF microfossil-bearing black cherts likely deposited in shallow to subaerial settings, although the three localities represent different environments including shallow marine and possible terrestrial hydrothermal setting. Additionally, the sites of deposition were much more sulfidic compared with the FQ black cherts, probably fueled by local hydrothermal activities. As the oldest estimates of the cyanobacterial evolution are 3.5 Ga or even 3.7 Ga ago, it is not entirely impossible that the SPF spheroids are cyanobacterial. Another possibility is that the SPF spheroids represent anoxygenic phototrophs utilizing hydrogen sulfide (H_2S) or elemental sulfur (S^0), which may be more plausible, considering their assumed sulfide-rich habitats. Recently described species of sulfur bacteria are morphologically similar to some of the microfossil assemblages of the the Strelley Pool Formation (Salman et al., 2011, 2013). Thus, one possible evolutionary scenario is that the SPF spheroid microbes with function of phototrophic sulfur oxidation had evolved to oxygenic phototrophs represented by the FQ spheroid microbes.

Flexible-walled large spherical spheroids of the Farrel Quartzite can be extracted by acid maceration and thus called acritarchs. Similar acritarchs have been reported from the 3.2 Ga-old Moodies Group, South Africa and the 2.5 Ga-old Gamohaan Formation, South Africa. These simple-shaped spheroid acritarchs have been abundantly described from the Proterozoic and the Phanerozoic (Figure 13.9a). They are called *Leiosphaeridia*, which may include cyanobacterial and eukaryotic affinities. However, due to their simple morphology and lack of well-identified excystment openings and ornamentations in many of specimens, their biological affinities have been poorly constrained. While spheroid acritarchs from the Moodies Group were abstemiously suggested to have been of possible cyanobacterial affinity (Javaux et al., 2010), equivalents from the Gamohaan Formation were implied to have been sulfur oxidizing bacteria (Czaja et al., 2016). The latter interpretation cannot be applied to the FQ populations, considering their habitat environment as discussed above. Their cyanobacterial affinity is worth to be considered, although further studies such as isotopic compositions of individual specimens, microstructures of vesicle and detection of biomarkers would be required.

13.4.2   BIOLOGICAL AFFINITY AND SURVIVAL STRATEGY OF LENTICULAR MICROBES

13.4.2.1   Possibility of Eukaryotic Affinities

Lenticular microfossils are morphologically unique, being characterized by spheroid central body surrounded by discoid flange. They range from 20 μm up to 100 μm across. I have described them from the ca. 3.0 Ga-old Farrel Quartzite and the ca. 3.4-old Strelley Pool Formation, Western Australia and the ca. 3.4 Ga-old Kromberg Formation, South Africa. As described earlier, morphologically similar microfossils are possibly present in the 3.2 Ga-old Fig Tree Group, South Africa and possibly in the 3.5 Ga-old Dresser Formation, Western Australia. Morphologically similar microfossils have not yet been reported from younger Archean successions, but are

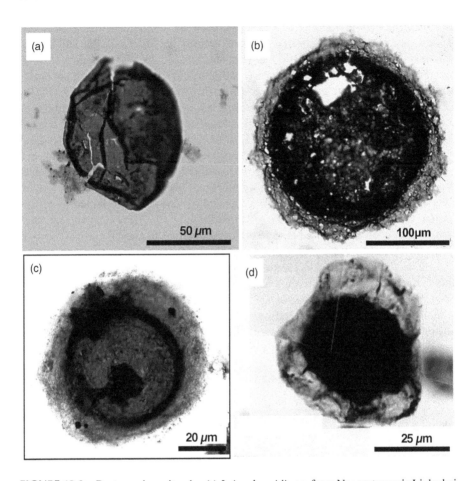

FIGURE 13.9 Proterozoic acritarchs. (a) *Leiosphaeridia* sp. from Neoproterozoic Liulaobei Formation, North China. (b) *Pterospermopsimorpha pileiform* is from the Mesoproterozoic Kaltasy Formation, Russia. (Reproduced from Sergeev et al. (2016), with permission from Elsevier.) (c) *Simia annulare*, from Neoproterozoic lower Shaler Supergroup, Arctic Canada. (Reproduced from Loron et al. (2019), with permission from Elsevier.) (d) *Pterospermella* spp. from the Proterozoic Thule Supergroup, Northwest Greenland. (Reproduced from Samuelsson et al. (1999), with permission from Elsevier.)

relatively common in the late Paleo- to Neoproterozoic successions. Such spheroidal large acritarchs with flange or double wall include three genera *Pterospermella*, *Pterospermopsimorpha*, and *Simia* (e.g., Samuelsson, 1997; Samuelsson et al., 1999; Sergeev et al., 2016; Vorob'eva et al., 2015; Tang et al., 2013; Miao et al., 2019) (Figure 13.9b–d). The two genera *Pterospermella* and *Pterospermopsimorpha* were named after mature phycoma with a flange-like ala of extant prasinophyte, *Pterosperma*. Strikingly well-preserved microfossils equivalent to phycoma of *Pterosperma* were reported from the Lower Devonian Rhynie chert (Kustatscher et al., 2014).

I was once impressed with similarity between lenticular microfossils and mature phycoma of *Pterosperma* and thus considered their possible phylogenic relationship. This consideration was later encouraged by identification of colonies composed of lenses and spheroids and specimen containing small spheroids inside, based on which *Pterosperma*-like lifecycle was reconstructed (Figure 12.1d). However, the assignment to *Pterosperma* now seems to be facile, although the eukaryotic affinity of lenticular microfossils can be still worth to be considered. The reason for this is as follows. Lenticular microbes are likely to have been planktonic, whereas phycoma of *Pterosperma* evolves after losing motility and resting on the sediment surface. *Pterosperma*-like life cycle and related reproduction style of lenticular microbes merely represent one of their diverse life cycles. Lenticular microbes appear to have reproduced by various ways, including binary fission, budding, and multiple fissions, in addition to *Pterosperma*-like reproduction. Reproduced daughter cells were not occasionally liberated to have comprised chain- or sheet-like colony.

Despite the assignment to *Pterosperma* being unlikely, the phylogenic relationship of the Archean lenticular microfossils with the Proterozoic spheroid microfossils *Pterospermella*, *Pterospermopsimorpha*, and *Simia* may be worth to be considered. Although their biological affinities have not yet been firmly constrained, eukaryotic or cyanobacterial affinity of distinctively double-walled *Pterospermopsimorpha* and *Simia* has been suggested (e.g., Sergeev et al., 2016; Vorob'eva et al., 2015, and reference therein). Compared these two genera, *Pterospermella* characterized by equatorial flange, the most similar to Archean lenticular microfossils, is morphologically simpler. However, Samuelsson et al. (1999) suggested that these two genera *(Pterospermella and Pterospermopsimorpha)* are congeneric with each other (Figures 13.9b and d).

The possibility of eukaryotic affinity of lenticular microfossils can be independently considered, based on criteria listed in the earlier section. Lenticular microfossils discovered from the Kaapvaal Craton, South Africa and the Pilbara Craton, Western Australia are large (>20 μm and up to 300 μm) and have acid-resistant vesicles. Although regularly arranged micron-scale ornamentation (polygonal plate comprising cell wall) cannot be identified, textures such as reticulation and striations are found on flange. Flange is far simpler than protrusions of Proterozoic eukaryotic microfossils but can be interpreted as a sort of appendage. Reticulation and striation can be sorts of ornamentation. These textures and structures are believed to have been constructed under a tight control of cytoskeleton. In conclusion, while their eukaryotic affinity cannot be asserted, its possibility should not be discarded. Of course, cyanobacterial affinity of the FQ population cannot be ruled out. Their "potential" affinity to eukaryotic algae or cyanobacteria is not inconsistent with their $\delta^{13}C$ values (Chapter 11, and also see Figure 7.2).

Speaking a little more about the eukaryotic affinity of lenticular microfossils, its possibility should be higher in the FQ population compared with the SPF one, because the depositional basin was likely oxygen oasis in which aerobic respiration was possible (Delarue et al., 2018). This seems to be critical, because regulation of large cells with a sort of protrusion requires a lot of energy, which can be obtained by aerobic respiration. Alternatively, nitrate respiration might have been an alternative. Indeed, some modern unicellular eukaryotes can utilize NO_3^- as an electron acceptor under anerobic conditions (e.g., Merz et al., 2020; Stein, 2018). This may be more realistic, because O_2 produced by cyanobacteria in the FQ "oxygen oasis (e.g., Olson et al., 2013)" was possibly readily utilized by ammonia-oxidizing (nitrifying) bacteria to convert NH_4^+ to NO_2^- as suggested from nitrogenisotopic composition (Chapter 4). Nearly 30 years ago, Finlay et al. (1983) suggested the possibility that primitive eukaryotes employed NO_3^- respiration.

Like spheroid microfossils, the possibility of sulfur bacterial affinity for lenticular microfossils appear to be higher in the SPF population. However, when considering this possibility, we need to mention to what electron acceptors were available. To my knowledge, modern sulfur bacteria utilize O_2 and NO_3^- (as electron acceptor). Their presence was highly possible in the FQ "oxygen oasis" as describe earlier, whereas less expected in the SPF habitats: I have no idea at present how this contradiction can be resolved. This issue should be reminded also for the other morphotypes, though not mentioned in the section 13.4.1.

13.4.2.2 Survival Strategy and Evolutionary Ecology of Lenticular Microbes

As suggested by, e.g., Catling and Clare (2005) and Cockell and Raven (2007), UV radiation was likely much higher in the Archean than today. Organisms must have avoided the risk of gene damage by this harmful radiation. One strategy was to explore subsurface and deep ocean habitats (e.g., Furnes et al., 2004; Homann et al., 2016). It is obvious from thick mafic to ultramafic successions preserved in greenstone belts that volcanic activity during the Paleoarchean was much more active (Lowe, 1999; Van Kranendonk et al., 2006), being attributed to higher mantle temperature (e.g., Herzberg et al., 2010). Surface habitats were often agitated by intense volcanic activities and related hydrothermal activities. Organisms had been repeatedly subjected to physical, temperature, and pH stresses. Identification of spherule beds suggests that the asteroid impact event was more frequent than in later periods (e.g., Lowe et al., 2014; Glikson et al., 2016). It is assumed that fracturing of crusts, huge tsunamis, and seawater boiling and evaporation had been induced by large asteroid impacts (e.g., Lowe, 2013; Lowe and Byerly, 2015). Despite such violent environments, it is evident from studies on biosignatures accumulated in these 50 years that diverse microbes had flourished into shallow to possibly even subaerial environments in addition to deep and subsurface environments during the early Archean (this book; also see Homann, 2019; Lepot. 2020 for review). This fact indicates that the Archean microbes had already evolved various strategies for survival and even prosperity in the harsh environment. In this section, survival strategy and ecological evolution of lenticular microbes are discussed.

To date, lenticular microfossils have been discovered from the 3.0 to 3.4 Ga-old sedimentary successions in the Pilbara and Kaapvaal cratons (Pflug, 1967; Walsh, 1992; Sugitani et al., 2007, 2010, 2013; Oehler et al., 2017). This occurrence implies that the lenticular microbes had an extremely long (at least 400 million years) lifetime as a group and that they were cosmopolitan. On the early Earth where morphologically diverse microorganisms with supposedly diverse metabo-lisms had already inhabited, the lenticular microbes might represent one of the most successful organisms. This success could be understood in the context of planktonic mode of life, large cell size, and robust envelope that characterize len-ticular microfossils.

A planktonic mode of life of microorganisms seems more advantageous for their spreading and exploring new places to inhabit, resulting in establishment of spatially separated populations. It is likely that the planktonicity had provided len-ticular microbes with chances of survival as species and higher taxonomic groups. Furthermore, opportunities of diversification as a result of adaptation to different environments and of geographic separation could have been given. Suspension in a photic zone was favorable or rather prerequisite for planktonic photoautotrophy, although some chemoautotrophs are also planktonic (e.g., Hadas and Pinkas, 2001). In photoautotrophy, organisms could utilize virtually infinite energy from the sun. On the other hand, the large cell size of lenticular microbes had some disadvantages. As predicted from Stoke's law, smaller cells tend to sediment at lower velocity than large cells (e.g., Sciascia et al., 2013). Namely, smaller cells could be suspended longer within photic zone. In addition, small cells could utilize nutrients more effec-tively than larger cells by diffusion {small package effect; Raven (1998)}, though this effect cannot be applied to eukaryotes. As demonstrated by Peperzak et al. (2003), however, the sedimentation rates of phytoplankton cells and colonies could not be constrained exclusively by their size. Density and morphological complexity are also important factors controlling sedimentation velocity. Rather, density is the most important factor (Peperzak et al., 2003; Padisák et al., 2003). Disadvantage from large cell size could be compensated by several ways. One way is to have vacuole that stores nutrients or metabolic products. If stored materials are of low density, as represented by gas, compared with cytoplasm, the overall density would be lowered, providing buoyancy. Another way is departure from sphericity (morphological com-plexity), as demonstrated by our sedimentation simulations. In addition, the more complex the morphology is, the lager the surface to volume ratio (S/V) becomes. A larger S/V ratio of the cell is more effective for nutrient uptake through a membrane (e.g., Salman et al., 2011). From these perspectives, the ecological evolution of len-ticular microbe is discussed below.

Evolution of robust wall might have been key and fundamental adaptation of lenticular microbes to harsh environment. This evolution was however a double-edged sword, because the robustness likely increased the cell density. Cellulose is known to make up the rigid walls of plant cells with a density of 1.5 g/cm^3. Cell wall densities of some eubacteria range from 1.18 g/cm^3 to 1.55 g/cm^3 (Scherrer et al., 1977 and references therein). These values are significantly higher than protoplasm (1.02–1.06 g/cm^3) and cholesterol fraction of plasma membranes (1.05 g/cm^3). Such deficit could have been compensated by enlargement of cell size associated with vacuolation. Thus, development of vacuoles filled with fluids or gases could have increased buoyancy and at the same time possessed capacity of nutrient storage. Obviously, the morphology of lenticular microfossils is significantly departed from a complete sphere, resulting in an increase in their S/V ratios. This might have helped nutrient uptake. As described in detail in Chapter 12, our sedimentation simulation demonstrated that the flange works more effectively for larger cells in lowering sedimentation velocity. The ratio of sedimentation velocity of flanged spheroid with a projected area of 1,000 μm^2 against equivalent sphere is 0.931, whereas that with a projected area of 4,000 μm^2 is 0.924. Though small (0.6%), this difference should not be overlooked, because the population was so large that such small difference could have significantly increased their opportunity to explore new habitats.

In summary, the combination of robustness, large cell size, and flange possibly had provided lenticular microbes with an ideal architecture for prosperity and survival. While physical protection against environmental stress could have been provided by the robustness of cell envelope, the lenticular microbes retained their planktonicity suitable for prosperity, escape from environmental crisis and possible photoautotrophy. This had been obtained by reduction of the cell density through cell enlargement and vacuolation in some case and evolution of flange. Finally, I would like to mention the most recent discovery of new morphotype of lenticular microfossil from the Panorama locality 2 in the Strelley Pool Formation (Delarue et al., 2021). The specimens have tail-like appendage and are suggested to have had active motility. Some species of lenticular microbes likely had already evolved a new tool for escaping from the harsh environment and exploring new habitats.

13.5 FUTURE DIRECTIONS OF THE PILBARA MICROFOSSILS

As shown in this chapter, the biological affinities of the Pilbara microfossils have been still poorly constrained, despite over 15 years studies into which over 20 researchers from six countries have been involved. In addition, there have been several unresolved problems. In this section, some of the unresolved problems are presented for future studies.

13.5.1 REEXAMINATION OF VERTICALLY TO SUBVERTICALLY ORIENTED COLUMNAR GIANT CRYSTALS

As described in Chapter 5, silicification is known to be a widespread phenomenon observed in Archean volcanic sedimentary successions, and in fact, I have observed many examples of silicification of minerals, clasts, and rocks in the Pilbara Cratons, some of which have long bothered me. Susceptivity to silicification is various depending on mineral species, and soluble evaporitic minerals are the most vulnerable to silicification. Thus, extensively silicified minerals are likely evaporitic in origin. The identification of their original mineralogy is indispensable, because evaporitic minerals could be environmental proxies. I have identified original mineralogies of some silicified crystals in the Farrel Quartzite based on occurrences, crystal habits, and/or analyses of inclusions. Based on data of interfacial angles (n = 715), vertically to subvertically oriented giant (~30 cm long) columnar crystals in CE1 at Mount Goldsworthy (Figure 9.4a) were identified as nahcolite ($NaHCO_3$) (Sugitani et al., 2003), which appeared to be consistent with the occurrence of nahcolite within the Barberton Supergroup, South Africa (Lowe and Fisher-Worrell, 1999).

I have also suggested that morphologically similar silicified crystals in beds closely associated with microfossil-bearing carbonaceous cherts within CE2 were nahcolite in analogy. However, a different interpretation (aragonite) has been given for similar-looking silicified giant crystals in, e.g., the 3.2 Ga Cleaverville Formation (Kiyokawa et al., 2006) in the Pilbara Craton. More recently, Otálora et al. (2018) examined similar-looking crystal pseudomorphs and crystal casts from the 3.48 Ga-old Dresser Formation in the Pilbara Craton, Western Australia using X-ray microtomography, energy-dispersive X-ray spectroscopy, and crystallographic methods and suggested that they were most likely originally aragonite. This is probably the case, and similar analyses may need to be applied to the FQ materials, particularly for those in CE2 hosting microfossil-bearing black cherts. This reexamination is preferably operated in the basin scale, because different evaporitic minerals could occur at different sites in the same basin and the combination of evaporitic minerals could give crucial constraints on the chemistry of water mass from which they were precipitated.

13.5.2 POTENTIAL OF BIOMARKER ANALYSES OF MICROFOSSILS

The detection of a biomarker that can specify biological affinity is the most coveted. As can be seen from arguments on syngenicity of biomarkers such as hopanes and steranes from 2.7 Ga-old shales (French et al., 2015 and references therein), it is hard to prove the syngenicity of biomarkers even though they are detected. In particular, biomarkers have generally been thought to be very trace in Paleo- to Mesoarchean rocks, which is inferred from low concentrations of organic C typically lower than 0.1 wt.% and from generally higher thermal maturation due to metamorphism up to greenschist facies. However, as suggested from the recent study of Alleon et al. (2018), better preservation of nitrogen- and oxygen-containing OM in the 3.4 Ga-old microfossils, compared with the 1.9 Ga-old Gunflint microfossils, should not be overlooked. This was explained by very early silicification (particularly entombment type) and lower permeability of host cherts. It was speculated that timing and extensive permeability loss might be a more critical factor than temperature for the molecule-level preservation of OM comprising microfossils (Alleon et al., 2018). In addition, as can be seen from macerated specimens, the preservation status of microfossils is various. Though rare, hyaline-walled microfossils are present. Thus, in my opinion, there still remains the chance that we could detect some syngenetic biomarkers, although we need a breakthrough of the technical side of things.

13.5.3 TAXONOMY OF THE LENTICULAR MICROFOSSILS AND LIFE HISTORY OF LENTICULAR MICROBES

As shown in Chapter 12, morphometry-based taxonomy was successfully done for the SPF lenticular microfossils (Sugitani et al., 2018). A similar taxonomy based on qualitative analyses of morphology has been coveted for the FQ lenticular microfossil population, although it has not yet been fully achieved. This is not because of my procrastination but because FQ population exhibit far more complex occurrences of colonies than the SPF lenticular populations (Sugitani, 2012, also see Chapter 12). This complexity likely reflects various life cycles. If the reconstruction of life cycle is combined with morphometry, taxonomy can be proposed in higher level of reliability. Unfortunately, maceration indispensable for precise morphometric analyses could destroy structures of colonies that are composed of detached individuals; it is difficult to relate morphometric data of individuals to types of colonies. However, as revealed by recent macerations, connected types of colony can be extracted without destruction (Figure 12.6a and c). Combination of morphometry of individual lenticular microfossils (oblateness, whole area and flange ratio) and the structures of such colonies could provide us with a firm basis for taxonomy of these types of lenticular microfossils. Furthermore, if the primary origins of various flange textures (hyaline, striated, fibrillar, and reticulated) would be successfully proved, these traits could be involved in the analyses, enabling us to establish the sophisticated taxonomy.

Along with proposing the taxonomy and its refinement, it is desirable to make clear whether some lenticular structures described by Wacey et al. (2018) from the 3.48 Ga-old Dresser Formation, Western Australia are really volcanic vesicles or microfossils. Furthermore, rediscovery of lenticular microfossils of the 3.2 Ga-old Fig Tree Group and the 3.4 Ga-old Kromberg Formation, South Africa is indispensable. The former was reported by Pflug (1967) more than 50 years ago. Published images are very poor, but they must be lenticular microfossils. The latter, which was first described by Walsh (1992), was recently rediscovered by Oehler et al. (2017) and by the author. However, all the identified specimens are contained in detrital phases and thus allochthonous, and their colonial occurrences have not yet been described. The discovery of autochthonous specimens is needed. Although the age of "the Farrel Quartzite" still has some ambiguities, it is possible that the life history of lenticular microfossils had continued at least 500 million years, or more, as discussed above. The whole picture of lenticular microfossils could only be correctly understood by revealing its long history.

COLUMN: ENDOSYMBIOTIC THEORY

Endosymbiotic theory is a theory proposed by Lynn Margulis to explain the origin of eukaryotic cells in 1970, although before her proposal, some researchers had already suggested that mitochondria and chloroplast, had originated from symbionts. It is now widely accepted that these organelles had been originated from free-living aerobic α-proteobacteria and cyanobacteria. This is well evidenced by the facts that these organelles have their own genes on a single circular DNA like prokaryotes and proliferate semiautonomously by division. Furthermore, chloroplasts in eukaryotic cells and cyanobacteria possess the fundamental similarities in their photosystems that are composed of Photosystems I and II. Obviously, oxygenic photosynthesis is limited to cyanobacteria and eukaryotes, although photosynthesis is distributed among wide clades of prokaryotes. Every organelle of the eukaryotic cells have once been suggested to be originated from endosymbiosis, including, e.g., flagellum, peroxisome, and even nucleus (Martin et al., 2015 and references therein). At present, however, this notion has been taken to be invalid (Archibald, 2015).

In the strict sense, this endosymbiosis should be called primary endosymbiosis. Secondary endosymbiosis, i.e., engulfment of a eukaryotic cell by another eukaryotic cell, is thought to have occurred several times and contributed to diversification of eukaryotes (Bhattacharya et al., 2004). Flagellate protist named *Hatena* is known to represent unique possibly on-going secondary endosymbiosis (Okamoto and Inouye, 2005). Almost all the cells of its natural populations have "a green plastid", which however would be inherited by only one daughter cell. Another daughter cell newly develops an apparatus for feeding and engulfs a free-living unicellular alga, which would retain some organelles such as nucleus, mitochondria, and plastid after engulfment, whereas lose flagella, cytoskeleton, and endomembrane system, becoming "a green alga".

REFERENCES

Alleon, J., Bernard, S., Le Guillou, C., Beyssac, O., Sugitani, K., Robert, F. 2018. Chemical nature of the 3.4 Ga Strelley Pool microfossils. *Geochemical Perspectives Letters* 7, 37–42. doi:10.7185/geochemlet.1817.

Altermann, W., Kazmierczak, J. 2003. Archean microfossils: A reappraisal of early life on Earth. *Research in Microbiology* 154, 611–617.

Altermann, W., Schopf, J.W. 1995. Microfossils from the Neoarchean Campbell Group, Griqualand West Sequence of the Transvaal Supergroup, and their paleoenvironmental and evolutionary implications. *Precambrian Research* 75, 65–90.

Amard, B., Bertrand-Sarfati, J. 1997. Microfossils in 2000 Ma old cherty stromatoltes of the Franceville Group, Gabon. *Precambrian Research* 81, 197–221.

Archibald, J.M. 2015. Endosymbiosis and eukaryotic cell evolution. *Current Biology* 25, R911–R921.

Awramik, S.M., Schopf, J.W., Walter, M.R. 1983. Filamentous fossil bacteria from the Archean of Western Australia. *Precambrian Research* 20, 357–374.

Bengston, S., Rasmussen, B., Ivarsson, M., Muhling, J., Broman, C., Marone, F., Stampanoni, M., Bekker, A. 2017. Fungus-like mycelial fossils in 2.4-billion-year-old vesicular basalt. *Nature Ecology & Evolution* 1, 0141.

Bhattacharya, D., Yoon, H.S., Hackett, J.D. 2004. Photosynthetic eukaryotes unite: endosymbiosis connects the dots. *BioEssays* 26, 50–60.

Blank, C.E., Sánchez-Baracaldo, P. 2010. Timing of morphological and ecological innovations in the cyanobacteria – a key to understanding the rise in atmospheric oxygen. *Geobiology* 8, 1–23.

Brocks, J.J., Logan, G.A., Buick, R., Summons, R.E. 1999. Archean molecular fossils and the early rise of eukaryotes. *Science* 285, 1033–1036.

Brocks, J.J., Love, G.D., Summons, R.E., Knoll, A.H., Logan, G.A., Bowden, S.A. 2005. Biomarker evidence for green and purple sulphur bacteria in a stratified Palaeoproterozoic sea. *Nature* 437, 866–870.

Brown, I.I., Bryant, D.A., Casamatta, D., Thomas-Keprta, K.L., Sarkisova, S.A., Shen, G., Graham, J.E., Boyd, E.S., Peters, J.W., Garrison, D.H., McKay, D.S. 2010. Polyphasic characterization of a thermotolerant siderophilic filamentous cyanobacterium that produces intracellular iron deposits. *Applied and Environmental Microbiology* 76, 6664–6672.

Butterfield, N.J. 2009. Modes of pre-Ediacaran multicellularity. *Precambrian Research* 173, 201–211.

Catling, D.C., Claire, M.W. 2005, How Earth's atmosphere evolved to an oxic state: A status report. *Earth and Planetary Science Letters* 237, 1–20

Celler, K., Koning, R.I., Koster, A.J., van Wezel, G.P. 2013. Multidimensional view of the bacterial cytoskeleton. *Journal of Bacteriology* 195, 1627–1636.

Cockell, C.S., Raven J.A. 2007, Ozone and life on the Archaean Earth. *Philosophical Transactions of The Royal Society A* 365, 1889–1901.

Couradeau, E., Benzerara, K., Gérard, E., Moreira, D., Bernard, S., Brown Jr., G.E., López-García, P. 2012. An early-branching microbialite cyanobacterium forms intracellular carbonates. *Science* 336, 459–462.

Czaja, A.D., Beukes, N.J., Osterhout, J.T. 2016. Sulfur-oxidizing bacteria prior to the Great Oxidation Event from the 2.52 Ga Gamohaan Formation of South Africa. *Geology* 44, 983–986.

Delarue, F., Robert, F., Sugitani, K., Tartèse, R., Duhamel, R., Derenne, S. 2018. Nitrogen isotope signatures of microfossils suggest aerobic metabolism 3.0 Gyr ago. *Geochemical Perspectives Letters* 7, 32–36.

Delarue, F., Bernard, S., Sugitani, K., Robert, F., Tartèse, R., Albers, S.-V., Duhamel, R., Pont, S., Derenne, S. 2021. Microfossils with tail-like structures in the 3.4 Gyr old Strelley Pool Formation. *Precambrian Research* 358, 106187.

Demoulin, C.F., Lara, Y.J., Cornet, L., François, C., Baurain, D., Wilmotte, A., Javaux, E.J. 2019. Cyanobacteria evolution: Insight from the fossil record. *Free Radical Biology and Medicine* 140, 206–223.

El Albani, A., Bengtson, S., Canfield, D.E., Bekker, A., Macchiarelli, R., Mazurier, A., Hammarlund, E.U., Boulvais, P., Dupuy, J.-J., Fontaine, C., Fürsich, F.T., Ganthier-Lafaye, F., Janvier, P., Javaux, E., Ossa, F.O., Pierson-Wickmann, A.-C., Riboulleau, A., Sardini, P., Vachard, D., Whitehouse, M., Meunier, A. 2010. Large colonial organisms with coordinated growth in oxygenated environments 2.1 Gyr ago. *Nature* 466, 100–104.

El Albani, A., Bengtson, S., Canfield, D.E., Riboulleau, A., Bard, C.R., Macchiarelli, R., Pemba, L.N., Hammarlund, E., Meunier, A., Mouele, I.M., Benzerara, K., Bernard, S., Boulvais, P., Chaussidon, M., Cesari, C., Fontaine, C., Chi-Fru, E., Garcia-Ruiz, J.M., Gauthier-Lafaye, F., Mazurier, A., Pierson-Wickmann, A.C., Rouxel, O., Trentesaux, A., Vecoli, M., Versteegh, G.J.M., White, L., Whitehouse, M., Bekker, A. 2014. The 2.1 Ga old Francevillian Biota: Biogenicity, taphonomy and biodiversity. *PLoS One* 9, e00438. doi:10.1371/journal.pone.0099438.

El Albani, A., Mangano, M.G., Buatois, L.A., Bengtson, S., Riboulleau, A., Bekker, A., Konhauser, K., Lyons, T., Rollion-Bard, C., Bankole, O., Baghekema, S.G.L., Meunier, A., Trentesaux, A., Mazurier, A., Aubineau, J., Laforest, C., Fontaine, C., Recourt, P., Fru, E.C., Macchiarelli, R., Reynaud, J.Y., Gauthier-Lafaye, F., Canfield, D.E. 2019. Organism motility in an oxygenated shallow-marine environment 2.1 billion years ago. *Proceedings of National Academy of Sciences of the United States of America* 116, 3431–3436.

Falcon, L.I., Magallon, S., Castillo, A. 2010. Dating the cyanobacterial ancestor of the chloroplast. *The ISME Journal* 4, 777–783.

Finlay, B.J., Span, A.S.W., Harman, J.M.P. 1983. Nitrate respiration in primitive eukaryotes. *Nature* 303, 333–336.

Flannery, D.T., Walter, M.R. 2011. Archean tufted microbial mats and the Great Oxidation Event: new insights into an ancient problem. *Australian Journal of Earth Sciences* 59, 1–11. doi:10.1080/08120099.2011.607849.

French, K.L., Hallmann, C., Hope, J.M., Schoon, P.L., Zumberge, J.A., Hoshino, Y., Peters, C.A., George, S.C., Love, G.D., Brocks, J.J., Buick, R., Summons, R.E. 2015. Reappraisal of hydrocarbon biomarkers in Archean rocks. *Proceedings of the National Academy of Sciences of the United States of America* 112, 5915–5920.

Furnes, H., Banerjee, N.R., Muehlenbachs, K., Staudigel, H., de Wit, M. 2004. Early life recorded in Archean pillow lavas. *Science* 304, 578–581.

Glikson, A., Hickman, A., Evans, N.J., Kirkland, C.L., Park, J.-W., Rapp, R., Romano, S. 2016. A new ~3.46 Ga asteroid impact ejecta unit at Marble Bar, Pilbara Craton, Western Australia: A petrological, microprobe and laser ablation ICPMS study. *Precambrian Research* 279, 103–122.

Golubic, S., Hofmann, H.J. 1976. Comparison of holocene and mid-precambrian entophysalidaceae (cyanophyta) in stromatolitic algal mats: cell division and degradation. *Journal of Paleontology* 50, 1074–1082.

Golubic, S., Sergeev, V.N., Knoll, A.H. 1995. Mesoproterozoic *Archaeoellipsoides*: akinetes of heterocystous cyanobacteria. *Lethaia* 28, 285–298.

Gueneli, N., Mckenna, A.M., Ohkouchi, N., Boreham, C.J., Beghin, J., Javaux, E.J., Brocks, J.J. 2015. 1.1-billion-year-old porphyrins establish a marine ecosystem dominated by bacterial primary producers. *Proceedings of National Academy of Sciences of the United States of America* 115, E6978–E6986.

Hadas, O., Pinkas, R. 2001. High chemoautotrophic primary production in Lake Kinneret, Israel: A neglected link in the carbon cycle of the lake. *Limnology and Oceanography* 46, 1968–1976.

Han, T.M., Runnegar, B. 1992. Megascopic eukaryotic algae from the 2.1-billion-year-old Negaunee Iron-Formation, Michigan. *Science* 257, 232–235.

Herzberg, C., Condie, K., Korenaga, J. 2010. Thermal history of the Earth and its petrological expression. *Earth and Planetary Science Letters* 292, 79–88.

Hofmann, H.J. 1976. Preambrian microflora, Belcher Islands, Canada: significance and systematics. *Journal of Paleontology* 50, 1040–1073.

Homann, M. 2019. Earliest life on Earth: Evidence from the Barberton Greenstone Belt, South Africa. *Earth-Science Reviews* 196, 102888.

Homann, M., Heubeck, C., Bontognali, T.R.R., Bouvier, A.S., Baumgartner, L.P., Airo, A. 2016. Evidence for cavity-dwelling microbial life in 3.22 Ga tidal deposits. *Geology* 44, 51–54.

House, C.H., Oehler, D.Z., Sugitani, K., Mimura, K. 2013. Carbon isotopic analyses of ca. 3.0 Ga microstructures imply planktonic autotrophs inhabited Earth's early oceans. *Geology* 41, 651–654.

Javaux, E.J., 2007. The Early Eukaryotic Fossil Record. In *Eukaryotic Membranes and Cytoskeleton. Advances in Experimental Medicine and Biology* 607, 1–19. Springer, New York. doi:10.1007/978-0-387-74021-8_1.

Javaux, E.J., Knoll, A.H., Walter, M.R. 2001. Morphological and ecological complexity in early eukaryotic ecosystems. *Nature* 412, 66–69.

Javaux, E.J., Knoll, A.H., Walter, M.R. 2004. TEM evidence for eukaryotic diversity in mid-Proterozoic oceans. *Geobiology* 2, 121–132.

Javaux, E.J., Marshall, C.P., Bekker, A. 2010. Organic-walled microfossils in 3.2-billion-year-old shallow-marine siliciclastic deposits. *Nature* 463, 934–938.

Javaux, E.J., Lepot, K. 2018. The Paleoproterozoic fossil record: Implications for the evolution of the biosphere during Earth's middle-age. *Earth-Science Reviews* 176, 68–86.

Kazmierczak, J., Altermann, W. 2002. Neoarchean biomineralization by benthic cyanobacteria. *Science* 298, 2351.

Kaźmierczak, J., Kremer, B., Altermann, W., Franchi, I. 2016. Tubular microfossils from ~2.8 to 2.7 Ga-old lacustrine deposits of South Africa: A sign for early origin of eukaryotes? *Precambrian Research* 286, 180–194.

Kaźmierczak J., Kremer, B. 2019. Pattern of cell division in ~3.4 Ga-old microbes from South Africa. *Precambrian Research* 331, 105357.

Kiyokawa, S., Ito, T., Ikehara, M., Kitajima, F. 2006. Middle Archean volcano-hydrothermal sequence: Bacterial microfossil-bearing 3.2 Ga Dixon Island Formation, coastal Pilbara terrane, Australia. *Geological Society of America Bulletin* 118, 3–22.

Klein, C., Beukes, N.J., Schopf, J.W. 1987. Filamentous microfossils in the early Proterozoic Transveaal Supergroup: Their morphology, significance, and paleoenvironmental setting. *Precambrian Research* 36, 81–94.

Knoll, A.H., Barghoorn, E.S. 1975. Precambrian eukaryotic organisms: a reassesment of the evidence. *Science* 190. 52–54.

Knoll, A.H., Strother, P.K., Rossi, S. 1988. Distribution and diagenesis of microfossils from the Lower Proterozoic Duck Creek Dolomite, Western Australia. *Precambrian Research* 38, 257–279.

Knoll, A.H., Javaux, E.J., Hewitt, D., Cohen, P. 2006. Eukaryotic organisms in Proterozoic oceans. *Philosophical Transactions of the Royal Society B* 361, 1023–1038.

Kremer, B., Kaźmierczak, J. 2017. Cellularly preserved microbial fossils from ~3.4 Ga deposits of South Africa: A testimony of early appearance of oxygenic life? *Precambrian Research* 295, 117–129.

Kumar, S. 1995. Megafossils from the Mesoproterozoic Rohtas Formation (the Vindhyan Supergroup), Katni area, central India. *Precambrian Research* 72, 171–184.

Kustatscher, E., Dotzler, N., Taylor, T.N., Krings, M. 2014. Microfossils with suggested affinities to the Pyramimonadales (Pyramimonadophyceae, Chlorophyta) from the Lower Devonian Rhynie chert. *Acta Palaeobotanica* 54, 163–171.

Lamb, D.M., Awramik, S.M., Chapman, D.J., Zhu, S. 2009. Evidence for eukaryotic diversification in the ~1800 million-year-old Changzhougou Formation, North China. *Precambrian Research* 173, 93–104.

Lepot, K., Compère, P., Gérard, E., Namsaraev, Z., Verleyen, E., Tavernier, I., Hodgson, D.A., Vyverman, W., Gilbert, B., Wilmotte, A., Javaux, E.J. 2014. Organic and mineral imprints in fossil photosynthetic mats of an East Antarctic lake. *Geobiology* 12, 424–450.

Lepot, K. 2020. Signatures of early microbial life from the Archean (4 to 2.5 Ga) eon. *Earth-Science Reviews* 209, 103296.

Lepot, K., Williford, K.H., Ushikubo, T., Sugitani, K., Mimura, K., Spicuzza, M.J., Valley, J.W. 2013. Texture-specific isotopic compositions in 3.4 Gyr old organic matter support selective preservation in cell-like structures. *Geochimica et Cosmochimica Acta* 112, 66–86.

Li, H., Lu, S., Su, W., Xiang, Z., Zhou, H., Zhang, Y. 2013. Recent advances in the study of the Mesoproterozoic geochronology in the North China Craton. *Journal of Asian Earth Sciences* 72, 216–227.

Loron, C.C., Rainbird, R.H., Turner, E.C., Greenman, J.W., Javaux, E.J. 2019. Organic-walled microfossils from the late Mesoproterozoic to early Neoproterozoic lower Shaler Supergroup (Arctic Canada): Diversity and biostratigraphic significance. *Precambrian Research* 321, 349–374.

Lowe, D.R. 1999. Geologic evolution of the Barberton greenstone belt and vicinity. *Geological Society of America Special paper* 329, 287–312. doi:10.1130/0-8137-2329-9.287.

Lowe, D.R., Fisher-Worrell, G. 1999. Sedimentology, mineralogy, and implications of silicified evaporites in the Kromberg Formation, Barberton Greenstone Belt, South Africa. *Geological Society of America Special Paper* 329, 167–188. doi:10.1130/0-8137-2329-9.167.

Lowe, D.R. 2013. Crustal fracturing and chert dike formation triggered by large meteorite impacts, ca. 3.260 Ga, Barberton greenstone belt, South Africa. *Geological Society of America Bulletin* 125, 894–912.

Lowe, D.R., Byerly, G.R. 2015. Geologic record of partial ocean evaporation triggered by giant asteroid impacts, 3.29–3.23 billion years ago. *Geology* 43, 535–538.

Lowe, D.R., Byerly, G.R., Kyte, F.T. 2014. Recently discovered 3.42–3.23 Ga impact layers, Barberton Belt, South Africa: 3.8 Ga detrital zircons, Archean impact history, and tectonic implications. *Geology* 42, 747–750.

Lyons, T.W., Reinhard, C.T., Planavsky, N.J. 2014. The rise of oxygen in Earth's early ocean and atmosphere. *Nature* 506, 307–315.

Martin, W., Garg, S., Zimorski, V. 2015. Endosymbiotic theories for eukaryote origin. *Philosophical Transactions of the Royal Society of London B* 370, 20140330.

Merz, E., Dick, G.J., de Beer, D., Grim, S., Hübener, T., Littmann, S., Oslen, K., Stuart, D., Lavik, G., Marchant, H.K., Klatt, J.M. 2020. Nitrate respiration and diel migration patterns of diatoms are linked in sediments underneath a microbial mat. *Environmental Microbiology* 23, 1422–1435. doi:10.1111/1462–2920.15345.

Miao, L., Moczydłowska, M., Zhu, S., Zhu, M. 2019. New record of organic-walled, morphologically distinct microfossils from the late Paleoproterozoic Changcheng Group in the Yanshan Range, North China. *Precambrian Research* 321, 172–198.

Nath, D. 2010. The prokaryotic cytoskeleton. *Nature Reviews Molecular Cell Biology* 9, s19.

Oehler, D.Z., Walsh, M.M., Sugitani, K., Liu, M.-C., House, C.H. 2017. Large and robust lenticular microorganisms on the young Earth. *Precambrian Research* 296, 112–119.

Okamoto, N., Inouye, I. 2005. A secondary symbiosis in progress? *Science* 310, 287.

Olson, S.L., Kump, L.R., Kasting, J.F. 2013. Qualifying the areal extent and dissolved oxygen concentrations of Archean oxygen oasis. *Chemical Geology* 362, 35–43.

Otálora, F., Mazurier, A., Garcia-Ruiz, J.M., Van Kranendonk, M.J., Kotopoulou, E., El Albani, A., Garrido, C.J. 2018. A crystallographic study of crystalline casts and pseudomorphs from the 3.5 Ga Dresser Formation, Pilbara Craton (Australia). *Journal of Applied Crystallography* 51, 1050–1058.

Padisák, J., Soróczki-Pintér, É., Rezner, Z. 2003. Sinking properties of some phytoplankton shapes and the relation of form resistance to morphological diversity of plankton – An experimental study. *Hydrobiologia* 500, 243–257.

Pang, K., Tang, Q., Schiffbauer, J.D., Yao, J., Yuan, X., Wan, B., Chen, L., Ou, Z., Xiao, S. 2013. The nature and origin of nucleus-like intracellular inclusions in Paleoproterozoic eukaryote microfossils. *Geobiology* 11, 499–510.

Pang, K., Tang, Q., Chen, L., Wan, B., Niu, C., Yuan, X., Xiao, S. 2018. Nitrogen-fixing heterocystous cyanobacteria in the Tonian period. *Current Biology* 28, 616–622.

Peng, Y., Bao, H., Yuan, X. 2009. New morphological observations for Paleoproterozoic acritarchs from the Chuanlinggou Formation, North China. *Precambrian Research* 168, 223–232.

Peperzak, L., Coljin, F., Koeman, R., Gieskes, W.W.C., Joordens, J.C.A. 2003. Phytoplankton sinking rates in the Rhine region of freshwater influence. *Journal of Plankton Research* 25, 365–383. doi:10.1093/plankt/25.4.365.

Pflug, H.D. 1967. Structured organic remains from the Fig Tree Series (Precambrian) of the Barberton Mountain Land (South Africa). *Review of Paleobotany and Palynology* 5, 9–29.

Planavsky N.J., Asael, D, Hofmann, A., Reinhard, C.T., Lalonde, S.V., Knudsen, A., Wang, X., Ossa, F.O., Pecoits, E., Smith, A.J.B., Beukes, N.J., Bekker, A., Johnson, T.M., Konhauser, K.O., Lyons, T.W., Rouxel, O.J. 2014. Evidence for oxygenic photosynthesis half a billion years before the Great Oxidation Event. *Nature Geoscience* 7, 283–286.

Ponce-Toledo, R.I., Deschamps, P., López-García, P., Zivanovic, Y., Benzerara, K., Moreira, D. 2017. An early-branching freshwater cyanobacterium at the origin of plastids. *Current Biology* 27, 386–391.

Raven, J.A. 1998. The twelfth Tansley Lecture. Small is beautiful: the picophytoplankton. *Functional Ecology*, 12, 503–513.

Rosing, M.T. 1999. ^{13}C-depleted carbon microparticles in >3700-Ma sea-floor sedimentary rocks from West Greenland. *Science* 283, 674–676.

Rosing, M.T., Frei, R. 2004. U-rich Archaean sea-floor sediments from Greenland – indications of > 3700 Ma oxygenic photosynthesis. *Earth and Planetary Science Letters* 217, 237–244.

Salman, V., Amann, R., Girnth, A.-C., Polerecky, L., Bailey, J.V., Høgslund, S., Jessen, G., Pantoja, S., Schultz-Vogt, H.N. 2011. A single-cell sequencing approach to the classification of large, vacuolated sulfur bacteria. *Systematic and Applied Microbiology.* 34, 243–259.

Salman, V., Bailey, J.V., Teske, A. 2013. Phylogenetic and morphologic complexity of giant sulphur bacteria. *Antonie van Leeuwenhoek* 104, 169–186.

Samuelsson, J. 1997. Biostratigraphy and paleobiology of Early Neoproterozoic strata of the Kola Peninsula, Northwest Russia. *Norsk Geologisk Tidsskrift* 77, 165–192.

Samuelsson, J., Dawes, P.R., Vidal, G. 1999. Organic-walled microfossils from the Proterozoic Thule Supergroup, Northwest Greenland. *Precambrian Research* 96, 1–23.

Sánchez-Baracaldo, P. 2015. Origin of marine planktonic cyanobacteria. *Scientific Reports* 5, 17418.

Satkoski, A.M., Beukes, N.J., Li, W., Beard, B.L., Jonson, C.M. 2015. A redox-stratified ocean 3.2 billion years ago. *Earth and Planetary Science Letters* 430, 43–53.

Scherrer, R., Berlin, E., Gerhardt P. 1977. Density, porosity, and structure of dried cell walls isolated from *Bacillus megaterium* and *Saccharomyces cerevisiae*. *Journal of Bacteriology* 129, 1162–1164.

Schirrmeister, B.E., Antonelli, A., Bagheri, H.C. 2011. The origin of multicellularity in cyanobacteria. *BMC Evolutionary Biology* 11, 45. http://www.biomedcentral.com/1471-2148/11/45.

Schirrmeister, B.E., Gugger, M., Donoghue, P.C.J. 2015. Cyanobacteria and the Great Oxidation Event: Evidence from genes and fossils. *Palaeontology* 58, 769–785.

Schopf, J.W. 1993. Microfossils of the Early Archean Apex Chert: New evidence of the antiquity of life. *Science* 260, 640–646.

Schopf, J.W., Blacic, J.M. 1971. New microorganisms from the Bitter Springs Formation (late Precambrian) of the north-central Amadeus Basin, Australia. *Journal of Paleontology* 45, 925–960.

Schopf, J.W. Walter, M.R. 1983. Archean microfossils: new evidence of ancient microbes. In J.W. Schopf Ed. *Earth's Earliest Biosphere, its origin and evolution.* pp, 214–239. Princeton University Press, New Jersey.

Sciascia, R., De Monte, S., Provenzale, A. 2013. Physics of sinking and selection of plankton cell size. *Physics Letters A* 377, 467–472. doi:10.1016/j.physleta.2012.12.020.

Sergeev, V.N., Knoll, A.H., Vorob'eva, N.G., Sergeeva, N.D. 2016. Microfossils from the lower Mesoproterozoic Kaltasy Formation, East European Platform. *Precambrian Research* 278, 87–107.

Sharma, M., Shukla, B. 2019. Akinetes from late Paleoproterozoic Salkhan Limestone (>1600 Ma) of India: A proxy for understanding life in extreme conditions. *Frontiers in Microbiology* 10, Article 397. doi:10.3389/fmicb.2019.00397.

Shih, Y.-L., Rothfield, L. 2006. The bacterial cytoskeleton. *Microbiolgy and Molecular Biology Reviews* 70, 729–754.

Stein, L.Y. 2018. Eukaryotic evolution: An ancient breath of nitrate. *Current Biology* 28, R875–R877.

Sugitani, K. 2012. Life cycle and taxonomy of Archean flanged microfossils from the Pilbara Craton, Western Australia. *Proceedings of the 34th International Geological Congress (IGC)*: AUSTRALIA 2012, 17.3, #257.

Sugitani, K., Mimura, K., Suzuki, K., Nagamine, K., Sugisaki, R. 2003. Stratigraphy and sedimentary petrology of an Archean volcanic–sedimentary succession at Mt. Goldsworthy in the Pilbara Block, Western Australia: implications of evaporite (nahcolite) and barite deposition. *Precambriam Research* 120, 55–79.

Sugitani, K., Grey, K., Allwood, A., Nagaoka, T., Mimura, K., Minami, M., Marshall, C.P., Van Kranendonk, M.J., Walter, M.R. 2007. Diverse microstructures from Archaean chert from the Mount Goldsworthy—Mount Grant area, Pilbara Craton, Western Australia: Microfossils, dubiofossills, *or pseudofossils? Precambrian Research* 158, 228–262.

Sugitani, K., Lepot, K., Nagaoka, T., Mimura, K., Van Kranendonk, M., Oehler, D.Z., Walter, M.R. 2010. Biogenicity of morphologically diverse carbonaceous microstructures from the ca. 3400 Ma Strelley Pool Formation, in the Pilbara Craton, Western Australia. *Astrobiology* 10, 899–920.

Sugitani, K., Mimura, K., Nagaoka, T., Lepot, K., Takeuchi, M. 2013. Microfossil assemblage from the 3400 Ma Strelley Pool Formation in the Pilbara Craton, Western Australia: Results from a new locality. *Precambrian Research* 226, 59–74.

Sugitani, K., Kohama, T., Mimura, K., Takeuchi, M., Senda, R., Morimoto, H. 2018. Speciation of Paleoarchean life demonstrated by analysis of the morphological variation of lenticular microfossils from the Pilbara Craron, Australia. *Astrobiology* 18, 1057–1070. doi:10.1089/ast.2017.1799.

Summons, R.E., Jahnke, L.L., Hope, J.M., Logan, G.A. 1999. 2-Methylhopanoids as biomarkers for cyanobacterial oxygenic photosynthesis. *Nature* 400, 554–557.

Tang, Q., Pang, K., Xiao, S., Yuan, X., Ou, Z., Wan, B. 2013. Organic-walled microfossils from the early Neoproterozoic Liulaobei Formation in the Huainan region of North China and their biostratigraphic significance. *Precambrian Research* 236, 157–181.

Tomitani, A., Knoll, A.H., Cavanaugh, C.M., Ohno, T. 2006. The evolutionary diversification of cyanobacteria: Molecular-phylogenetic and paleontological perspectives. *Proceedings of National Academy of Sciences of the United States of America* 103, 5442–5447.

Van Kranendonk, M.J., Hickman, A.H., Smithies, R.H., Williams, I.R., Bagas, L., Farrell, T.R. 2006. Revised lithostratigraphy of Archean supracrustal and intrusive rocks in the northern Pilbara Craton, Western Australia. *P.57. Geological Survey of Western Australia, Record* 2006/15.

Vorob'eva, N.G., Sergeev, V.N., Petrov, P.Y. 2015. Kotuikan Formation assemblage: A diverse organic-walled microbiota in the Mesoproterozoic Anabar succession, northern Siberia. *Precambrian Research* 256, 201–222.

Wacey, D., Noffke, N., Saunders, M., Guagliardo, P., Pyle, D.M. 2018. Volcanogenic pseudofossils from the ~3.48 Ga Dresser Formation, Pilbara, Western Australia. *Astrobiology* 18, 539–555.

Walsh, M.W. 1992. Microfossils and possible microfossils from the early archean Onverwacht group, Barberton mountain land, South Africa. *Precambrian Research* 54, 271–293.

Walter, M.R., Oehler, J.H., Oehler, D.Z. 1976. Megascopic algae 1,300 million years old from the Belt Supergroup, Montana: a reinterpretation of Walcott's *Helminthoidichnites. Journal of Paleontology* 50, 872–881.

Walter, M.R., Du Rulin, R., Horodyski, R.J. 1990. Coiled carbonaceous megafossils from the middle Proterozoic of Jixian (Tianjin) and Montana. *American Journal of Science* 290-A, 133–148.

Wang, Y., Wang, Y., Du, W. 2016. The long-ranging macroalga *Grypania spiralis* from the Ediacaran Doushantuo Formation, Guizhou, South China. *Alcheringa: An Australasian Journal of Palaeontology* 40,303–312.

Waterbury, J.B., Stanier, R.Y. 1978. Patterns of growth and development in pleurocapsalean cyanobacteria. *Microbiology and Molecular Biology Reviews* 42, 2–44.

Xiao, S., Knoll, A.H., Kaufman, A.J., Yin, L., Zhang, Y. 1997. Neoproterozoic fossils in Mesoproterozoic rocks? Chemostratigraphic resolution of a biostratigraphic conundrum from the North China Platform. *Precambrian Research* 84, 197–200.

Yin. L., Yuan, X., Meng, F., Hu, J. 2005. Protists of the upper mesoproterozoic Ruyang Group in Shanxi Province, China. *Precambrian Research* 141, 49–66.

Yin, L., Meng, F., Kong, F., Niu, C. 2020. Microfossils from the Paleoproterozoic Hutuo Group, Shanxi, North China: Early evidence for eukaryotic metabolism. *Precambrian Research* 342, 105650.

Index

Note: **Bold** page numbers refer to tables and *italic* page numbers refer to figures.

Milton Keynes UK
Ingram Content Group UK Ltd.
UKHW022037141024
449569UK00014B/637